World Internet Development Report 2017

Chinese Academy of Cyberspace Studies
Editor

World Internet Development Report 2017

Translated by Peng Ping

 Springer

Editor
Chinese Academy of Cyberspace Studies
Beijing, China

ISBN 978-3-662-57523-9   ISBN 978-3-662-57524-6   (eBook)
https://doi.org/10.1007/978-3-662-57524-6

Jointly published with Publishing House of Electronics Industry, Beijing, China

The print edition is not for sale in China Mainland. Customers from China Mainland please order the print book from: Publishing House of Electronics Industry.

Library of Congress Control Number: 2018950206

This Springer imprint is published by the registered company Springer-Verlag GmbH, DE part of Springer Nature
The registered company address is: Heidelberger Platz 3, 14197 Berlin, Germany

Foreword

A new-round technical and industrial revolution is emerging represented by information technology, which is a strong drive for the economic and social development. Meanwhile, Internet development has given rise to new challenges to all countries' sovereignty, security, and interests. It is the common goal and vision of the international community to promote the sustaining development of the Internet so that all nations can share the Internet development achievements. The goal requires joint effort and participation of all countries.

Against this backdrop, we decided to compile the *World Internet Development Report 2017* ("the *Report*" for short hereinafter) to study and analyze the world Internet development, summarize the past experience, analyze the present status, and manifest the Chinese academic circles' knowledge and thinking of the world Internet development, providing new reference and intelligent support for the global Internet development. Three efforts have been made in the Report as follows:

1. We always take President Xi Jinping's thoughts on international cyberspace governance as the theoretical basis for exploring the Chinese solution to the world Internet development and governance.

The global Internet governance system has entered a critical period, in which ideas on international governance of Internet coexist, competing and communicating with each other, and there is a rising voice for an inclusive solution. Mr. Xi Jinping, President of the People's Republic of China, has illustrated international Internet governance on many important international occasions, especially the "four principles" on promoting the global Internet governance system reform and the "five proposals" on building a community of shared future in cyberspace, which have won extensive praise and universal recognition from the international community and provided the Chinese solution to the world Internet development and governance. The *Report* will take President Xi's thought on international cyberspace governance as the theoretical basis and main thread running through, comprehensively illustrate his thoughts, ideas, and proposals on cyberspace governance and objectively and comprehensively interpret the Chinese solution to Internet governance. It is aimed to provide a new choice for the global Internet development and governance and contribute to promoting the building of a community of shared future in cyberspace.

2. We always take the practices of the world Internet development as the research basis for compiling the *Report* into a mini-encyclopedia about the Internet.

Since the beginning of the Internet, information technology has been developing fast and has comprehensively been integrated into social production and life, with great impact on the pattern of global economy, interests, and security. Major countries all take the Internet as their strategic focus and priority of development and speed up its construction. They have had new practices, measures, and achievements in the areas from technology to application and from Internet development to cyberspace governance. The *Report*, focusing on the status quo and trend of the world Internet development as well as the practices and achievements of all countries' Internet development, especially the new situation and new progress in 2017, reviews and analyzes the development of key Internet areas such as information infrastructure, information technology, cyber security, digital economy, e-government, network media, and international cyberspace governance. It is a panoramic research achievement covering all areas of the world Internet development.

3. We try to make everything comprehensive, accurate, and objective and to construct the assessment system of global Internet development research.

The *Report* is aimed to be international, authoritative, and accurate as well as instructive, theoretical, and general. We try to interpret the global situation from the Chinese perspective to predict the future Internet development. In particular, we set up and release the World Internet Development Index, which is an unprecedented one both at home and abroad. The index helps to measure the general situation of the world Internet development from the Chinese perspective, paying equal attention to both aggregate and per capita indicators, redressing the neglect of aggregate ones. It is beneficial for comprehensive assessment of both aggregate and per capital indicators. As the first comprehensive assessment of Internet development of major countries, it showcases the strengths and advantages of these countries.

The *Report* is a tentative study of Chinese academic circles in providing original and theoretical reference for the world Internet governance. We will continue to pay attention to the world Internet development and keep proposing our analysis and ideas to help all countries to benefit from the world Internet development and to contribute Chinese wisdom and effort to the building of a community of shared future in cyberspace.

Beijing, China   Chinese Academy of
December 2017   Cyberspace Studies

Contents

Overview

The Internet, as an important achievement of the development of the world's civilization, is becoming an important drive for innovative development and social progress, benefiting the whole humankind. This report, with reference to the existing research results from both China and other countries, tries to establish the Global Internet Development Index to make a comprehensive assessment and quantitative assessment of the Internet development of major countries, so that it can be used as a reference for all countries in promoting their Internet development. Generally speaking, the explosive growth of the global mobile Internet is coming to an end and the world's Internet connection scale growth is entering a power conversion period. The Internet development is seeing a transfer from "person-to-person connection" to "connection of all things." Rising network information technologies like artificial intelligence have become the new highlands and digital economy, the new power for all countries' economic growth. Cyberspace has become the new area of the global governance system revolution; smart society, the new social formation of people's production and life; and the Internet, the common home connecting you and me. All countries and the international community should take the well-being of humankind as the fundamental consideration and adhere to the philosophy of built, shared and governed by all. We should make the global Internet governance system more equitable and reasonable and work together to build a community of shared future in cyberspace featured equality, respect, innovative development, openness, sharing spirit, security, and order.

I. General Situation of the World Internet Development

In today's world, the information technology revolution represented by the Internet is still in progress. It has become the locomotive of innovation-driven development, fundamentally changing people's life and work and pushing the society forward. By

June 2017, the total number of Internet users worldwide had reached 3.89 billion, with a coverage rate of 51.7%[1]. Almost all major countries are speeding up informatization and reaping digital bonus to cope with the sluggish growth in the post-financial crisis era, to realize inclusive and sustainable growth and to enhance their international competitiveness. In the Internet sector, they will accelerate infrastructure deployment, foster innovation capacity, promote the development of the industry, tap the potential of its applications, spare no effort to safeguard cyber security, and direct the global governance system toward a more just and equitable future.

1. Globalization has entered the new era of being driven by data, and the Internet sector is now the strategic focus and development priority across the world.

At present, as the world is undergoing re-balance of power and readjustments, globalization is at a new turning point: It was driven first by international trade in version 1.0 and later by international finance in version 2.0, and now it is driven by data in its latest updated version 3.0. Major economies in the world are riding the new tides of information technology revolution and accelerating the development and use of the Internet to promote their own economic and social development and maintain or enhance their overall strength and international competitiveness. At present, 100 countries and regions have promulgated or updated national strategies for cyber security or informatization. For instance, the USA is stepping up its efforts to deploy next-generation network facilities, big data, advanced manufacturing, and artificial intelligence to consolidate its technological and industrial leadership; the EU is trying to remove the digital trade barriers among its member states and accelerate the formation of a unified digital market; Germany has adopted "Industry 4.0" as a national strategy to consolidate its status as an industrial power; Russia has issued the "Roadmap for the Development of Information Technology Industry" to enhance its overall advantages in its information industry; India is promoting the building of a "digital India" by speeding up its development of optical network, software, and cyber security; China is trying to build itself into an Internet power by promoting the integration of the Internet, big data, artificial intelligence, and real economy and enhancing its economic innovation capacity and competitiveness. Both emerging and developing countries are seeking to go digital and strengthen their competitiveness by developing the Internet.

2. The energization level of the Internet is producing increasing influence, and informatization has become the great engine of economic development and social progress.

At present, the energization level of the Internet is playing an increasingly important role in driving and leading economic and social development. It is embodied in four "new things," namely new production elements, new infrastructure, new economy, and new governance. First, data as a new production element are playing an obvious role and the opening up, sharing, and application of

---

[1]Data source: www.internetworldstats.com.

data can help to optimize the allocation and use efficiency of traditional elements and thus to improve the total factor productivity of resources, capital, and talents. In his book *Powershift: Knowledge, Wealth, and Violence at the Edge of the 21st Century*, Alvin Toffler, a futurist, points out, "The world has bid farewell to the era dominated by violence and money, and the magic of the world's future politics will be in the hand of 'information as power'." The more data are developed, the more valuable they are. It is estimated by 2019, the global data flow will have increased by 66 times in comparison with that in 2005. Secondly, the Internet has become the new infrastructure in economic and social development. From 2013 to 2017, there has been an increase of 196 Tbps in the bandwidth of the Internet, and now, the bandwidth is 295 Tbps, with an annual increase of 30%[2]. A research by the World Bank shows that every 10 percent increase in the broadband popularity can help the GDP to increase by 1.38%. The Internet is becoming the kind of infrastructure like water and electricity. It is the "main artery" of economic and social development. The energization efficiency of "Internet+" is emerging, and intelligent medical care, transportation, and education are thriving. Thirdly, to develop digital economy has become the common option of major countries and regions in enhancing their global competitiveness. G20 released the *G20 Digital Economy Development and Cooperation Initiative* at the Hangzhou Summit 2016, which is a centralized demonstration of the consensus of all participating countries on knowing and developing digital economy. Research shows that 22% of GDP of the world is closely related to digital economy covering techniques and capital. The application of digital techniques and technologies will help the world's economy to witness an added value of two trillion US dollars by 2020, and half of the increment of the total global economic gross will come from digital economy by 2025[3]. Fourthly, information promotes the modernization of the national governance system and governance capability. With the increase of the technical level of infrastructure, computer, Internet, and program development software and hardware, e-government is witnessing the shift from electronic stage to network, data, and intelligence stages. A survey made by the UN Department of Economic and Social Affairs shows that from 2003 to 2016, all countries were promoting e-government construction, and the e-government index rose from 0.402 in 2003 to 0.4922 in 2016, with an annual steady increase of about 1.6%. Developed countries from Europe and America are building the "online government" to boost delayering, precise, and standard national governance. Emerging countries in the Asian and Pacific region are also promoting modernization of government through informatization and have made great achievements.

3. The revolution of information technologies represented by the Internet is progressing, and new technologies like artificial intelligence are the new height of global innovation.

Information technologies are the area with the most intensive R&D, the most active innovation, the most extensive application, and the greatest influence in the world.

---

[2] Data source: Telegeography.
[3] Accenture. Digital Disruption: The Growth Multiplier.

A report by the World Intellectual Property Organization (WIPO) shows that in the past two decades, among the top 30 companies of the world with patent registration, 80% are Internet-related companies. At present, there is another round of innovation in information technologies. Following the three waves of innovation, namely PC, communication equipment and Internet, and intelligent terminal, the fourth wave keeps updating day by day, represented by cloud computing, big data, Internet of things, and artificial intelligence. These technologies are witnessing upgrading, overall penetration, and accelerated innovation. According to the analysis by the World Semiconductor Trade Statistics Organization (WSTS) and Ccid think tank, the sales volume of semiconductor all over the world increased from 298.3 billion US dollars in 2011 to 360.9 billion US dollars in 2017. The year 2016 marked the entry of the global chip manufacturing techniques into post-Moore's law period. Intel, Samsung, and Taiwan Semiconductor Manufacturing Company (TSMC) have all announced that they have realized the mass production of 10-nanometer chips. The USA, Japan, Republic of Korea, and China are competing in chip industry and technology. Data from Gartner show that cloud computing has become the focus of Internet giants, with Amazon, Microsoft, and Alibaba as the top three on the global cloud computing market in 2016, accounting for 50% of the market share. New technologies like 5G, quantum communication, and satellite communication are accelerating, with the mature standard of 5G to be completed and officially used for commercial purposes in 2020. In the new technical revolution and industrial revolution, artificial intelligence in multiple disciplines including integrated mathematics, statistics, mathematical logic, computer science, and neuroscience is playing an increasingly important role. According to the prediction by the International Data Center (IDC), the compound annual growth rate of the income from artificial intelligence from 2016 to 2020 will be 54.4%, which is expected to boost the market value of trillions of US dollars and to bring about new revolutions in its industry and other industries.

4. The Internet is becoming the spiritual and cultural home, so the governance and order of cyberspace are becoming ever more important.

The Internet has totally broken the traditional information dissemination way, and cyberspace is becoming the major center for people to exchange their ideas and cultures. According to statistics from Global Mobile Suppliers Association (GSA), by June 2017, the number of global mobile users had reached 7.72 billion and the cell phone has become the most frequently used Internet terminal, followed by the PC. The development of mobile Internet has led to the rapid expansion of social media like Facebook, Twitter, and WeChat. By June 2017, the number of active users of social media in the world had amounted to 3.028 billion, and that of active users of mobile terminals to 2.78 billion. With news push, payment transaction and live video show being added to social media, the mobilizing competence and public opinion influence of social media are being enhanced. In particular, as populism is spreading worldwide and pan-ideologization tends to rise, there are greater challenges to the governments of different countries in safeguarding the order of cyberspace and governing the content in it. In recent years, many countries have made investigations into unfair competition and privacy invasion of Google, Amazon, and Facebook. All countries have carried out comprehensive administration over the cyberspace by means of legislative governing, administrative supervision, and technical monitoring to prevent the emergence and spread of bad information in cyberspace. The USA has set up Social Network Monitoring Center and launched the Social Network/Media Capacity and other projects to monitor the social media like Facebook, Twitter, MySpace, and YouTube. Russia has set up the world's leading cyber data monitoring system—"Action Monitoring System" to completely monitor the telephone, voice over Internet phone, and Internet communication within its border. Large social media like Facebook, Twitter, YouTube, and WeChat are controlling harmful information with advanced technologies like artificial intelligence.

5. There are obvious non-traditional security problems, so cyber security is becoming the priority of priorities of national security.

At present, there are frequent threats to cyberspace information security. There are more obvious security problems such as cyber attacks, cyber crimes, and privacy leak. Cyber security is an important issue concerning the security of all countries and regions. According to the European Network and Information Security Agency (ENISA), 78% of Web sites have security holes, 15% of which are serious ones. According to FBI, the loss caused by bot net to the USA has amounted to 9 billion US dollars, that to the world has amounted to 110 billion US dollars, with about 500 million mainframes harmed in the world every year. *Global Phishing Report* shows that 2016 witnessed the highest number of phishing attacks and the highest number of domain names used for phishing, with at least 255,065 phishing attacks, an increase of over 10% in comparison with that in 2015. Blackmail attack number is explosively growing worldwide. Research by Kaspersky shows that by the end of 2016, 44,000 blackmail attack software variants had been discovered and at least 114 countries and regions had been affected by blackmail attack events. With the intelligence and cyber development, more and more physical devices are connected, so they are subject to attacks and their cascade amplification, resulting in domino effects. Statistics from HIS, a market survey organization, show that by 2025, the sell-in of relevant devices will have increased to 1.2 billion, but there is insufficient protection or no protection for the devices of Internet of things. Especially, key information infrastructure cyber security is being threatened. According to statistics, since 2016, many different sectors have been attacked on the cyberspace by hackers or terrorists, for instance, governmental sectors like Irish Government, Indian Ministry of Foreign Affairs and military network, NASA, United States Department of Defense, and US National Security Agency Data Center; financial sectors like Japanese banking; energy sectors like Ukrainian Grid, Israeli Electricity Authority, and German nuclear power stations; medical and health sectors like Australian Ministry of Health, American medical care institutions; hydraulic sectors like the server of American Hydro and sewage treatment plants; and transportation sectors like the Utah Airport Website. Besides, because some countries keep improving their cyber military strength, there seem to be cyber wars coming. The USA set up its cyberspace command in 2009 and announced in August 2017 that this command would be upgraded to be the highest level United Combatant Commands and the country is planning to build 133 cyber armies with synthetic battle capability before 2018, including 40 attacking ones. It is also developing cyber weapons, for instance, wireless intrusion, that can break physically insulated measures. Military or economic powers like Germany, Israel, Republic of Korea, and Japan have all channeled cyberspace into their military security area. Their cyberspace military transformation is inevitable since they are setting up cyber armies to develop their capacity of cyber military fighting.

6. Global cyberspace governance has entered the multilateral and multi-party governance stage.

The governments of all countries are more and more dominant in the cyberspace governance. Four special agencies and mechanisms like the UN group of governmental experts on cyber security and its framework and regional multilateral organizations represented by Economic Cooperation Organization (ECO) and APEC, and bilateral dialogues between China and the USA, China, and Russia, and the USA and Russia provide the basic framework for the inter-governmental cooperation in global cyberspace governance. The voice advocating the cyberspace sovereignty principle and multilateralism becomes louder, forming the situation that can contend with multi-stakeholders, and there have been substantial results in ICANN mechanism reform. There is also rising voice for surpassing the confrontation between two patterns and seeking for inclusive solutions. Traditional and emerging powers and inter-governmental organizations like the UN, Internet communities, and relevant research institutions are promoting the formulation of rules on global cyberspace. Great progress has been made in the formulation of basic principles on global cyberspace governance, and Internet technical standards and norms, as well as in cyber crime punishment and digital economy development.

II. Status Quo of Some Major Countries' Internet Development

This section, centering on the Global Internet Development Index that includes six dimensions (infrastructure, innovation capacity, industry development, Internet application, cyber security, and Internet governance), through synthesized and quantified assessments, provides a comprehensive, scientific, and objective analysis of the status quo of the Internet development in some major countries and the overall trend of the world's Internet development.

(I) Main Considerations of the Global Internet Development Index System

1. The Global Internet Development Index is the specification, standardization, and indexation of President Xi Jinping's thought on global Internet governance.

In the past few years, Chinese President Xi Jinping elaborated on global Internet governance ideas on several important international occasions. In his keynote speech delivered at the opening ceremony of the Second World Internet Conference, he put forward four principles for the transformation of global Internet governance system (*The four principles are: respect for cyber sovereignty, maintenance of peace and security, promotion of openness and cooperation, and cultivation of good order*) and five proposals on building a community of shared future in cyberspace (*First, speed up the building of global Internet infrastructure and promote inter-connectivity. Second, build an online platform for cultural exchange and mutual learning. Third, promote innovative development of cyber economy for common prosperity. Fourth, maintain cyber security and promote orderly development. Fifth, build an Internet governance system to promote equity and justice.*), winning wide applaud and recognition from the international community. These proposals were lauded as China's intelligent contribution to global Internet governance. The Global Internet Development Index is the specification, standardization, and indexation of President Xi Jinping's thought on global Internet governance, aiming to provide quantifiable, referable, and comparable data and evidence for countries to assess their own status quo of their Internet development so that they can make good plans and take corresponding measures.

2. As an assessment of the overall status of the Internet development across the world is missing, the Internet Development Index will fill this gap.

The attempts to assess Internet development level are long-standing; the United Nations, the International Telecommunication Union, and the World Economic Forum have all made such attempts: the Global E-Government Development Index, the Global ICT Development Index, and the Networked Readiness Index have gradually been accepted by the international community. They have helped countries to know about their current status of Internet development, the stage they are in, and their strengths and weaknesses in that aspect. These indices, however, are mostly special ones to measure such specific dimensions as informatization or cyber security. They fail to reflect the big picture and the overall trend of Internet development across the world. The Global Internet Development Index System contains 6 first-level indicators, 12 second-level ones, and 32 third-level ones, concerning six dimensions, namely infrastructure, innovation capacity, industry development, Internet application, cyber security, and Internet governance. They are adopted to comprehensively analyze the development of the Internet in various countries. It will help countries learn from each other and make progress together to promote the development of the Internet in their regions and benefit the people and the society at large.

3. The Global Internet Development Index involves both per capita and aggregate
   indicators of Internet development level and thus provides a more comprehensive, accurate, and scientific system for assessment.

To assess the Internet development of a country, per capita indicators are, without a
doubt, important; aggregate indicators such as technological input, number of talents, industry development, and scale of application are also valuable tools.
According to Metcalfe's Law, the value of a telecommunications network is pro-
portional to the square of the number of connected users of the system, indicating
that aggregate index is crucial in measuring Internet development level. To redress
the neglect of aggregate indicators in existing systems, the establishment of this
system is aimed to evaluate both aggregate and per capita indicators and therefore is
more comprehensive, objective, and accurate.

(II) Specific Content and Assessment Results of the Global Internet Development Index

Given the experimental nature of this newly established index and the limited
availability of data, the 2017 Global Internet Development Index has involved 38
countries, covering major economies of five continents and countries with better
Internet development. It can basically reflect the latest development of the Internet
across the world. These 38 countries include:

America: the USA, Canada, Brazil, Argentina, Mexico, and Chile;
Asia: China, Japan, Republic of Korea, Indonesia, India, Saudi Arabia, Turkey,
UAE, Kazakhstan, Malaysia, Singapore, Thailand, and Vietnam;
Europe: the UK, France, Germany, Italy, Russia, Denmark, Estonia, Finland,
Netherlands, Norway, Portugal, Spain, Sweden, and Switzerland;
Oceania: Australia;
Africa: South Africa, Egypt, Nigeria, and Kenya.

The index (see Fig. 1 and Table 1) contains six dimensions: infrastructure,
innovation capacity, industry development, Internet application, cyber security, and
Internet governance. The infrastructure section mainly refers to each country's
broadband construction level; innovation capacity, to the capability of innovating,
including technology, human resources, investment; the industry development
section, to the development level of the Internet industry in these countries; the
application section, to their Internet application level from three dimensions: personal application, e-commerce, and e-government; the security section, to their
cyber security level; the governance section, to their managerial expertise and
participation in domestic and international cyberspace governance.

By giving these indicators reference values, we now have the Internet
Development Index scores of 38 countries (see Table 2). It is clear that developed
economies score the highest in terms of average Internet development level. They
are followed by some European countries and emerging Asian countries.

Fig 1 Global Internet
Development Index system

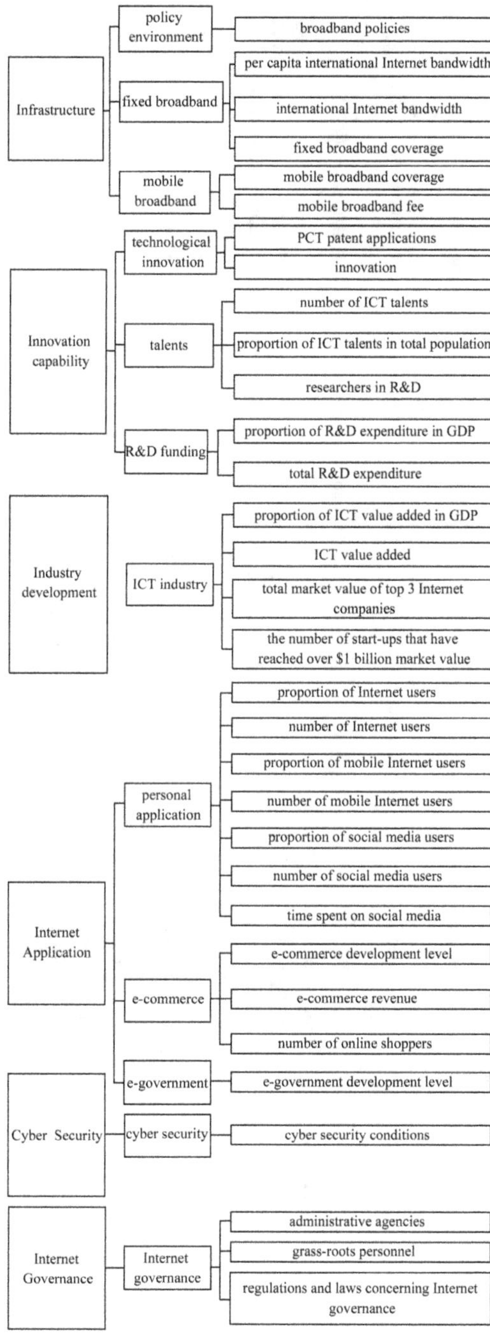

Table 1 Global Internet Development Index system

First-level indicator	Second-level indicator	Third-level indicator	Specification	Data source
1. Infrastructure	1.1 Broadband environment	1.1.1 Broadband policies	Importance attached to broadband development	ITU databases (2015)
	1.2 Fixed broadband	1.2.1 Per capita international Internet bandwidth	Development level of international Internet	ITU databases (2015)
		1.2.2 International Internet bandwidth	Coverage of International Internet	Per capita international Internet bandwidth × the number of population (2015)
		1.2.3 Fixed broadband coverage	Construction level of fixed broadband	ITU databases (2015)
	1.3 Mobile broadband	1.3.1 Mobile broadband coverage	Construction level of mobile broadband	ITU databases (2015)
		1.3.2 Mobile broadband fee	Mobile broadband fee	ITU databases (2015)
2. Innovation capacity	2.1 Technical innovation	2.1.1 PCT patent applications	Innovation level and ability in ICT industry	World Development Indicators from the World Bank (2014–2015)
		2.1.2 Innovation	Innovation environment for ICT industry	WIPO (2016)
	2.2 Talents	2.2.1 Number of ICT talents	ICT talents	ILO databases (2015–2016)
		2.2.2 Proportion of ICT talents in total population	Proportion of ICT talents in total population	ILO databases (2015–2016)
		2.2.3 Researchers in R&D	Researchers in R&D	World development indicators from the World Bank (2013–2015)

(continued)

Table 1 (continued)

First-level indicator	Second-level indicator	Third-level indicator	Specification	Data source
2. Innovation capacity	2.3 R&D funding	2.3.1 Proportion of R&D expenditure in GDP	Proportion of R&D expenditure in GDP	WIPO statistics (2013–2015)
		2.3.2 Total R&D expenditure	R&D input	Proportion of R&D expenditure in GDP $\times$ GDP (2013–2015)
3. Industry development	3.1 ICT industry	3.1.1 Proportion of ICT value added	ICT value added	The UN databases (2016)
		3.1.2 ICT value added	ICT value added	ICT value added $\times$ GDP (2016)
		3.1.3 Total market value of top 3 Internet companies	Total market value of top 3 Internet companies	*World Startup Wiki* statistics (2014)
		3.1.4 The number of start-ups with over $1 billion market value	The number of start-ups with over $1 billion market value	*The Washington Times* statistics (2014–2016)
4. Internet application	4.1 Personal application	4.1.1 Proportion of Internet users	Proportion of Internet users of the total population	ITU databases (2016)
		4.1.2 Number of Internet users	Total number of Internet users	Proportion of Internet users of the total population $\times$ total population (2016)
		4.1.2 Proportion of mobile Internet users	Proportion of mobile Internet users of the total population	Global Web index statistics (2015)
		4.1.3 Number of mobile Internet users	Total number of mobile Internet users	Proportion of mobile Internet users of the total population $\times$ total population (2015)

(continued)

Table 1 (continued)

First-level indicator	Second-level indicator	Third-level indicator	Specification	Data source
4. Internet application	4.1 Personal application	4.1.4 Proportion of social media users	Proportion of social media users of the total population	Global Web index statistics (2015)
		4.1.5 Number of social media users	Total number of social media users	Proportion of social media users of the total population × total population (2015)
		4.1.6 Time spent on social media	Time spent on social media	Global Web index statistics (2015)
	4.2 E-commerce	4.2.1 E-commerce environment	E-commerce environment	UNCTAD statistics (2014)
		4.2.2 E-commerce revenue	E-commerce revenue	Statista Inc. statistics (2017)
		4.2.3 Number of online shoppers	Number of online shoppers	Government Web sites statistic (2016–2017)
	4.3 E-government	4.3.1 e-government development level	E-government development level	The UN statistics (2016)
5. Cyber security	5.1 Cyber security	5.1.1 Cyber security conditions	Cyber security level	ITU statistics statistic (2016)
6. Internet governance	6.1 Internet governance	6.1.1 Administrative agencies	Central-level Internet administrative agencies	Government Web sites statistic
		6.1.2 Administrative personnel	Internet administrative personnel	Government Web sites statistic
		6.1.3 Regulations and laws concerning Internet governance	Regulations and laws concerning Internet governance, including content management, privacy protection, market order and crackdown on cyber crimes, violence, or terrorism	Government Web sites statistic

Table 2 Global Internet Development Index

Ranking	Country	Score
1	The USA	57.66
2	China	41.80
3	Republic of Korea	38.86
4	Japan	38.11
5	The UK	37.85
6	Singapore	37.71
7	Sweden	36.64
8	Finland	35.72
9	France	35.39
10	Germany	35.22
11	Australia	35.21
12	Canada	34.63
13	Netherlands	34.60
14	Estonia	34.59
15	Norway	34.25
16	Switzerland	33.84
17	Denmark	31.47
18	Russia	30.52
19	Italy	30.09
20	Spain	29.96
21	Malaysia	28.83
22	United Arab Emirates	28.18
23	Brazil	27.87
24	Portugal	27.57
25	India	26.72
26	Mexico	26.68
27	Thailand	25.83
28	Argentina	25.70
29	Turkey	23.84
30	Egypt	23.39
31	Saudi Arabia	22.96
32	Chile	21.88
33	Indonesia	21.69
34	South Africa	21.30
35	Kenya	20.33
36	Vietnam	19.96
37	Kazakhstan	19.33
38	Nigeria	16.27

Developing countries in Latin America and sub-Saharan Africa are stepping up their efforts to catch up.

1. Progress has been made in infrastructure, but its development is still quite uneven (see Table 3).

The number of Internet users has increased from one billion in 2005 to 3.89 billion today, two-thirds of which are found in developing countries. Mobile Internet usage has surpassed fixed broadband Internet usage. In addition, the number of mobile phone users in developing countries is 5.5 billion, while the figure in developed countries is merely 1.5 billion[4]. Such a large user base ensures promising future for the development of the Internet in developing countries.

Asia is home to 49.7% of the world's total Internet users, but its Internet coverage rate is merely 46.7%, lower than the world's average. Fixed broadband access in East Asia and Northeast Asia is as high as 74.89% and that in the Middle East is 58.7%, but in other Asian regions, it is less than 10%[5]. Among them, Singapore tops the list in terms of infrastructure and Internet coverage rate. Its mobile broadband coverage rate is as high as 142.2%, with 4G coverage reaching 82% and speed reaching 46 Mbps, the fastest in the world[6].

North America and Europe rank first and second in terms of Internet coverage rate, with 88.1% and 80.2%, respectively. In Europe, 218 million (99.9%) households have access to fixed or mobile Internet: Fixed broadband coverage in Finland, Sweden, Estonia, Norway, and the UK has reached 97.5%; mobile Internet coverage has generally exceeded 95%, with the exception of Germany (92.1%) and Slovakia (90.7%).

As a less populous continent, Oceania is home to only 0.7% of Internet users, but its Internet coverage rate is as high as 69.6%. The rate in Latin American region is 62.4%, close to the global average. In recent years, the development of the Internet is gaining momentum in Brazil, Chile, and other countries.

The Internet sector in most African countries is relatively backward, and the Internet coverage rate in Africa lags behind that in all other regions. As the second largest economy on the continent, South Africa enjoys the highest level of Internet development that the continent has seen. By November 2015, 71.4% of African population still had no access to the Internet.

2. To cultivate Internet innovation capacity, a favorable environment must be created.

Innovation capacity is a major indicator of the development potential of a country (see Table 4). North America generally attaches great importance to technical

---

[4]Harnessing the digital economy for developing countries, http://www.keepeek.com/Digital-Asset-Management/oecd/development/harnessing-the-digital-economy-for-developing-countries_4adffb24-en#page13.

[5] http://www.unescap.org/resources/state-ict-asia-and-pacific-2016-uncovering-widening-broadband-divide.

[6]ITU IDI 2016.

Table 3 Infrastructure

Ranking	Country	Score
1	Singapore	3.37
2	Finland	2.93
3	Sweden	2.82
4	Estonia	2.75
5	The UK	2.69
6	Norway	2.58
7	Republic of Korea	2.51
8	Denmark	2.26
9	Switzerland	2.21
10	Australia	2.20
11	The USA	2.20
12	Netherlands	2.20
13	Germany	2.15
14	Japan	1.90
15	France	1.89
16	Spain	1.70
17	Canada	1.67
18	Russia	1.65
19	Italy	1.61
20	Portugal	1.61
21	United Arab Emirates	1.59
22	Saudi Arabia	1.55
23	Brazil	1.43
24	Chile	1.33
25	Kazakhstan	1.29
26	Malaysia	1.27
27	China	1.23
28	Argentina	1.21
29	Thailand	1.16
30	Turkey	1.15
31	South Africa	1.05
32	Mexico	0.89
33	Egypt	0.80
34	Vietnam	0.72
35	Indonesia	0.66
36	India	0.39
37	Nigeria	0.37
38	Kenya	0.37

Table 4 Innovation capacity

Ranking	Country	Score
1	The USA	8.91
2	Republic of Korea	6.87
3	Japan	6.73
4	China	6.35
5	Sweden	5.85
6	Germany	5.80
7	Switzerland	5.65
8	Denmark	5.40
9	Finland	5.20
10	The UK	4.96
11	Netherlands	4.77
12	Singapore	4.67
13	France	4.65
14	Australia	4.46
15	Norway	4.31
16	Canada	4.24
17	Estonia	3.80
18	India	3.70
19	Spain	3.43
20	Italy	3.37
21	Portugal	3.20
22	Malaysia	2.91
23	Russia	2.73
24	Brazil	2.61
25	United Arab Emirates	2.45
26	Turkey	2.41
27	Mexico	2.21
28	Saudi Arabia	2.10
29	Thailand	2.06
30	Vietnam	1.94
31	South Africa	1.88
32	Argentina	1.86
33	Chile	1.79
34	Indonesia	1.63
35	Egypt	1.52
36	Kazakhstan	1.32
37	Kenya	1.16
38	Nigeria	0.72

innovation. The USA has been generously supporting research projects of cutting-edge information technology through government funding, and hence, such famous Internet companies as Google have risen as giants. In Latin America, although Argentina, Mexico, Brazil, and other countries lag slightly behind in innovation capacity, governments of these countries are actively introducing top-level design and providing tax and financial incentives to encourage the development of high-tech industries.

Europe enjoys a better start for Internet development. To stay ahead, high-income economies such as Sweden, Switzerland, Denmark, Germany, Finland, the UK, and the Netherlands are committed to improving their innovation environment, human capital, and R&D investment, hence higher innovation level.

In Asia, Republic of Korea and Japan are top innovators in the world, and they stay far ahead in patent and R&D investment. China is also catching up, and it is relatively strong in patent registration.

The foundation for innovation in Africa is weak; African countries, however, are making great efforts to create a favorable environment for innovation by actively cultivating Internet talents and enhancing digital literacy of the public.

3. The development of the Internet industry assumes various forms and corporate performance is a key factor.

Enterprises are the main force in promoting the development of the Internet industry, and the main bodies in scientific innovation and industrial transformation, supporting the development of digital economy in a country. Generally speaking, Internet companies are actively expanding the scope of their business and rushing to secure their places in emerging high-tech industries, spawning new technologies, new businesses, new models, and new services (see Table 5).

North America leads the world in Internet industry. The USA is home to a large number of leading Internet companies: Such Internet giants as Apple, Google, Microsoft, and Amazon are trendsetters of the global Internet scene. The rapid development of the Internet industry in Latin America provides a good investment environment for foreign Internet companies.

Europe's great language diversity and small average population are adverse to the making of Internet giants. For this reason, by far, none of the world's top 20 Internet companies in terms of market value is from Europe.

The development of the Internet industry in Asia is extremely uneven. Internet companies from China, India, Japan, and Republic of Korea are major players. It is particularly worth mentioning that Internet start-ups in China are growing rapidly. The Billion Dollar Start-up Club now has eight Chinese members (accounting for 11% of the total) and four Indian members. Companies from Republic of Korea and Malaysia have also made the list[7]. In the Arab world, Internet companies can provide a full range of services including online music, e-commerce, e-payment, fashion, travel, job hunting, cab hailing, education, and social commerce. Souq, an

---

[7] http://www.cankaoxiaoxi.com/finance/20150225/678201.shtml.

Table 5 Industry development

Ranking	Country	Score
1	The USA	15.14
2	China	5.96
3	The UK	2.28
4	Japan	2.11
5	India	1.81
6	Sweden	1.63
7	Republic of Korea	1.55
8	Malaysia	1.49
9	Germany	1.47
10	Estonia	1.43
11	Finland	1.40
12	Canada	1.35
13	France	1.35
14	Netherlands	1.24
15	Denmark	1.14
16	Switzerland	1.12
17	Spain	1.07
18	Singapore	1.04
19	Indonesia	0.98
20	Norway	0.97
21	Italy	0.91
22	Brazil	0.90
23	Chile	0.83
24	Australia	0.82
25	Portugal	0.78
26	Kazakhstan	0.73
27	South Africa	0.67
28	Turkey	0.67
29	Mexico	0.57
30	Kenya	0.27
31	Russia	0.25
32	Vietnam	0.21
33	Argentina	0.12
34	United Arab Emirates	0.06
35	Nigeria	0.06

Note Thailand, Saudi Arabia, and Egypt are not listed due to lack of data.

e-commerce company, and Careem, a transportation network company from the United Arab Emirates, are regional leaders.

Although Africa has not yet been able to turn out a world-famous Internet company, its cooperation with ICT giants such as Google, Microsoft, Huawei, and others has attracted growing foreign investment in science and technology into Africa, with South Africa, Nigeria, and Kenya being the biggest receivers. This inflow of investment will promote the development of local technology start-ups in Africa.

4. The Internet has huge potential in its application and can lead to profound social and economic changes.

The Internet directly impacts economy, society, and the general public through its application (see Table 6). In Asia, thanks to their wider Internet coverage, China, Republic of Korea, and Singapore have higher public Internet application level. China has become the largest B2C e-commerce market in the world. In Korea, most of online time is spent on games and social media, and therefore, the two sectors are exceptionally well developed. Singapore has become a model smart city-state. Mobile TV and mobile wallet have been widely used in Japan.

The overall development of the Internet and digital economy in Europe is relatively high as e-government and industrial digitization in the region have drawn worldwide attention. As part of its post-Brexit plans, the UK Government launches the Digital Strategy to maintain its position as a world leader in digital government built on the concept of "Government as a Platform." Germany, proceeding from Industry 4.0, vigorously promotes the digital transformation of its industrial production. The European Union also presented "digitizing of European Industry Plan"[8] to help European industry, SMEs, researchers, and public authorities make the best use of new technologies.

By the end of 2016, 4G coverage in Latin America reached 68% of the total population, with smartphones accounting for 55% of total mobile connections in the region. Thanks to the development of the mobile Internet, Latin America and the Caribbean regions have higher rate of mobile payment than other parts of the world, with commercial payments done via mobile means accounting for 57% of the total number of transactions (the global average is 5%) and bulk electronic payments accounting for nearly 7% (the global average is just over 2%)[9].

With better access to the Internet, Africa has also seen the Internet being more widely applied in various sectors: communication, social networks, e-commerce, entertainment, and media. Notable examples include M-Pesa, a mobile phone-based money transfer service, and Ushahidi, an open-source crisis response site in Kenya[10].

---

[8] Commission sets out path to digitise European industry, http://europa.eu/rapid/press-release_IP-16-1407_en.htm?locale=en.

[9] http://www.businesswire.com/news/home/20170302005560/zh-CN.

[10] http://mgafrica.com/article/2016-04-29-africa-connectivity.

Table 6 Internet application

Ranking	Country	Score
1	The USA	13.22
2	China	13.20
3	Republic of Korea	11.40
4	The UK	11.34
5	Australia	10.74
6	Singapore	10.64
7	Japan	10.62
8	France	10.45
9	Canada	10.39
10	Sweden	10.32
11	Germany	10.26
12	Netherlands	10.25
13	Denmark	10.24
14	Finland	10.03
15	Spain	10.03
16	United Arab Emirates	10.03
17	Norway	9.98
18	Argentina	9.45
19	Estonia	9.31
20	Italy	9.23
21	Russia	9.21
22	Switzerland	8.90
23	Brazil	8.75
24	Chile	8.51
25	Malaysia	8.49
26	Portugal	8.40
27	India	8.24
28	Mexico	8.16
29	Turkey	8.05
30	Saudi Arabia	7.72
31	Thailand	7.53
32	Kazakhstan	7.23
33	South Africa	6.92
34	Vietnam	6.64
35	Indonesia	5.94
36	Egypt	5.60
37	Kenya	4.55
38	Nigeria	3.93

5. Cyber security is universally stressed, but tactics and capacity against cyber attacks are to be improved.

Cyber threats and hazards have become a problem that concerns all countries. Although cyber security is an issue of national strategic significance for almost all countries, currently only 38% have released relevant strategies and 12% are formulating strategies. This means that 50% of the world's countries have no clearly defined strategies against cyber threats[11] (see Table 7).

Cyber security is uneven in Asia. Singapore, as a sovereign city-state, tops the list thanks to its high level of digitalization[12]. Malaysia is a dark horse as it actively advocates strengthening cyber security. China, India, and Thailand are in the middle, with room for improvement.

American countries, such as the USA and Canada, generally stay ahead. Compared with other countries, the USA has the largest number of and the most comprehensive laws and regulations concerning cyber security. Its level of cyber security is second only to that of Singapore.

Europe is quite strict with the protection of personal information. The promulgation of *General Data Protection Regulation* extends the scope of the EU data protection law to all non-European companies processing data of EU residents.

The situation of cyber security in Africa is grave: disconnections occur from time to time, hampering the development of the Internet. Cooperation with other countries, however, has been stepped up to crack down on cyber crimes that are increasingly rampant.

6. The importance of Internet governance has been recognized, and the governance system in all countries is in urgent need of improvement.

The sound development of the Internet cannot be achieved without the active participation and effective governance of all countries. In recent years, all countries are exploring Internet governance models that suit their own conditions with such concerns as privacy protection, digital trade regulation, Web content administration, competition regulation, and crackdown on cyber crimes. Generally speaking, in a region where the Internet sector is more developed, the Internet governance is better (see Table 8). For instance, in the USA and Canada, policies and regulations pertaining to the Internet are relatively well developed and strictly enforced, and specialized agencies and personnel concerned with the Internet have been set up.

Given the fact that Europe is an early adopter of the Internet, its Internet users are better educated and its related laws and regulations are relatively sound; most European countries, such as the UK, Germany, and France, usually have their Internet sector supervised by self-regulatory organizations under the guidance of the government and in accordance with laws and regulations or voluntary agreements.

---

[11] Global Cyber Security Index 2017 released by ITU, http://wemedia.ifeng.com/21651005/wemedia.shtml.

[12] http://www.ccpit.org/Contents/Channel_4126/2017/0905/872246/content_872246.htm.

Table 7 Cyber security

Ranking	Country	Score
1	Singapore	9.25
2	The USA	9.19
3	Malaysia	8.93
4	Estonia	8.46
5	Australia	8.24
6	France	8.19
7	Canada	8.18
8	Russia	7.88
9	Japan	7.86
10	Norway	7.86
11	The UK	7.83
12	Republic of Korea	7.82
13	Egypt	7.72
14	Netherlands	7.60
15	Finland	7.41
16	Sweden	7.33
17	Switzerland	7.27
18	Thailand	6.84
19	India	6.83
20	Germany	6.79
21	Mexico	6.60
22	Italy	6.26
23	China	6.24
24	Denmark	6.17
25	Brazil	5.93
26	Turkey	5.81
27	Kenya	5.74
28	Saudi Arabia	5.69
29	Nigeria	5.69
30	United Arab Emirates	5.66
31	Spain	5.19
32	Portugal	5.08
33	South Africa	5.02
34	Argentina	4.82
35	Indonesia	4.24
36	Chile	3.67
37	Kazakhstan	3.52
38	Vietnam	2.45

Table 8 Internet governance

Ranking	Country	Score
1	The USA	9.00
2	Japan	8.90
3	France	8.85
4	Estonia	8.85
5	China	8.80
6	Canada	8.80
7	Russia	8.80
8	The UK	8.75
9	Germany	8.75
10	Australia	8.75
11	Finland	8.75
12	Singapore	8.75
13	Italy	8.70
14	Republic of Korea	8.70
15	Sweden	8.70
16	Switzerland	8.70
17	Netherlands	8.55
18	Norway	8.55
19	Spain	8.55
20	Portugal	8.50
21	United Arab Emirates	8.40
22	Brazil	8.25
23	Argentina	8.25
24	Mexico	8.25
25	Indonesia	8.25
26	Kenya	8.25
27	Thailand	8.25
28	Vietnam	8.00
29	Egypt	7.75
30	Denmark	6.25
31	Saudi Arabia	5.90
32	India	5.75
33	South Africa	5.75
34	Turkey	5.75
35	Chile	5.75
36	Malaysia	5.75
37	Nigeria	5.50
38	Kazakhstan	5.25

Emerging countries in Asia have formed two types of Internet governance models: One is government-regulating (China and Singapore), as the government retains high regulatory authority over Internet governance; the other is self-regulating (e.g., Japan), as regulatory power mainly rests with associations, societies, and other non-governmental organizations while government interference is limited.

As the Internet sector is booming in Africa, governments of African countries, recognizing the importance of Internet governance, are also working to speed up its institutional design in the hope of blazing a trail that befits their own development. By the end of 2016, African Internet Governance Forum had been successfully held for five sessions, a testament to the commitment of African countries.

III. Internet Development of Some Representative Countries

By comparing the scores of 38 countries, we see the developed economies are the best, and then come European countries and emerging economies of Asia, while Latin America and Sub-Saharan Africa are making their efforts. Developed economies like the United States, Republic of Korea, the United Kingdom, Japan and Singapore, and BRICS (China, Russia, Brazil, India and South Africa) see most representative Internet development, and they are the leaders of the global Internet development at different stages. Here are the introductions of the above-mentioned 10 countries.

(I) Internet Development of Five Developed Economies

1. The United States is still the leader of the global Internet development.

The USA, the origin of the Internet, is one of the five largest Internet markets. It ranks first in Internet Development Index ranking, and it is one of the leaders in innovation capacity, industry development, Internet application, and Internet governance. Its cyber security level is only behind that of Singapore, but it ranks 11th in terms of Internet infrastructure due to its vastness and hence the big digital gap (see Fig. 2).

The US Government attaches importance to the construction of Internet infrastructure, so they have launched many a strategy, including the *National Information Infrastructure Plan* (*Information Highway*), *National Broadband Plan*, stressing that every American can have access to broadband.[13]

In terms of technical innovation, the USA is the leader of the world. The government has launched many a policy to boost the R&D of the new-generation

---

[13] http://intl.ce.cn/zhuanti/it/zt/kdjh/201007/22/t20100722_21643519.shtml.

Fig. 2 Internet Development
Index of the USA

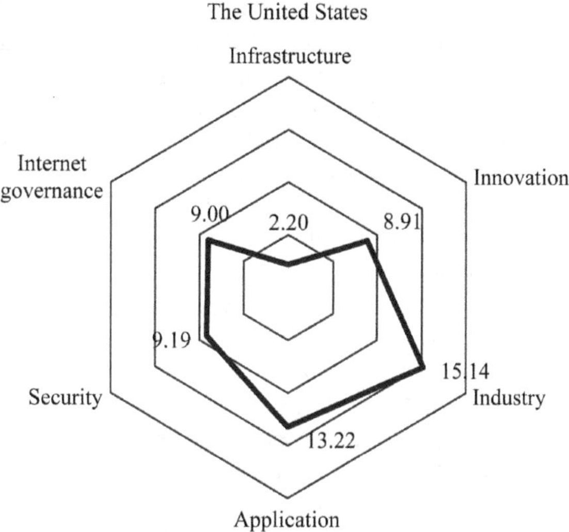

information technology. For instance, the Network Information Technology Research and Development (NITRD) provides financial aid and fund for basic R&D through organizational cooperation, providing cooperative platforms for federal agencies. Besides, Defense Advanced Research Projects Agency is the important promoter for the R&D of cutting-edge technology of network information. Famous Internet companies like Google have developed on the basis of technical R&D results funded by the government.

In terms of industry development, the USA is ahead of other countries. Internet giants represented by Apple, Google, Microsoft, Amazon, Facebook, IBM, Intel, and Cisco almost determine the development focus and direction of the global Internet business. Thanks to its developed Internet, American digital economy plays an important role in economy. According to statistics released by American Department of Commerce, American online service trade volume accounts for half of the country's total service trade volume, and one-sixth of the country's total export volume (including goods trade and service trade)[14].

In terms of Internet application, the USA takes the lead in e-government development. Since the 1990s, the country has released E-government Strategy— Simplified Delivery of Services to Citizens, *E-government Act of 2002, E-government Strategy of 2003, Federal Mobile Government Strategy and Digital Government Strategy: Building a Century Platform to Better Serve the American* to foster the e-government. The USA also attaches importance to the open utilization of government data, having released *Transparency and Open Government Memorandum, Executive Order—Making Open and Machine Readable the New Default for Government Information, Open, Public, Electronic, and Necessary*

---

[14] http://www2.caict.ac.cn/zscp/ictdc/xxhyg/201612/t20161214_2184664.html.

Government Data Act, and Open Government Initiative 2016, and launched data.gov, thus leading the global open government development.

The American Government stresses the domestic cyber security capacity building. Totally, the country has formulated over 140 laws and regulations on the Internet, covering network infrastructure, freedom in speech, press and publication, personal privacy, protection of minors, intellectual property rights, cyber crime, and national security and other areas of national and social life. Therefore, the USA has a complete Internet legislation system and its measures are borrowed or imitated by other countries.

In terms of Internet governance, the US Government has, based on three documents, namely *The National Strategy to Secure Cyberspace (The National Strategy for short), International Strategy* for cyberspace (International Strategy for short) and The Comprehensive National Cybersecurity Initiative (The Initiative for short), set up a complete and coordinated national and international cyberspace governance framework. *The National Strategy and The Initiative* focuses on the domestic Internet governance. For instance, they are used to prevent key American infrastructure from being attacked from online to reduce the damage that the country may suffer from cyber attacks. *The International Strategy* focuses on international cooperation and contributes to the establishment of an effective international Internet governance framework involving different partners.

2. The Internet of Republic of Korea is developed in general.

Republic of Korea ranks third in the Internet development. It ranks second in terms of innovation capacity, third in Internet application, and sixth (tied) in Internet governance. But it is not so good in network infrastructure (ranking 7th), industry development (ranking 7th), and cyber security (ranking 12th) (see Fig. 3).

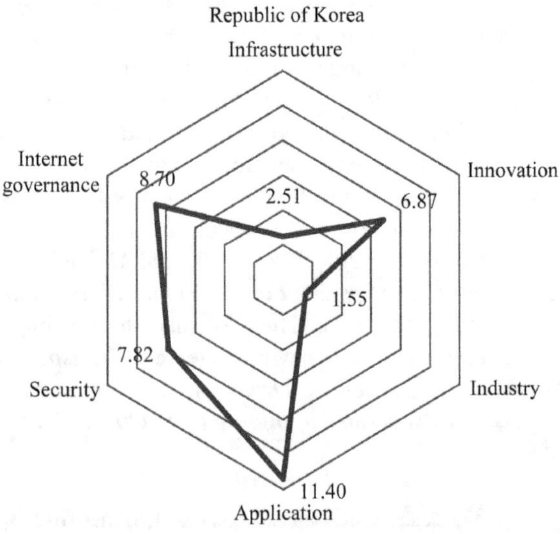

Fig. 3 Internet Development Index of Republic of Korea

The Internet in Republic of Korea enjoys wide coverage and high speed, ranking seventh in terms of network infrastructure. By January 2017, Republic of Korea's Internet user coverage rate had reached 90%; that of mobile Internet coverage rate, 115%; the coverage ratio of 3G and 4G, 99%, with cafés, parks, subways, and airports all having free WiFi. As one of the countries with the fastest Internet access, Republic of Korea has an Internet speed twice of the American and nearly four times of China's. However, because of the high consumption, Republic of Korea has the highest communication fee.

In terms of infrastructure construction, Republic of Korea is developing the 5G network and announces that it will provide the world's first 5G network for commercial purpose in 2019. It is estimated KRW 1.6 trillion (approximately RMB 9.5 billion yuan) will be spent on it, and the network speed at that time will be 1000 times that of today, about 20 Gbps.

In terms of the Internet application, Internet users of Republic of Korea spend the most time on online games and social media, accounting for more than one-third of the Internet users' time online on average. Besides, the country enjoys high-level e-government. Because of the implementation of national strategies like *e-KOREA Vision 2006*, *E-government Planning 2003–2007*, and *E-government Planning 2008–2012*, Republic of Korea is among the top countries in terms of e-government. In June 2013, keeping up with the mobile Internet development, the government released *Basic Plan 3.0 for Government Building* to provide tailored service for the people and establish the people-centered and service-oriented government.

Republic of Korea attaches great importance to Internet governance. It is the first country to make laws on Internet censorship and to establish the Internet censorship organizations. In 1995, the world's first Internet censorship organization— Information and Communication Ethics Committee—was founded in Republic of Korea and *Business Law on Electronic Communication* was released there at the same time. The law authorizes the Information and Communication Ethics Committee to monitor the cyberspace, to make arbitration on cyberspace conflicts, to close illegal or unhealthy Web sites, and to shield overseas offensive ones. In 2008, the Government of Republic of Korea set up the Broadcast and Communication Review Committee to shoulder the above-mentioned responsibilities. In July 2000, Cybercrime Response Center was established in the Police Bureau of the country and the center accepts accusations, makes investigations, develops new technologies, and conduct international cooperation to fight cyber crimes. In 2010, to respond to frequent cyber attacks, the Ministry of Defence of Republic of Korea set up the Information Security Command to safeguard national cyber security. Besides, the country has established complaints channels like the Center for Law-breaking and Harmful Information Reporting to monitor online information. To guarantee the cyber security, regulate the cyber activities, and protect the users' rights and interests, the Government of Republic of Korea has formulated and improved a series of laws on cyberspace administration, such as *Basic Law on Telecommunication*, *Basic Law on National Informationization*, *Law on Promoting Information and Communication Cyberspace Use and Protecting*

Information, Basic Protective Law on Information and Telecommunication, Regulations on Cyber Security, and *Law on Communication Confidentiality Protection*. In January 2017, the Government of Republic of Korea submitted to the National Assembly *the National Cyber Security Act* to further improve the legislation on the national cyber security. The country has set up the National Cyber Security Committee, raised the cyber security administration to a high level, and enhanced the cyber security guarantee.

3. Japan has strong overall momentum in the Internet development.

Japan has been attaching importance to the Internet development. In the Internet Development Index, Japan ranks 4th, with Internet governance (ranklng second), innovation capacity (ranking third), industry development (ranking fourth), Internet application (ranking seventh), and cyber security (ranking ninth) among the top indicators of the world (see Fig. 4).

Japan enjoys high-level development of its network infrastructure. It has released *IT Basic National Strategy, IT Basic Law, E-Japan Strategy, u-Japan Strategy*, and *i-Japan Strategy*, which have helped Japan to build an environment in which anyone can access to the Internet at any place at any time. By January 2017, the Internet coverage rate of the country had reached 93%; the mobile Internet coverage rate, 147%; the 3G and 4G coverage rate, 98%. Due to Japan's developed economy, the broadband fee there is higher than in other countries.

Japan has been the leader of the world in innovation. For years, it has attached importance to the development of robots and artificial intelligence. In the 1980s, industrial robots produced in Japan began to be used widely in the automobile industry and electronic and electric industries, taking the lead in the world's industrial robot development. The use of robots was soon extended into the service

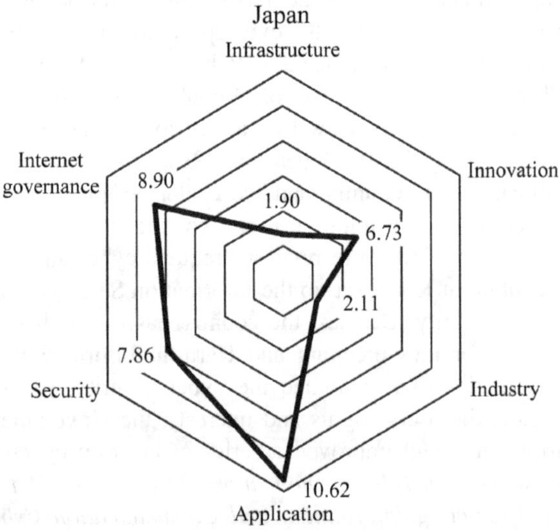

Fig. 4 Internet Development Index of Japan

industry. In recent years, Japan has maintained its advantages in robot R&D, production, application, and key parts supply, and thus its position as a leading country of robots. It enjoys more than 90% of market shares of the world not only in the field of complete appliance but also in the field of key parts of robots, for instance, precision retarding mechanism, servo, and other sensors. The government offers support in policies for the industry. In January 2015, Japan released *New Strategies for Robots* in hope of maintaining its position as the biggest country of robots and making the country into "the world's robot innovative base," "the world's largest country of robot application," and "the country leading the world into a new robot era." In January 2016, a Japanese cabinet meeting reviewed and passed the *Fifth Issue of Basic Plan for Science and Technology* (2016–2020), according to which Japan will take manufacturing as the core, flexibly use information and communication technology and build a world-leading "super-smart society" (Society 5.0) based on the Internet and Internet of things.

In terms of industry development, Japan has failed to make good use of the Internet era since there is no famous Internet company in the country, but there is a renowned Internet investor—SoftBank, who has invested in Yahoo!, Alibaba, ARM, and Boston Power as well as Indian online retailer Snapdeal, Paytm (a digital payment company), and all rivals of Uber in Southeast Asia. Last year, SoftBank acquired ARM, the British chip giant.

Japan saw early application of the Internet and the mobile Internet. In 2013, the mobile Internet there already saw its maturity, with mobile TV and mobile wallet widely used. LINE, the largest and most popular mobile phone social platform in Japan, actively explored the overseas markets after it was founded in the country. At present, LINE is mainly used in Asia, with the number of its users being maintained at 200 million. Japan used to be the leader of computer games in the world, but with the development of smartphones and the mobile Internet, industrialized and standardized European and American development has replaced closed Japanese development, but there have emerged some famous Japanese developers of mobile games. For instance, Puzzle & Dragons, a representative game developed by GungHo, has been issued in 33 countries and regions. Among Video apps, Youtube is the most popular among Japanese users.

Japan attaches importance to cyber security, taking "safeguarding the country through cyber security" as a national strategy. In June 2013, the country officially released *Japanese Cyber Security Strategy*, which proposed the objective of "establishing the strong and lively world-leading cyber space." In January 2015, it set up Internet Security Strategy Headquarters made up of cabinet members and formulated the new *Cyber Security Strategy* in May of the same year. In February 2017, the country decided to revise the *Cyber Security Strategy* to ensure the cyber security during the 2020 Tokyo Olympics and Paralympics and to set up the Coordination and Response Center for the Olympics and Paralympics as the coordinator between infrastructure operators and competent authorities.

Japan has a developed cyber administration system in terms of Internet governance, having formed a pattern with the government as the leader, and the network service providers as the core and with the involvement of multiple subjects. Specifically, the Japanese Government does not directly intervene in the Internet governance, but non-governmental organizations like the Internet Industry Association acts as the coordinator. Japanese industries are highly self-disciplined, and Internet companies there have strong sense of responsibility, so they set up different associations to formulate rules and regulations that they strictly observe. Internet users are also self-disciplined. Attaching importance to the roles that laws play in Internet government, Japan has formulated systematic and comprehensive laws. Almost all Internet problems can be solved on the basis of different laws, so that the government can have the Internet under its control with the aid of a series of relevant laws.

4. The Internet of the United Kingdom is developing steadily.

In the Internet Development Index System, the UK ranks fifth. The country has good foundation for the Internet development and attaches great importance to the sustainable development capacity, remaining one of the top countries in the world. The UK released *Digital Economy Strategy 2015–2018* in 2015 and *UK Digital Strategy* after Brexit, which shows that the British Government has always taken the Internet development as a prior social development philosophy. The country enjoys high-level development in terms of industry development (ranking third), Internet application (ranking fourth), infrastructure (ranking fifth), and Internet governance (tied for fifth place), but there is space for improvement in terms of innovation capacity (ranking tenth) and cyber security (ranking 11th) (see Fig. 5).

The UK enjoys good network infrastructure, but different regions do not develop at the same level. According to statistics from Ofcom, the competent authority of British communication, the networking rate of British families reached 86% in

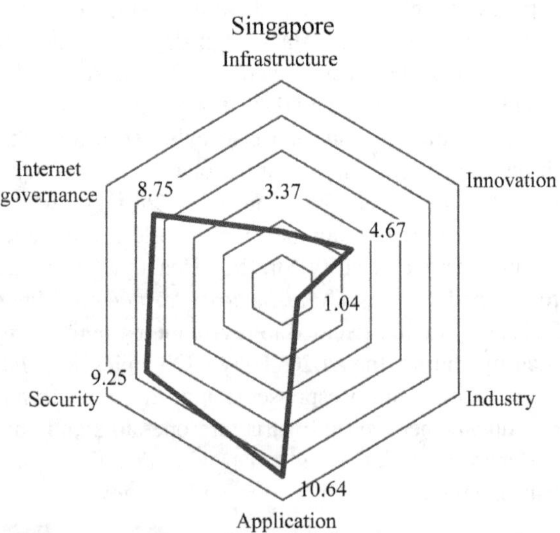

Fig. 5 Internet Development Index of the United Kingdom

2016, in which the broadband access rate was 79% and mobile Internet access rate was 66%[15].

In terms of Internet application, the UK ranks first in Europe, with 173.7 billion euros of income from e-commerce in 2016, accounting for 37.1% of the European total from e-commerce and contributing 7.16% to the national GDP, the highest in Europe[16]. The country's sharing economy, for which the British Government offers support in policies because they are determined to make the country the "Global Center for the Sharing Economy." For that purpose, the British Government has set up a new special trade body—Sharing Economy UK. The UK ranks first in the world in terms of e-government, having formulated UK Digital Strategy and taking digitalization as prior means of service. To implement the strategy, the country has established the Government Digital Service (GDS) and promoted GOV.UK as the single domain name of the government and the only platform of information release and service provision by the Central Government, relevant departments, and non-governmental organizations.

While developing the Internet, the UK stresses cyber security. The country issued *National Cyber Security Strategy 2016–2021* at the end of 2016 and invested about 1.9 billion pounds (approximately 2.3 billion US dollars) in enhancing cyber security and capacity to ensure the economic prosperity in the digital era.

Attaching importance to the domestic Internet governance, the UK is one of the first countries that have established the supervision system. The indirect administration pattern adopted by the UK in the Web content administration has become the benchmark of the international institutionalization. The procedure is as follows: The Internet Watch Foundation (IWF) filters bad or illegal information in accordance with the industrial voluntary agreements, for instance, R3 *Safety-Net Agreement*, and the government only supervises the important cyber information concerned with social stability and economic development.

5. Singapore takes the lead in the world in terms of Internet development foundation.

Singapore is among the leading countries in the world in terms of Internet development, ranking sixth of the 38 assessed countries. In terms of infrastructure and cyber security, it ranks first while in terms of industry, it ranks 18th since there are no powerful Internet giants in the country (see Fig. 6).

Singapore is among the top countries in terms of network infrastructure. Since the early 1980s, the country has been formulating and implementing the national blueprints for the Internet development, for instance, the *National Computerization Plan (1980–1985), National IT Plan (1986–1991), IT 2000, iN2015,* and *iN2025*. The Internet coverage rate in the country is high, and the mobile broadband coverage rate is as high as 142.2%. The 4G coverage rate is 82%, and Singapore ranks first with its 4G network speed as high as 46 Mbps[17].

---

[15] Ofcom's Communications Market Report 2016: Chart, Page 5.
[16] Ecommerce Foundation: United Kingdom B2C Ecommerce Report 2017—Light.
[17] ITU IDI 2016.

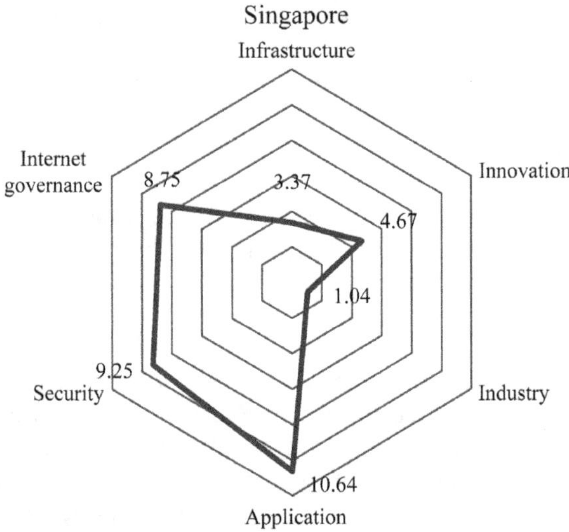

Fig. 6 Internet Development Index of Singapore

High Internet coverage rate lays a solid foundation for Singapore's Internet application. Take its e-government as an example. The country has released *e-Government Action Plan*, *Singapore Second e-Government Action Plan*, *iGov2010*, and *eGov2015* to optimize the governmental service flow and has established one-stop public service websites. It is building the cooperative government with interaction from the people and with joint innovation, and developing mobile public service. In transportation, Singapore has launched many intelligent transportation systems; in medical care, it has developed the comprehensive medical care information platform; in education, it has launched the Future Schooling Program to enhance interaction and involvement of students in their learning process.

Singapore has set up a perfect cyber security protection system, so it has strong capacity of cyber security. Since 2005, it has formulated three masterplans for information security. The first two are called *Infocomm Security Masterplan*, focusing on public areas' capacity of resisting cyber threats and construction of important information and communication infrastructure. The third is *The National Cyber Security Masterplan 2018*, providing comprehensive guarantee for the security of the environment of governmental, corporate and personal information and communication. In 2016, the country released Singapore Cybersecurity Strategy to urge all stakeholders including governmental sectors, Internet industry, scholars, and service providers to make joint efforts to combat cyber crime. To keep up with the above-mentioned planning, Singapore has established special supervision organization. For instance, it organized the National Cyber Incident Response Team (NCIRT) in 1997, Information and Communication Technology Security Administration in 2009, and Cyber Security Agency (CSA) in April 2015 to coordinate cyber security of all governmental departments. Singapore has signed with Australia, France, India, Netherlands, the UK, the USA, and Germany

Memorandums of Understanding or letters of intent to strengthen cyber security cooperation. Global Complex for Innovation, a special organization combating cyber crime set up by TERPOL, is located in Singapore.

Singapore has discovered an Internet governance pattern with the governmental intervention playing the dominating role. In 1999, it set up the Info-communications Development Authority (IDA) to make strategic planning and give political guidance for the increasingly intensive information and communication market of the country. In January 2003, it set up the Media Development Authority (MDA) in charge of cyberspace content administration. In August 2016, the two organizations merged into Info-communications Media Development Authority (IMDA), and at the same time, the Government Technology Agency (GovTech) was established. The former is in charge of information spread and content, while the latter, the construction of the "Intelligent Nation" of IDA. A series of laws formulated by Singapore, such as the *Broadcasting Act, Internal Security Act, Internet Code of Practice, Internet Behavior Act, Spam Control Act, Personal Data Protection Act, Computer Misuse and Cybersecurity Act*, and *Electronic Transactions Act*, all play regulatory roles in Internet governance. While setting up special organizations and laws and regulations, Singapore encourages its Internet companies to establish their self-disciplinary system and has formulated *Internet Code of Practice*, a resource nature self-disciplinary regulation within the industry, which defines the rules on fair competition, self-monitoring, and user service. To sum up, Singapore has a strong demand to maintain social stability by strictly regulating the Internet, so it has the control much stricter than western countries over the cyberspace content. The combination of governmental control and industrial disciplining embraces the government's will to control the Internet.

(II) Internet Development in BRICS

1. China is a latecomer but a winner in the Internet development.

China's Internet development level is only behind that of the USA, with Internet governance (ranking 4th), Internet application (ranking 2nd), industry development (ranking 2nd), and innovation capacity (ranking 4th) among the top. However, the country has to improve its infrastructure (ranking 24th) and cyber security (ranking 23rd) (see Fig. 7).

China's network infrastructure keeps being improved. With the strategy of "broadband China" and the policy of "facilitating faster and more affordable Internet connection" being implemented, the country is speeding up the coverage of optical fiber broadband, with obvious effect. The total number of IPv4 and IPv6 addresses is among the top ones of the world, and the number of optical fiber access users has amounted to 255 million, accounting for 80.2% of the national total number of fix broadband users. On the other hand, due to the vastness of the territory, especially the complex landforms in rural areas and remote areas, China is

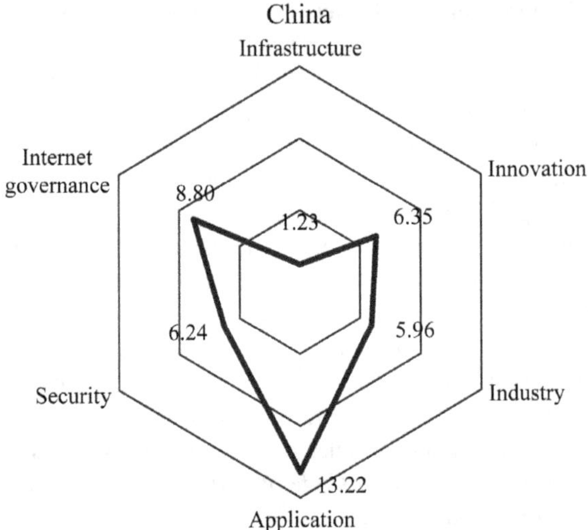

Fig. 7 Internet Development Index of China

faced with difficulties in broadband network construction and a big digital gap between areas.

China is among the top countries in innovation capacity. Especially in terms of PCT patent registration, it is far ahead of other countries, which indicates the rapid improvement of the country's innovation capacity. In recent years, the Chinese Government has been stressing the self-innovation capacity and has provided sufficient financial support for its companies' R&D.

China's Internet industrial strength has been booming, and the Internet innovation and entrepreneurship are more and more active. A large number of unicorn companies with estimated value exceeding one billion US dollars have emerged, totally 103, accounting for 40% of the world's total. Among these unicorn companies, there are seven with their respective estimated value exceeding 10 billion US dollars and a total value of 213.7 billion US dollars.

As the country with the largest population, China is witnessing the Internet becoming something common in people's daily life. E-commerce of the country is thriving, and it has become the largest market of e-commerce, the transaction volume of which accounts for 40% of the world's total. Mobile payment scale is increasing, and digital contents like online games, webcast, and online video are very popular. Besides, China's sharing economy is developing fast.

Since 2016, China's Internet governance has been centered on the improvement of the governance awareness, clarification of platform responsibility, regulation of content making, and cyber crime combating. As the Internet governance organization, the Cyberspace Administration of China launched 16 special actions in 2016, doing one after another cleaning for the cyberspace. The governance covered Web portals, search engines, Web site navigation, Weibo and WeChat, mobile terminals and cloud disks, and the content subject to governance included all illegal texts, pictures, and audio and video information. A series of policies and laws and regulations have been released, for instance, *Regulations on Internet Information Searching Service, Regulations on Mobile Internet Application Program Information Service*, and *Regulations on Webcast*, which have made cyberspace supervision standard and refined.

2. Russia is gaining momentum in its Internet development.

As a mid-level performer, Russia ranks 18th in the Internet Development Index. In sub-indexes, Russia ranks 18th in infrastructure, 23rd in innovation capacity, 31st in industry development, 21st in Internet application, eighth in cyber security, and the fourth in Internet governance together with China and Canada (see Fig. 8).

Among BRICS countries, Russia is obviously in its upward trend in Internet development. With a complete system centering on national informatization strategy supported by regulations in informatization sub-sectors, the government has introduced a host of policies and programs, including *the Russian Federation Information Society Development Strategy, Russian Federation Information Society National Program 2011–2020, Russian Federation Information Society Development Strategy 2017–2030*, to lead the country's Internet development.

In terms of infrastructure, by 2017 nearly 35,000 km of optical fiber lines have been completed within the territory of the Russian Federation, and the number of settlements that have over 10,000 inhabitants but are denied fiber access has also been reduced from 32 in 2012 to four. Mobile broadband has a much higher coverage rate than fixed broadband in Russia. In 2015, out of every 100 citizens, only 18.8 were fixed broadband users but 71.3 were active mobile broadband users. In that year, Russia had 160 mobile phones per 100 citizens, surpassing the figure of such Nordic countries as Denmark, Finland, and Sweden. Over the same period, Russia's average fee for mobile broadband (based on prepaid handheld terminals, 500 Mb) was only 0.3% of the country's per capita GNI.

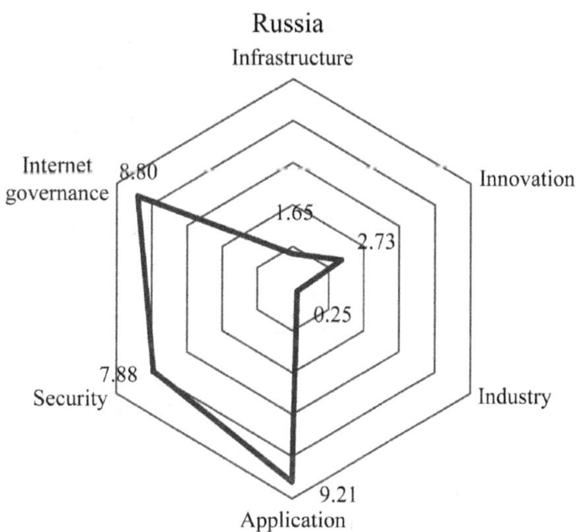

Fig. 8 Internet Development Index of Russia

In terms of R&D of information technology, the Russian Government has taken a series of measures to revitalize its IT R&D capabilities since 2007. For instance, it has funded the Elbrus microprocessor, Russian equivalent of the Windows operating system, RoMOS mobile operating system, Rupad Tablet PC, and other new products. To achieve technological breakthroughs and seize the high ground of international R&D competitions, in October 2012, the Russian Government approved the establishment of the Future Research Fund, analogous to America's Defense Advanced Research Projects Agency, to promote promising military research and development projects that are selected through competitions. The Strategy for Developing the Information Technology Sector in the Russian Federation in 2014–2020 and until 2025, approved in November 2013, singled out six long-term R&D priorities in the IT sector in the next 10–15 years and a set of short-term research priorities. The report *Russia 2030: Science and Technology Foresight*, approved in January 2014, listed seven priority fields and their expected results.

Backed by the government, export of Russia's information products and services is growing rapidly. The *Development of the Information Technology Industry Action Plan* (Roadmap) approved in July 2013 set the goal of doubling the volume of exported hi-tech products and services by 2018 and corresponding measures to achieve it. To promote the market share of local software industry and replace key systems with homegrown ones, in June 2015, Russian President Putin signed the bill to launch the Russian Computer and Database Program and instructed the Ministry of Communications and Mass Media to register domestic software after expert reviews. In January 2017, Medvedev, Prime Minister of Russia, signed a government decree about creation of Fund of Research of Information Technology (FRIT). Russian enterprises, after passing the relevant financial audit, can obtain preferential loan support from the fund for exporting products and homegrown replacement products.

In terms of cyber security and regulation, on the basis of its own conditions, Russia has created a unique theoretical system of "information security" in close connection with the country's sovereignty. "Information space" can be divided into three levels: the hardware, the software, and the human, and cyber space concerns the first two levels only. The *Federal Law on Security of Critical Russian Federation Network Infrastructure* signed in July 2017 sets out the basic principles for ensuring security of Russia's critical network infrastructure and defines the powers of state bodies for relevant matters. The Yarovaya Law passed in July 2016 requires telecom providers to store (the content of data shall be stored for 6 months, and the metadata on them for 1 year) and disclose these data and metadata to authorities on request. In July 2017, VPN Law was promulgated to prohibit VPN services in Russia so as to block Internet users from visiting banned Web sites.

3. Internet applications are increasingly active in Brazil.

As the largest and the most populous country in South America, Brazil ranks 23rd in the Internet Development Index. In sub-indexes, Brazil ranks 23rd in infrastructure, 24th in innovation capacity, 22nd in industry development, 23rd in

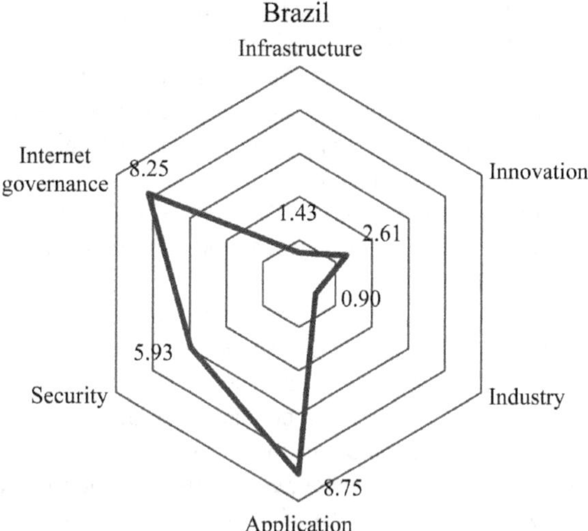

Fig. 9 Internet Development Index of Brazil

application, 25th in cyber security, and mid level in Internet governance (see Fig. 9).

Brazil's fiber optic network coverage is not high, but the mobile Internet coverage is as high as 100%. To boost its broadband coverage, the country launched the National Network of Education and Research (RNP) dedicated to providing high-speed Internet connections for research and higher education communities as early as 1989. By the end of 2015, the Network, which runs 22,000 km, had been able to provide Internet services to 1219 research, medical, and higher education institutions and 3.5 million users. To further reduce the cost of Internet access and increase public Internet usage, the Brazilian Government launched *National Broadband Plan* (PNBL) in 2010. In the following 5 years, Brazil spent BRL 12.8 billion (about RMB 43.7 billion yuan) in expanding broadband network coverage and reduced its monthly fee to BRL 15–35. Brazil has also launched the Special Taxation Regime of the Broadband National Program (REPNBL) to provide tax relief to buyers of certain products and services in the hope of stimulating private sector's investment in the network.

On the strength of Brazil's favorable Internet conditions, social media thrive in Brazil. Social media users account for 87.6% of total Internet users and 92% of young people. On average, they spent 3.3 h daily on social media and most of them are users of Youtube, Facebook, and WhatsApp. Brazil hence is referred to as "the Social Media Capital of the Universe" by the *Wall Street Journal* and "the Future of Social Media" by the *Forbes* magazine.

The Brazilian Government supports the development of the information industry, with particular emphasis on the development of the export of software. Thanks to the promulgation of *Growth Promotion Plan*, *National Broadband Plan*, and *Plan of Boosting Information Industry* and a set of tax relief policies, Brazil's software export is growing by leaps and bounds. Today, the country has grown to be Latin America's largest and the world's fifth largest e-commerce market. Statistics from Internet Retailer show that out of the top 500 e-commerce companies, 299 are from Brazil. According to statista.com, a world-renowned data Web site, Brazil's e-commerce market reached 16.5 billion US dollars in revenue in 2016 and is expected to grow at an average annual rate of 11.9% (CAGR 2016–2021) to 29 billion US dollars by 2021[18]. In tune with the rapid development of its e-commerce, the country is also strong in online payment. Its online payment ratio has been increasing year by year and its mobile payment ratio jumped from 0.3% in 2011 to 21.5% in 2016[19].

Internet governance in Brazil dates back to the 1990s, and therefore, the country is an early actor among the developing countries. In December 2008, the National Defense Strategy (NDS) was issued. The Strategy singled out Internet technology as strategic to the Brazilian defense and called for the establishment of an institution dedicated to strengthening industrial and military Internet capabilities to achieve technology self-sufficiency, especially information technology used in submarines and weapon systems. Brazil plans to master key technologies through capacity building in educational and military communities, and enhances intercommunication among forces. The Institutional Security Cabinet of Brazil is responsible for public cyber security. In 2010, the Cabinet created the Cyber Defense Center that would coordinate actions to protect the country's vital military, government, and information infrastructure. In 2012, the center received 4.5 million US dollars from the government for its activities. As part of the Cyber Defense Center, the Integrated Electronic Warfare Center (CIGE as its Portuguese acronym) has purchased Cyber Operations Simulator (SIMOC as its Portuguese acronym) to conduct training based on real scenarios of cyber catastrophes. The Integrated Electronic Warfare Center is headed by a Lieutenant General to thwart attacks to Brazil's Army network. Disclosure of PRISM in 2013 led Brazil to openly denounce the USA on several occasions, even at the UN General Assembly. *The Brazilian Civil Rights Framework for the Internet* was approved by the Brazilian Congress in March 2014, marking a new stage of Internet development and governance of the country.

4. India performs exceptionally well in Internet industry development.

India ranks 25th in the Internet Development Index. In sub-indexes, India ranks 36th in infrastructure, 18th in innovation capacity, 5th in industry development, 27th in application, 19th in cyber security, and relatively lagged behind in Internet governance (see Fig. 10).

India's infrastructure still has much room for improvement. According to its real-time Internet monitor, the country had 460 million Internet users in 2016, second only to China. As the second largest Internet market in the world, India, however, has only the Internet coverage rate of merely 34.8%. For cultural reasons,

---

[18] http://www.ccpit.org/Contents/Channel_4114/2017/0824/866119/content_866119.htm.
[19] https://www.baijingapp.com/article/11744.

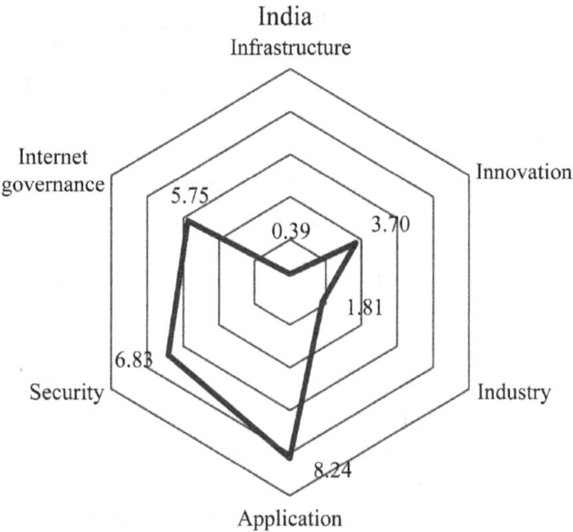

Fig. 10 Internet Development Index of India

it has far less female Internet users. According to data from Internet and Mobile Association of India, only about 30% of Internet users are female.

Over the years, India has focused on information technology outsourcing in the development of the information industry. Its software outsourcing industry takes the lead in the world and has the largest IT outsourcing market in the world. Since 2012, the Indian Government has formulated a series of strategies and plans to make up for its shortcomings in the Internet industry, such as *India's Five-Year Plan for Telecommunication Industry* (2012–2017), *National Optical Fibre Network*, and *Digital India*. The initiative of "Make in India" launched by Prime Minister Narendra Modi has included a number of policies to encourage and protect foreign investment and improve India's manufacturing capabilities.

India's large population forms a strong user base for Internet applications. E-commerce, cab hailing, and online food delivery services are growing by leaps and bounds. With the rapid development of mobile payment, telecommunication and e-commerce enterprises are pouring into India's blue ocean of mobile payment. Waves of entrepreneurship spawned by mobile technology are gaining momentum. Mobile apps in the country are predominantly the products of the USA, India, and China.

In terms of cyber security, India surpassed the USA in 2012 to become the world's largest spam source. Indian governmental agencies and financial centers are major targets of hackers. In recent years, the government has taken a series of measures to strengthen cyber security regulation. In 2000, they promulgated the *Information Technology Act*. In 2011, they introduced the National Cyber Security Strategy (Draft) to boost the development of homegrown IT products and reduce possible threats posed by imported high-tech products to national security. In 2013, the Ministry of Communications and Information Technology released the *National Cyber Security Policy*, which set out the goals and action plans for the development of cyber security in India in the next 5 years in an attempt to establish an overall network security framework. In order to establish a cyber security system under its administration, the Indian Government has set up the National Cyber Coordination Centre, which is given unlimited access to all online accounts.

In terms of Internet governance, *India's Information Technology Act* provides legal framework for Internet application, e-commerce, and cyber security, with recognition of the validity of electronic records and digital signatures; defines cyber crimes and prescribed penalties for them; authorizes the government to establish a Cyber Appellate Tribunal; requires Internet companies to remove content that is defamatory, hateful, unsuitable for children or that infringes copyright within 36 h after being informed by the authorities; authorizes the government to shut down illegal Web sites and filter illegal content; and penalizes posting of provocative or offensive content. In general, the Indian Government exercises great authority over Internet content and has the power of interception or monitoring or decryption of any information. Any person who is deemed to have committed a crime will be sentenced to 7 years of imprisonment. In addition, all licensed ISPs are required to sign agreements that authorize the government to obtain users' data.

5. South Africa is the leader of Africa in Internet development.

South Africa is one of the most developed countries of Africa. In the Internet Development Index, it ranks 34th, with its infrastructure ranking 31st, innovation capacity 31st, industry development 27th, Internet application 33rd, and cyber security 33rd (see Fig. 11).

The penetration of Internet in South Africa is higher than in most countries of Africa. The number of Internet users increased from 2.4 million in 2000 to 21 million in 2016, which resulted from the improvement of network infrastructure and loading capacity. In 2011, South Africa released its broadband strategy, according

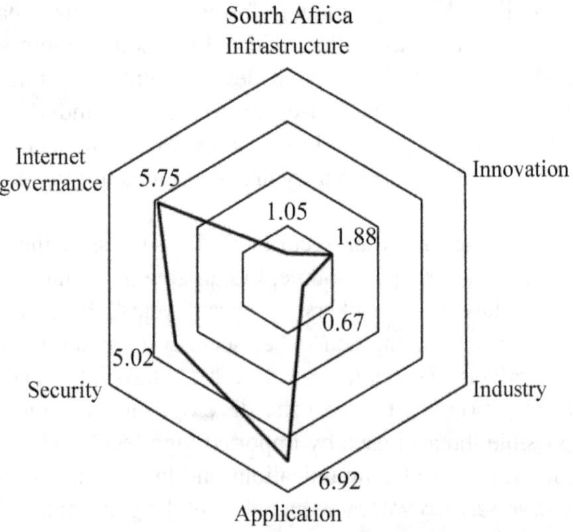

Fig. 11 Internet Development Index of South Africa

to which the country would appropriate ZAR 450 million for the broadband infrastructure and service in areas with poor Internet service and in rural areas. In 2013, the country released a new strategy with the theme "South Africa Connection," according to which a series of actions and projects would be launched to improve the broadband of the country and to change the broadband speed. The strategy also proposed that there would be a review every year.

With the construction of the network, there has been increasing Internet application in South Africa, in which contact, communication, and social networks are more frequently seen. Besides, e-commerce is increasing at a speed of 10% every year. Online entertainment and media are also developing fast.

South Africa has no eminent Internet company, so the country attaches importance to its cooperation with Google, Microsoft, and Huawei. Microsoft used to launch the initiative "White spaces," in which the company provided low-cost wireless Internet access service for five schools in Limpopo of South Africa by means of white TV spaces and solar energy base stations. BroadReach and United States Agency for International Development (USAID) jointly launched the "90-90-90 Challenge Project" to help South Africa to prevent AIDS from expanding, aiming to help 90% of HIV carriers to know about their health, to help 90% of AIDS-infected people to receive antiretroviral treatment, and to help 90% of the AIDS patients receiving treatment to reduce the effect of the virus. Google has also launched a series of projects. For instance, it released the indoor maps of airports, department stores, casinos, conference halls, and universities of South Africa. The Government of South Africa has also cooperated with NVIDIA to provide 28,000 schools and over 12 million students with computers and Internet access service, so that every seven students can share one computer. The project has till now covered 105 schools in North West of South Africa.

The present status of cyber security in South Africa is not consistent with its Internet development, since there are frequent cyber attacks there. Most enterprises in South Africa, especially banks, are faced with serious threats from cyber crime. In recent years, with most enterprises and financial institutions strengthening their cyber security facilities construction, cyber criminals are shifting their focus to the weakest in the cyber links—Internet users. In its cyberspace governance, South Africa has neglected the legislation on cyber security, which is an important reason why the country is a place with frequent cyber law breaches and cyber crimes. To deal with the cyber security problems, the country has formulated *E-communication and Transaction Law*, *Cyber Security Policy Framework*, and *Cyber Crime and Cyber Security Act* to discover a practical pattern of cyber governance. Besides policies and laws, the Government of South Africa has established South African Computer Security Incident Response Team (ECS-CSIRT) to discover, prevent, tackle, and manage cyber threats and thus to enhance the domestic cyber security.

IV. Trend of the World Internet Development and Suggestions on Policies in this Regard

Generally, the explosive expansion of the Global Mobile Internet is coming to an end and the growth of the global Internet connection scale is coming into the stage of power conversion. The Internet development is transferring from "person-to-person connection" to "Internet of everything." Artificial intelligence, block chain, and quantum communication are rising; Internet of things, cloud computing, and big data are developing fast, and the 5G era is drawing near. The benefits brought about by the digital era keeps emerging, the four factors boosting the global Internet evolution and innovation, namely capital, technology, data, and demand, are evolving, and there is more serious conflict between new and old systems. When we look into the future, we suggest that all the countries and the international community should take the well-being of the humankind as the fundamental consideration and contribute to the building of a community of shared future in cyberspace. We should make the global cyberspace governance more equitable and reasonable and strive to realize the goal of building the cyberspace with more equality, respect, innovation, openness and sharing, security and order.

1. We should promote innovations and creations in the field of the Internet and realize common development of all countries.

The Internet plays a leading role in promoting innovation-driven development. It is an important option for all countries to realize better economic development in the post-financial crisis era. All countries should encourage and support all kinds of innovations and creations based on the Internet, speed up the development of the new-generation information technology such as cloud computing, Internet of things, and artificial intelligence, as well as the cultivation of new technology, new application, and new type of business of the Internet. We should also accelerate the in-depth integration of the Internet and the real economy and promote the digitalization, network transformation, and intelligence transformation of traditional industries. All countries should deepen their cooperation in technical R&D, cross-border e-commerce, and innovation of small- and medium-sized enterprises on the principle of opening and coordination and thus create more interest convergences and new cooperation highlights in terms of cyberspace.

2. We should speed up cyber coverage to better benefit all the peoples.

The essence of the Internet is the interconnection, and the value of information lies in intercommunication. There remains a big digital gap across the world. Statistics released by the UN show that least developed countries will have had only 17.5% of Internet coverage by the end of 2017. With the acceleration of informatization, the Internet has become the big artery of economic and social development and the indispensable new type of infrastructure of modern society. The international community has the responsibility for boosting universally beneficial development of information technology worldwide and provides fund, technology, and talent support for developing countries and least-developed countries in enhancing their Internet development capacity and thus creates conditions for eradicating poverty and promoting common development. All countries should strengthen strategic link between countries and between regions, and their communication and cooperation in distant medical care, online education, e-government, and smart city building. We should try to discover new means of and approaches to making the Internet more convenient and beneficial to the people, so that the Internet development results will better benefit all nations.

3. We should expand humanity communication through cyberspace and promote exchanges between civilizations.

The Internet is an important carrier of excellent cultures and a tool of positive energy spread. By March 2017, the number of Internet users speaking top ten languages in the world had reached 2.9 billion, accounting for 78% of the total number of Internet users of the world, while the number of Internet users using other languages only accounted for 22%[20]. All countries should work together to construct cultural exchange platforms on the Internet to show the diversity of civilizations and to seek ways for common development. We should give full play to the role of the Internet in disseminating civilization, promote digitalization and network communication of excellent cultural products, and spread positive things like justice, kindness, liveliness, and goodness in the cyberspace. We should enhance the cyber ethic construction and Web site opening and visiting in good manners and thus reinforce the protection of juveniles.

4. We should face and handle together the challenges to cyber security and safeguard the order of the cyberspace.

To safeguard cyber security is the common responsibility of the international community, so all countries should have common and sustainable cyberspace philosophy of mutual trust and cooperation. We should not allow the situation to occur in which one country is safe while another is not. No country is allowed to seek its own security by sacrificing another country's security. All countries should strengthen communication and negotiation concerning cyber security and strategic mutual trust in this field. We should establish the normalized emergency responding mechanism and deepen cooperation in technical R&D, rule and regulation making, information sharing and talents cultivation, enhance key network infrastructure security protection, improve the awareness of protection and guarantee competence in global cyber security, and jointly tackle threats to it. We should enhance the punishment of cyber crimes, such as cyber attacks and cyber terrorism, protect intellectual property rights, personal privacy, national security and public interests, and thus jointly build peaceful and safe cyberspace.

---

[20] Data source: internetworldstats-www.internetworldstats.com/stats7.htm.

5. We should formulate complete cyberspace governance rules and promote the reform of the global governance system.

Cyberspace is the space of activity for all human beings, and its future should be controlled by all countries together. At present, the global Internet development governance is at a key turning point, so it requires all counties to be visionary, to treat one another equally and promote equity and justice. We should promote the establishment of a multilateral, democratic, and transparent global Internet governance system, taking respect for cyberspace sovereignty as a basic principle, safeguarding the equal right to develop, participate, and govern, reforming any unjust or unreasonable arrangement in the global Internet governance system and improving emerging market countries' and developing countries' right to represent and speak. No cyberspace hegemony is allowed, neither is single domination or only a few countries' decision. The governments, international organizations, Internet businesses, technical communities, private sectors, and individuals should all play their role through effective and constructive cooperation to build a community of shared future in cyberspace. We should work together to build a safer and healthier global cyberspace which is more peaceful, open, cooperative, and prosperous.

Chapter 1
Development of World Internet

Chinese Academy of Cyberspace Studies

Abstract Looking at the history of world civilizations, humanity has progressed along with agricultural, industrial and information revolutions. Each of these industrial and technological revolutions has had great and profound impact on our way of production and life. Today, information technologies represented by the Internet are experiencing rapid changes with each passing day. They have become the leading power for the innovation-driven economic and social development. The world Internet development history is a cross-language, cross-national and cross-cultural history, a history in which all countries make joint effort and work together to build a community of shared future in cyberspace, and a history the integration of information technology innovation with industries. It is an important part of the history of world civilizations. Comprehensively considering technical innovation, industrial development, number of Internet users and situation change, we can divide the world Internet development into five stages: the birth of the Internet (1960s–1992); expansion and application of the Internet (1993–2000); popularization of the Internet (2000–2007); mobile Internet (2008–2016); Internet of Everything (2016–present).

1.1 The First Stage: Birth of the Internet

The Internet is not a coincident invention in social development, but the inevitable achievement in scientific and technical development. It was not invented from nothing, but had been developed for hundreds of thousands of years. The Babbage Computer Machine, telegraph, Turing Machine, and Information Theory were all landmark achievements in the history of the Internet development, playing the role of theoretical guidance and technical foundation. They can all be traced in today's Internet.

In the Cold War period of the 20th century, the competition between the United States and the Soviet Union was not only reflected in military field, but also in the technical field. On October 4th, 1957, the Soviet Union launched the first

Chinese Academy of Cyberspace Studies
Beijing, China

Chinese Academy of Cyberspace Studies (ed.), *World Internet Development Report 2017*,
https://doi.org/10.1007/978-3-662-57524-6_1

man-made satellite Sputnik, which shocked the United States and in turn urged the latter to establish the Advanced Research Projects Agency (ARPA)[1] in the Department of National Defense. ARPA appropriated fund for basic research of universities and businesses to reestablish America's leadership in military technology application and development. The investment of the Department of National Defense had immense returns, a series of core technologies playing important roles in American technical innovation and economic prosperity, including ARPANET, the early form of the Internet.

1.1.1 Birth of ARPANET

1. **Hypothesis and proposal of the theory**

ARPANET was originally aimed to provide researchers with a network for time-sharing use of computer resources. In 1962, JCR Licklider, a computer scientist, became the first director of ARPA Information Processing Technology Office and hence the first man in charge of the computer research project. In April 1963, Licklider formed the earliest thought about the network or inter-network system —"Intergalactic Computer Network", i.e. constructing a communication network between computer users so that every user could access to the data and programs of any website through the computers. Packet Switching Network provided solutions for Licklider's Intergalactic Computer Network. In July 1961, Leonard Kleinrock from MIT published his paper in which he put forward the concept "Packet Switching Network" for the first time, i.e. transmitting data packet containing source and purpose from one position to another. In the same period, Paul Baran from NPL (the National Physical Laboratory) also put forward the concept "Packet switching". All these theories and tests laid foundation for the later construction of ARPANET.

2. **Launch of ARPANET**

In February 1966, Licklider's successor Bob Taylor successfully persuaded ARPA to provide fund for ARPANET. At the end of that year, Larry Roberts proposed the concept "computer network" and became in January 2017 the project manager of ARPA Information Processing Technology Office. Then he officially launched the ARPANET program. ARPANET's physical structure was started in 1968. After Roberts completed the general structure and standards of ARPANET, BW Company was authorized by ARPA to construct the network. Finally, in October 1969, ARPANET came into being, with the first four nodes being UCLA, The Augmentation Research Center at Stanford Research Institute, UCSB and The University of Utah School of Computing, at the speed of 2.4 kbps. ARPANET was not an Internet, but initial cyberspace. In the following years, there were other Packet Switching networks that came into being, such as Ethernet by Xerox PARC

---

[1]The acronym was modified for times into DARPA at last.

(Xerox Palo Alto Research Center) and DECNET by Digital Equipment Corporation and the SATNET between the United States and Europe (Norway). But these networks were not compatible or inter-connective.

3. Completion of TCP/IP symbolizing the birth of the Internet

After the ARPANET was completed, computers were soon connected into the network. ARPA began to develop hot-to-host protocol with complete functions to connect all systems. In December 1970, the Network Working Group (NGG) under the leadership of Steve Crocker completed the original ARPANET host Protocol, i.e. Network Control Program (NCP), which, however, was only applicable to special computers. Therefore, it was necessary to develop a new and more inclusive open-source protocol network. From 1973 to 1974, Robert E. Kahn, who had assisted in developing IMP at BBN, and Vinton Cerf, who had got involved in protocol development, developed TCP/IP (Transmission Control Protocol/Internet Protocol), making specific stipulations for how the data packet passed through the network and reached the destination. It became the standard protocol of ARPANET in 1983, marking the birth of the Internet.

1.1.2   Early Applications of the Internet

Birth of the Internet made people aware of the convenience, quickness and economy of the Internet in information exchange. The field of research was the first to realize the importance of making use of the Internet. In 1970s, ARPANET was expanded into universities and governmental organizations in the east of the United States and into Europe. The number of nodes rose to 23 in 1972. Not only was APRANET itself experiencing rapid changes, but also a number of innovations and applications were developed on the trunk. The applications in that period are as follows:

(1) E-mail

E-mail is regarded as a killer application. It was one of the key innovations in 1970s promoting the Internet development. After the birth of ARPANET, e-mail sending between different computer systems was continuously tested. In 1972, the BBN engineer Ray Tomlinson successfully developed the local user e-mail program SNSMSG (Send Message), which contained the experimental document transmission program, and chose "@" as the interval marking of the user name and address. The first cross-computer e-mail was sent on the ARPANET. With the popularization of the ARPANET, the service software and standard concerned with e-mail (Simple Mail Transfer Protocol, SMTP) emerged and the number of email users began to rise dramatically. By 1973, e-mail traffic accounted for 75% of the total traffic of ARPANET. E-mail not only expanded the use of early ARPANET, but also became a new communication model.

(2) Bulletin board and newsgroup

In 1978, two members of the Computer Communication Society of Chicago set up the first Bulletin Board System (BBS) in which online forums could be established and messages could be sent. It could also provide files and software. Any user could join the system. In 1979, based on the Unix-to-Unix Copy (UUCP), students from Duke University and University of North Carolina established the Usenet System with personal computers as the host ones and with the "newsgroup", i.e. a topic discussion forum, in it. In the "newsgroup", messages could be sent and answered. Usenet was an important part of the online community. By 1984, Usenet terminals were used in all universities and research institutes of the United States.

(3) Information retrieval

With the increase of information on the Internet and the standardization of FTP and Telnet, information retrieval application became possible and necessary. In 1989, students from McGill University of Montreal tried for the first time to set up the FTP site documentation "Archie" to search the Internet. Through the software, known open FTP websites could be periodically logged in, the documents could be listed and a searchable software index could be set up. In 1991, University of Minnesota developed a more friendly Internet menu search system named Gopher, which could help the user to search online and distribute files through the navigation based on the menu. The user only needed to input or click a number to choose the desirable menu. Gopher gained such popularity that there emerged over 10,000 Gopher systems worldwide.

1.1.3 *Establishment of NSFNET and Infancy of Internet Governance*

ARPANET was originally set up by the government for military purpose and then extensively used by computer research staff. From it, people realized the value of the Internet and the civilian demand kept increasing. In 1980s, the National Science Foundation Network (NSFNET) was established, pushing the Internet to be opened for civilian use. By 1989, 200,000 computers had been connected into the Internet.

1. Establishment of NSFNET

As PCs became cheaper and more popular, more people wished to access to the Internet, so the number of ARPANET users began to grow exponentially year by year. As more and more people from non-military fields began to use it, ARPANET was divided in September 1984 into two independent parts, namely, MILNET, used for military communication, and ARPANET, used for non-military purpose. In 1981, National Science Foundation of the United States financially supported the establishment of Computer Science Network (CSNET), the original form of NSFNET, which enabled computer departments of academic and research institutes limited by fund or authorization to access to ARPANET. CSNET was later connected with the computer science departments of Israel, Australia, Canada, France, Germany, Republic of Korea and Japan and finally connected with more than 180 organizations worldwide. In 1985, under the leadership of Dennis Jennings, the National Science Foundation set up NSFNET, which connected five supercomputer centers with National Center for Atmosphere Research (NCAR) of the United States to form a public research network. It chose to support the TCP/IP protocol and the existing basic facilities of ARPANET and developed into a backbone network in the U.S. Internet. Due to the rise of NSFNET, ARPANET was closed in 1990. The greatest contribution of NSFNET was to make the Internet available to the public, not just to research staff and governmental agencies. While the United States was developing NSFNET, other countries began to construct their wide area network (WAN). In Europe, EBONE, a pan-Europe backbone network, was launched in 1992, when UUCP was connected into the United Kingdom, Holland, Denmark and Sweden. In 1994, China Science and Technology Network was first connected into the Internet.

2. Early Internet governance

In the period of ARPANET, it was necessary to establish mapping relationship between the host computer name and Internet address when the user browsed the Internet for resources, but since there were not many host machines and users, the Stanford Research Institute Network Information Center (SRI-NIC) sponsored by the U.S. Government maintained the mapping relationship manually, but the rapid growth and diversity of host machine environment made the operation difficult. In 1984, Jonathan B. Postel and Joyce Reynolds co-designed DNS, which is still used today. The establishment of DNS made the Internet more unified and convenient. It became no longer necessary to remember the meaningless figures, but the website name or the name of the server with which the computer was connected. Postel was in charge of the management and allocation of the core resources in the early period of the Internet. In 1987, he set up seven root servers (later 13 such servers), all located in the United States. In 1988, the U.S. Government asked Postel to adopt more secure and reasonable measures to ensure the management and allocation of core resources of the Internet, so Postel set up Internet Assigned Numbers Authority (IANA), which was responsible for the allocation of Internet resources.

3. Founding of network technology communities

Network technology communities came into being with the birth of the Internet. Research staff of the ARPANET in the early period did their work through the communities. All the ARPA computer research projects were projects of collaboration of more than one contractor through all kinds of mechanism. For each project there was a working unit, which developed into a working unit of the Internet. In 1979, Vinton G. Cerf and other ARPANET research people set up an informal committee—Internet Configuration and Control Committee, which was restructured into the Internet Activities Board (IAB), later the Internet Architecture Board, which developed basic Internet protocols still used today. In 1986, the Internet Engineering Task Force (IETF) was established. After that, IAB made plans for the whole architecture and long-term development of the Internet and supervised the formulation of the Internet standards. IETF is an open standard-formulating organization, in charge of the development of the draft of Internet protocols. There was no formal member in IETF, and all the participants had no pay. They did their work in mailing list work groups and informal discussion work groups, and decided on whether the standards could be adopted on the principle of the "consensus and effective code". The file recording the standards was called RFC (request for comments). Actually, RFC occurred earlier than IETF's founding. RFC1, the earliest RFC, was the technical specification compiled by Stephen D. Crocker in 1969 and used to connect early ARPANET nodes. All RFC files could be obtained online and respective design and operation information could be shared through RFC files. RFC was the first application in the field of Internet information sharing, contributing to the later development of the Internet.

1.1.4 Origin and Development of Internet Culture

Early Internet culture originated from hacker subculture of the United States. Hackers originally referred to those who were fond of and good at computer science, programming and designing, coming from Tech Model Railroad Club (TMRC), the first hacker community, founded in MIT in 1959. It created a series of hacker terms producing impact on the whole Internet culture. Another important Internet culture group was Homebrew Computer Club formed in 1975 in the Menlo Park Garage of California. The club was the practitioner of new hacker ethics. In late 1970s, Internet culture developed from a technical community (in a narrow sense) into a subculture with extensive and diverse computer users. This subculture was called "Cyberpunk". One of the key figures was Stewart Brand, who connected counterculture with technical culture in a perfect way and introduced elite culture and popular culture into network community culture. Thus a totally new mixture of cultures was formed.[2] As early as in 1968, the *Whole Earth Catalog* founded by Brand was a biblical book of counterculture of the United States, which directly enlightened a new generation of IT stars represented by Steve Jobs and helped to form the basic values and spirit of the Silicon Valley. In 1985, Brand founded The Whole Earth 'Lectronic Link (WELL), the earliest network community, which attracted the most active Internet culture elites including some hackers.

---

[2]Fred Turner, *From Counterculture to Cyberculture: Stewart Brand, the Whole Earth Network, and the Rise of Digital Utopianism*, trans. Zhang Xingzhou, et al., Beijing: Publishing House of Electronics Industry, 2013.

1.2   The Second Stage: Expansion and Application of the Internet

Since the end of 1980s, the Internet has developed from a tool only used by research and technology staff into a necessity in common people's life. The birth of the World Wide Web (www) has promoted the expansion and application of the Internet, and a large number of technical innovations have stimulated the industrial growth. Social and institutional innovations concerning the Internet then emerged, boosting the investment in and the technical development of the Internet.

1.2.1   Birth of the World Wide Web

The World Wide Web (www), one of the most influential inventions in the history of the Internet, was launched in early 1990s. Tim Berners-Lee, a UK engineer from the European Organization for Nuclear Research (CERN), discovered that the combination of the hypertext and the Internet could help to establish the global connection. By the end of 1990, Berners-Lee had set up a whole set of tools to put his imagination into reality, including Hypertext Transfer Protocol (HTTP), Hypertext Markup Language (HTML), the first application software of the basic browser, the first server software and the first web page describing the project itself. In 1991, www was soon spread among other European research centers. In August of the same year, Berners-Lee published the www project introduction to seek for cooperators, marking that www was opened to a wider Internet world.

1.2.2   Application of the Internet and Rapid Development of Network Economy

Since it was originally sponsored by the government, the Internet was at the beginning used in research, education and government affairs. The commercial use was forbidden. This policy lasted till early 1990s, when independent commercial networks began to develop. In May 1995, the National Science Foundation of the United States stopped its financial support for the backbone networks of the Internet and cancelled the limit to the commercial use of the Internet, ushering in the development upsurge of the Internet and diversified application of browsers, search engines, Internet service provision, instant messaging, e-commerce and online subscription. The computer and Internet have been a key drive for the world's economic growth ever since.

1. Browsers

In 1993, Marc Andreessen and Eric Bina launched Mosaic website browser, the first widely-used network graphic browser. In 1994, Andreessen founded Mosaic Communications Corporation, later renamed Netscape Communications Corporation, which launched the first commercial network browser—Netscape Navigator (later renamed Netscape), which occupied 90% of the market within one year. Netscape removed technical barriers in Internet use, making Internet available to everyone. The browser thus became a key power promoting the development of www. In August 1995, Netscape was listed with NASDAQ at US$28 per share and on the very first day its market value amounted to US$ 2.7 billion, causing great sensation and a wave of entrepreneurship in the field of the Internet. In the same year, Microsoft bought the technology of Spyglass and shortly after that released Internet Explorer (IE), causing a new round of browser wars. Microsoft made IE the successful dominator of the browser market by bundling its software with its windows OS, with IE market occupation rate exceeding 95% in 2002, while Netscape was acquired by America Online (AOL) in 1998.[3]

2. Search engines

Yahoo! was the first search engine in the history of the Internet. In October 1994, there were already 3.864 million websites,[4] which led to the emergence of search engines. In 1995, Yang Zhiyuan and David Filo, two students from Stanford University, launched the website Yahoo!, on which the existing websites were catalogued in accordance with the pre-classification, which improved the practicability of the Internet. In 1998, Google launched by Larry Page and Sergey Brin took over the dominating position in search engines. It adopted PageRank Algorithm to mark the relevance or importance of a website, so it became a popular search engine. Besides website search, Google began to provide other services, such as image, newsgroup, map and video searching, and soon became the largest and most influential search engine. The company also became one of the largest ones of the world.

3. Instant messaging

The release of ICQ software marked the beginning of instant messaging. In 1996, Mirabilis of Israel released the original ICQ software, which became popular soon. In 1998, the number of its users reached 12 million[5] and ICQ became an alternative to the telephone. From the rapid growth of ICQ, Yahoo! and Microsoft, the two leaders in the field of the Internet, saw commercial opportunities, so they

---

[3]Ian Peter. *History of the Internet–the Browser wars*, http://www.nethistory.info/History%20of% 20the%20Internet/browserwars.html.

[4]Arun Rao & Piero Scarruffi, *A History of Silicon Valley: the Greatest Creation*, trans. Yan Jingli and Hou Aihua, Beijing: Post & Telecom Press, 2014.

[5]*ICQ Surpasses 50 Million Registered Users*. TimeWarner, Dec 1, 1999. http://www.timewarner. com/newsroom/press-releases/1999/12/01/icq-surpasses-50-million-registered-users.

respectively launched Yahoo! Messenger, MSN Messenger and MS Chat. The market of instant messaging began to grow, and its function was extended from instant text chatting to file transmission, audio/video chatting, game playing and online friends making.

4. E-commerce

The founding of eBay marked the beginning of the fundamental revolution in e-commerce. In 1995, Pierre Omidyar founded the auction website Auctionweb, which was renamed eBay in 1997, aimed to make transactions successful without the necessity for strangers to meet. By 1998, the total auction volume of eBay had reached US$740 million.[6] In 1995, Jeff Bezos, former manager of Hedge Fund in Wall Street, founded Amazon.com, the largest bookstore in the world. The company began as an Internet merchant of books and later sold a wide variety of products and services. Today, it is the world's largest Internet retailer.

1.2.3 Rudiments of Internet Governance Systems

With the commercialization and internationalization of the Internet, the exponential growth of the number of users and commercial activities inspired people to pay more attention to standards and domain name management systems of the Internet, which, in turn, led to the establishment of Internet Society (ISOC), World Wide Web Consortium (W3C) and Internet Corporation for Assigned Names and Numbers (ICANN), marking the rudiments of Internet governance systems with technical governance as the core.

1. Founding of the ISOC

In 1992, ISOC was founded under the leadership of Robert E. Kahn and Vint Cerf to undertake the social responsibilities and legal action related to the Internet standards resulting from the Internet development. ISOC, IETF and IAB constitute an umbrella structure in the field of the Internet. It is "governed by a diverse Board of Trustees that is dedicated to ensuring that the Internet stays open, transparent and defined by you," boosting the policies, technologies and standards of the Internet as well as its future development.[7] Another Internet standard formulating organization is W3C, founded in October 1994 by Tim Berners-Lee from MIT. It has the development of protocols and www as its mission, which is similar with that of IETF, i.e. coordinating the production and sharing of the Internet. W3C has formed a cooperative and supportive relationship with ISOC, IETF and IAB.

---

[6]Arun Rao & Piero Scarruffi, *A History of Silicon Valley: the Greatest Creation*, trans. Yan Jingli and Hou Aihua, Beijing: Post & Telecom Press, 2014.

[7]ISOC. *About Internet Society*, https://www.internetsociety.org/about-internet-society.

2. **Founding of ICANN**

In June 1998, the U.S. Government released *Management of Internet Names and Addresses,*[8] which proposed to found Internet Corporation for Assigned Names and Numbers (ICANN), a non-profit civilian organization to be in charge of the Internet names and address resources. It also promised to hand over the authority over IANA. ICANN was founded in Los Angeles in October 1998, governed by the laws of California. At the beginning, the Department of Commerce of the United States signed a contract with ICANN, according to which, the root servers managed by IANA should be properties of the U.S. Government, who commissioned ICANN to manage IANA via a contract of a limited period. The Department of Commerce should have the final right to manage IANA. However, other countries and Internet communities were dissatisfied for a long time with the U.S. Government's ultimate management authority over IANA, which later remained for a long time a key problem and dispute in the field of Internet governance.

1.2.4 Emergence of Internet Culture

In early 1990s, the launch of www and the National Information Infrastructure (NII), a new hi-tech program of the U.S. Government, helped to increase the popularity of the Internet, which, in turn, led to the emergence of Internet culture and its entry into the mainstream society.

1. *Wired*

In 1993, Louis Rossetto and Jane Metcalfe founded the renowned Internet culture magazine *Wired*, which, before the first wave of the Internet, participated in the development of the Internet by launching the first banner advertisement, marking the beginning of technical consumption media and promoting the spread of Internet culture.

2. **Electronic Frontier Foundation (EFF)**

In 1990, John Perry Barlow, an activist of WELL Community, and Mitchell Kapor, founder of Lotus, a software business, founded Electronic Frontier Foundation (EFF), a non-profit organization, which was aimed to settle the disputes between government and technology, and to protect hackers, who were a disadvantaged group at the that time, from being treated unfairly. Through legal action assistance, analysis of policies, grass-root movements and technical support, EFF advocated users' privacy, speech freedom and innovation to safeguard citizens' rights and

---

[8]ICANN. *White Paper.* https://www.icann.org/resources/unthemed-pages/white-paper-2012-02-25-en.

interests in cyberspace. In the following decades, EFF spoke for and launched right-defense movements for source software, intellectual property right, security research and file sharing tools.

1.3 The Third Stage: Popularization of the Internet

The popularization of the Internet and the excessive investment in it gave rise to Internet foam. From the end of 1990s to the early 21st century, the Internet witnessed rapid development and popularization. In 2005, the number of its users reached one billion for the first time. The growth attracted a large amount of capital and thus resulted in the foam of the Internet. In 1999, 457 IPOs in the United States were mostly made by hi-tech startups and 100 startups were directly related to the Internet.[9] In March 2000, when NASDAQ crashed, a large number of startups went bankrupt, resulting in a total IT industry loss of five trillion US dollars. On the other hand, a group of representative Internet businesses emerged, including Amazon, Google and Yahoo!. The gradual maturity of the Internet economy afterwards promoted the Internet development.

1.3.1 Birth of Web 2.0

Web 2.0 refers to World Wide Websites that emphasize user-generated content, usability and interoperability. The term was coined by Darcy DiNucci in 1999 and popularized at the O'Reilly Web 2.0 Conference in 2004. Web2.0 does not refer to an update of any technical specification, but innovation in the way web pages are designed and used. It may allow users to interact and collaborate with each other in a social media dialogue as creators of user-generated content.[10] Examples of Web 2.0 features include social blogs, Wikis and P2P.

1. **Blogs**

The first blog diary keeper was Justin Hall. In January 1994, he started his web-based diary *Justin's Links from the Underground*, recording his activities, thoughts, meditation and even secret content in the form of diary. In 1997, John Barger created the term "weblog", and later the short form "blog" became a daily expression still used today. Ev Williams launched a new blog script tool and the free software blogger, which made blogging simple, easy and popular. The occurrence of political blogs and news blogs led to the mainstreaming of blogs,

---

[9]Arun Rao & Piero Scarruffi, *A History of Silicon Valley: the Greatest Creation*, trans. Yan Jingli and Hou Aihua, Beijing: Post & Telecom Press, 2014.

[10]Web 2.0. Wikipedia. https://en.wikipedia.org/wiki/Web_2.0.

which became important tools for media, political advisors and even election candidates to promote and express their position.

2. Wikis

The earliest Wikis software was WikiWikiWeb developed by Ward Cunningham in 1995. The user could directly edit on the website using wiki software, and every historical version of the website could be saved. During this process, no supervisor or checking person pre-reviewed the change. Cunningham opened his own basic software to others for modification and use, which led to the occurrence of a series of Wiki websites, among which Wikipedia was the most famous. In 2000, Jimmy Wales and Lawrence Mark founded Wikipedia based on Wiki technology. It is by far the most popular wiki-based website and one of the most widely viewed sites of any kind in the world. As a multi-lingual online knowledge body which can be visited and edited freely, Wikipedia is edited through collaboration among Internet communities. It has been widely used in the world since it was launched.

3. P2P

Peer-to-peer (P2P) is a distributed application architecture that partitions tasks or workloads between peers, who share resources rather than use centralized management systems. In 1999, Shawn Fanning founded Napster, a music service website, the very first P2P service widely used since consumers worldwide can share the music files. Such a downloading way soon stirred a "tornado", producing a great impact on the way digital files are used. But a legal action was soon launched against Napster, which was forced to be closed in July 2001, but its occurrence led to the emergence of file sharing models.

1.3.2 Development of Social Media

The Internet has multi-functions, including technical, media, entertainment and application services, so it is of strong social property. Email and bulletin board, which occurred in the early period of Internet development, already embody the social property of the Internet. The popularization of social media resulted from Web2.0, and SNS formed by P2P. Early social network sites that were popular were Friendster founded in 2002 and MySpace founded in 2004. In February 2004, Mark Zuckerberg from Harvard University and his classmates launched the facebook.com, which became popular instantly. After two years of dramatic growth, Facebook was opened to the world in August 2006, and in the following decade it contributed to the network traffic and number of Internet users, producing great social impacts. Today, other worldwide social network sites include Instagram, Twitter, WhatsApp, LinkedIn, Snapchat, Google+, BBS and blog. Social applications in China include Weibo, WeChat and Tencent QQ, those in Republic of Korea include Line and KakaoTalk, and those in Russia include VK (former Vkontakte) and Odnoklassniki, which have developed fast.[11] The popularization of social networks has changed the way of communication, producing influence on social development.

1.3.3 World Summit on Information Society (WSIS) Marked the Beginning of Global Governance of the Internet

In the early 21st century, the rapid development of information and communication technologies has led to revolutions of modern society and rise of problems such as widening of the gap between the rich and the poor and digital divide. Therefore, the United Nations organized the WSIS to get international organizations, governments, civil organizations and private businesses involved in hope that the Internet will serve the development of the whole humankind. WSIS marked the beginning of global governance of the Internet.

1. Geneva Summit 2003

Over 10,000 stakeholders from 175 countries and regions participated in Geneva Summit held in December 2003. It was the first global discussion about Internet governance. The management right over ICANN became the focus. The Summit passed two resolutions: *Declaration of Principles* and the *Plan of Action*, two documents of policies that laid out the general plan of action and work objectives of WSIS in the future. The Working Group on Internet Governance (WGIG) was formed to define Internet governance and identify the public policies concerning it and the roles and responsibilities of all stakeholders.

2. Tunis Summit 2005

Tunis Summit held in November 2005 resulted in agreement on the *Tunis Commitment* and the *Tunis Agenda for the Information Society*. *Declaration of Principles* was reiterated and progress of all decisions was identified. The Internet Governance Forum (IGF) was established to have policy dialogues on Internet governance. At that summit, the Internet governance mechanism became the discussion focus and two opposite positions were formed, one advocating the maintenance of the existing system while the other advocating the establishment of new intergovernmental systems.

3. WSIS Forum

To monitor and review the progress of the two previous summit resolutions, WSIS decided to organize annual forums from 2006, advocating and promoting the implementation and development of *Geneva Plan of Action* and *Tunis Agenda*.

---

[11]ChinaLabs: *Research Report on Internet Development of G20 Countries*, 2016.

1.3.4   Movement of Internet Sharing Culture

In 2005, the number of Internet users in the world reached one billion, a milestone in the history of the Internet development. According to ITU statistics, in 2005, the number of worldwide Internet users was one billion and the total population of the world was approx. 6.5 billion, so the Internet popularization rate was 15.8%.[12] Among the Internet users, 60% lived in developed countries of Europe and North America. In the early 21st century, as the popularization of the Internet increased people's desire for knowledge, culture and information sharing on the Internet, the sharing culture movement emerged, manifested in online information sharing and copyleft movement. In 2001, Larry Lessing, a professor of law from Stanford University, founded Creative Commons advocating looser copyright systems. In 2002, Richard Stallman launched Free Software Movement (FSM), with the goal of promoting the freedom to run, duplicate, change and redistribute software. In 2003, Piratbyrån, a civil Swedish anti-copyright organization, founded the Pirate Bay, a point-to-point file sharing website, to promote awareness of Internet sharing. Adhering to the philosophy of freedom, equality and free information sharing, the Pirate Bay soon won support and response from millions of Internet users world-wide and then became the world's largest and most influential BT sharing website. On the other hand, the Pirate Bay, as a free and open online resource sharing platform, has been forced offline or been warned by different governments for times.

1.4   The Fourth Stage: Mobile Internet

With the rise of 3G mobile networks and popularization of the new-generation smart terminals represented by iPhone, the world entered a mobile Internet era. The number of smartphone users grew from nearly three billion in 2007 to nearly six billion in 2011, with the growth rate of 100% in five years, forming a strong drive for the Internet development. Super platforms have emerged worldwide and the global Internet governance system and network relations between major countries have undergone great changes.

---

[12]ITU. Time Series of ICT Data for the World 2005–2017, http://www.itu.int/en/ITU-D/Statistics/Pages/stat/default.aspx.

1.4.1 Development of Mobile Communication Technologies and Popularization of Smartphones

1. Development of mobile communication technologies

The maturity of mobile communication technologies is the basic guarantee of the development of the mobile Internet. In 2001, Japan launched 3G service while in May 2008, CDMA2000 of the United States, WCDMA of Europe and TD-SCDMA of China were acknowledged by ITU as the 3G mobile communication standards. In early 2012, LTE-Advanced and Wireless MAN-Advanced technical specifications were recognized as international 4G standards. The speed of 3G and 4G sound and data transmission has been greatly increased, so wireless roaming can be done worldwide, and image, music and video streaming can be processed better, which has promoted the rapid development of mobile applications and the mobile Internet.

2. Popularization of smartphones

By the end of 1990s, the mobile phone had become a widely-used communication device and the first smartphones were developed, represented by Blackberry produced by RIM of Canada. Blackberry smartphones monopolized the market at that time. In June 2007, Apple Inc. officially launched its smartphone iPhone, which supported emailing, mobile calling, messaging, website browsing and other wireless communication services. For iPhone's iOS operation system, multi-touch technology (touch-screen technology which originated in Europe) was adopted. The APP Store of iPhone was launched in 2008. Thanks to the great success of iPhone, the smartphone was nominated by *Time* as an innovation product of 2007. By August 2011, Apple's market value had exceeded that of Exxon Mobil, ranking first in the world. On the market of operation systems of mobile phones, iOS has the major rival—Android of Google, who acquired Android in 2005, a Linux system company founded by Andy Rubin, and thus expanded into the mobile phone operating system market. Different from Apple, Google takes Android as an open platform, whose source code can be used for free by hardware manufacturers on their mobile phones and tablet computers. In 2007, Google set up Android Open Handset Alliance by cooperating with scores of mobile operators, handset manufacturers and chip makers. In 2010, the Android mobile handset surpassed iPhone in shipments and became the world's largest mobile phone operating system.

1.4.2 Rise of Super Network Platforms

Since 2010, super network platforms including Apple, Google, Microsoft, Amazon and Facebook (FAMGA) have been top five businesses in terms of market value. China has entered a new stage at which super network platforms are dominant. Other super network platforms include Yandex and Mail.ru from Russia, Naver and

DaumKakao from Republic of Korea, Rakuten and Line from Japan, Infibeam from India and MercadoLivre from Brazil.

1. Super search engines

Google, benefiting from its huge profit on the online advertising market, has rapidly grown into the leading brand in the field of search engines. Other top ones are Baidu from China, Yandex from Russia and Yahoo! from Japan.

2. Super e-commercial platforms

So far, in the terms of market penetration rate of global online retailing and auction, Alibaba has been ranking first in the world as China's leading e-commercial platform. It also has the absolute market share in key niche fields of its industry by means of investment and merger and acquisition. In terms of sales volume and market value, Amazon is the world's largest online retailer, who has set up localized websites in more than 10 countries.

3. Super social platforms

In July 2017, Facebook had 2.2 billion active users from over 200 countries and region. It is the most widely-used social website in the world. Except China (WeChat) and Russia (Vkontakte), where the local social tools are in the dominant position, Facebook is the top social website in most countries.

These super network platforms represent the most cutting-edge achievements in technical innovation and progress, and they are becoming important information infrastructure, with strong capacity of mobilization and dominant position in their industry. They have, to a great extent, undertaken the public service function, which should have been undertaken by the State. On the other hand, they have caused problems in data security, private protection, social morality and legal order, which have attracted the attention of governments of all countries.

1.4.3   *Changes in the Global Internet Governance System*

The first global cyber war in 2007 and Snowden Incident in 2013 indicate that cyber security has become a big challenge that the world has to face, attracting the attention of all countries and producing great impact on the global Internet governance system and on relations between major countries.

1. Beginning of the formulation of cyber security norms

In April 2007, the first national cyber warfare broke out in Estonia, a country highly reliant on the Internet. As some people protested against Estonia's relocation of the Soviet Second World War memorial from Tallinn, the critical national information infrastructure of Estonian government, banks and media suffered three-week of DDoS (distributed denial of service) attacks, and a number of websites were forced to close or could not provide services. That was a large-scale cyber attack out of political purpose rather than economic, resulting in many political debates on cyber conflicts and wars and urging political and military organizations to rethink the importance of cyber security to the security of modern countries. Estonia had to formulate the *Cyber Security Strategy of the Republic of Estonia*, participate in international cooperation and organize the NATO Cooperative Cyber Defence Center of Excellence. Since 2010, it has hosted the Conference on Cyber Conflict every year. NATO Cooperative Cyber Defence Center of Excellence, founded in May 2008, released in 2013 *Tallinn Manual 1.0*, which, for the first time, provides global standard suggestions on cyber warfare. According to the *Manual*, a country has sovereignty over the Internet components within its territory. In 2017, the *Manual* was updated into *Tallinn Manual 2.0*.

2. **Snowden Incident**

In June 2013, Edward Snowden, a former employee of CIA and former contractor of NSA, disclosed in Hong Kong the secret program PRISM launched by NSA to *The Guardian* of the United Kingdom and *The Washington Post* of the United States, revealing the secret of the global surveillance programs launched by the United States, and hence causing great sensation worldwide. According to the disclosed documents, a great number of technical businesses have got involved in the PRISM Program, including Microsoft, Yahoo!, Google, Facebook, Paltalk, YouTube, Skype, America Online and Apple. Through the Program, NSA has obtained data including emails, video and speech, pictures, VoIP talk, file transmission and detailed information on social networks. Besides, the strong international surveillance network of the United States cooperates with security departments of other countries (for instance, the listening stations of the UK Government Communication Headquarters, GCHQ for short). Snowden Incident aroused the attention of the world to cyber security and privacy and boosted the debates on global Internet governance policies. The Incident also revealed the severe challenge to the reform of the global Internet governance system and to the relations between major countries. Brazil took diplomatic action after the Incident. It criticized the U.S. surveillance at the 68th Session of the United Nations General Assembly and launched with ICANN the NETmundial Initiative in October 2013 to discuss the solution to Internet surveillance. The relationship between Europe and the United States got into a ditch and EU began to worry and complain about the U.S. privacy protection. Max Schrems, an Austrian lawyer, accused Facebook of privacy infringement, which led to the termination of *SafeHarber*, an agreement on data transmission signed in 2000 by EU and the United States.[13] The disclosure impaired the moral basis of the United States in Internet governance policies, so the country was forced to transfer the DNS management right to ICANN. In March 2014, the National Telecommunications and

---

[13]*An Austrian Young Man Won the Lawsuit against Facebook*, October 07th, 2015. http://news. xinhuanet.com/world/2015-10/07/c_1116748332.htm.

Information Administration announced that the U.S. Government would give up the management right over critical Internet functions and transfer it to the multi-stakeholder community.

1.4.4   Diversification of Internet Culture

Seen from the content on the Internet, there exist great gaps in the usage of languages. According to W3Techs statistics, the number of English websites accounts of 51.2%,[14] so English is inevitably the top Internet language (See Fig. 1.1). By June 2017, the number of Internet users of top 10 languages had amounted to three billion, accounting for 77% of the total number (3.89 billion) of Internet users of the world, while the number of Internet users of other languages accounted for only 23%, so they are truly the "minorities". Languages have gone beyond borders but have their own borders on the other hand. The population of top ten languages is 5.1 billion, accounting for 68% of the total population (7.5 billion), or two thirds. Obviously, with the development of Internet and society, multilingualism becomes a new trend. With the change of the number of Internet users in different countries and the trend of multilingualism, the Internet culture will be diversified. Compensations and communications between languages, cultures and nations are inevitable in the development of Internet culture.

1.5   The Fifth Stage: The Era of Internet of Everything

With the occurrence of 5G technology and AI and their relevant applications, the era of Internet of Everything, which involves human beings, programs, data and things, is coming. 5G technology represents the new-generation mobile communication system oriented towards 2020, with super-high spectral efficiency and super-low energy consumption. The transmission rate, resource use, wireless coverage and user experience will all improve in comparison with that in the 4G era. The advent of 5G era will be a great opportunity to subvert the present situation of communication. In terms of national strategies, political support for 5G technologies and competition for its standard will become the key point of competition in the field of mobile communication of all countries. In terms of industry, development of 5G technologies is inevitable, so operators of capital and ICT, Internet businesses, device manufacturers and chip producers are turning to 5G industries. The official commercialization of 5G technologies will accelerate the society's entry into the era of Internet of Everything and will change people's lifestyle and

---

[14]*Usage of Content Languages for Websites.* W3Techs.com, Dec 9, 2017, https://w3techs.com/technologies/overview/content_language/all.

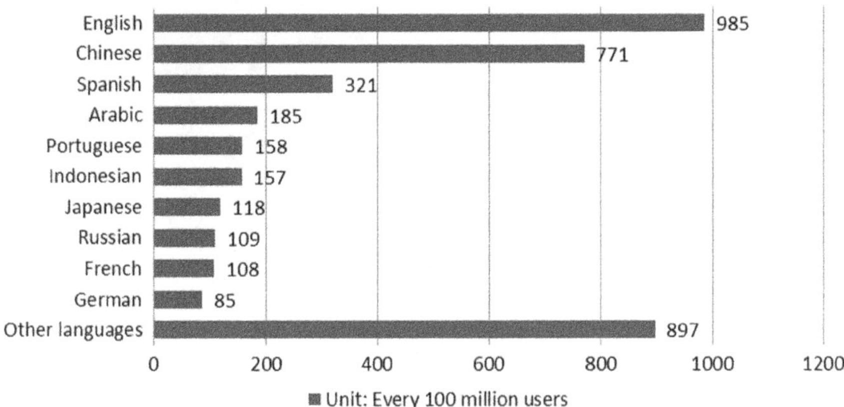

Fig. 1.1 Top 10 languages used on the internet (*Source* http://www.internetworldstats.com/stats7. htm) by June 30th, 2016, the total number of Internet users had amounted to 3.885567619 billion

industrial layout and help to reconstruct the civilization of post-industrial era. Internet's integration with politics, economy, culture, society and military affairs will make the whole society more reliant on the Internet whose governance, in turn, will no longer be only the governance of cyber technology and cyberspace, but an indispensable part of social, state and international governance. The change of the meaning and scope of international Internet governance will promote all countries to play an increasingly important role in Internet governance; otherwise, they will not be able to efficiently tackle cybercrime, terrorism and disorder. To be short, the government will inevitably play an important role in Internet governance.

By June 2017, the number of Internet users worldwide had reached 3.89 billion, marking the Internet popularization rate of 51.7%,[15] a half of the total population (7.5 billion). Among the future new 3.0 billion Internet users, 90% will come from developing countries, which indicates that the major field of Internet development and cyberspace governance will shift into developing counties. It is the mission of the Internet to enable everyone to know and change the world through it at the lowest cost but the highest efficiency. It is the common responsibility of the international community to enable the remaining 3.0 billion people to become Internet users. The world Internet should be developed for the wellbeing of human beings. In other words, Internet should be developed, used and governed for the purpose of benefiting society and its people and of realizing the people's yearning for a better life.

---

[15]*Internet Users in the World*. Internet World Stats. http://www.internetworldstats.com/stats.htm.

Chapter 2
Development of the World's Information Infrastructure

Chinese Academy of Cyberspace Studies

Abstract (1) With the evolution of fixed broadband network technology, fiber broadband access is becoming the mainstream and the number of its users keeps increasing. By the first quarter of 2017, the number of fixed broadband users in the world had reached 875.1 million, including over 500 million fiber users. (2) The commercial scale and number of users of 4G technology keep increasing rapidly and its application scenarios are enriched. Its technical standard is evolving toward 4G+ and 4.5G at a high speed. By June 2017, 192 countries and regions had opened 601 LTE (4G) commercial networks, with 7.72 billion mobile users and the 4G penetration rate of 28%. (3) The demand for WLAN keeps increasing and the throughput is being calculated in gigabit. By the end of 2016, there were 94 million public WLAN access points (AP) and the number is expected to increase by six times in 2021. (4) With the explosive growth of the demand for Internet addresses, IPv6 sees new development opportunities. The layout is being accelerated and the number of its users is increasing. By July 2017, the number of IPv6 users in the world had accounted for 11.25%. (5) Internet broadband is witnessing rapid growth and the undersea cable construction is booming. Google, Microsoft and Facebook and other Internet giants are important force in the construction. In 2017, the speed of Internet broadband reached 295 Tbps and 270 undersea cables had been put into use. (6) High-orbit broadband communication satellites are developing fast, so are satellite Internet systems. Communication technologies like hot-air balloons and drones are being tested. By the end of 2016, there were 59 high-orbit broadband satellites and the Internet access capacity was 600 Gbps. (7) Data centers and cloud computing platforms are under centralized construction, which is witnessing obvious synergy effect. By 2016, 45% of large data centers were located in the United States. Content distribution networks (CDNs) are developing at a high speed and they can bear higher traffic. In 2016, the global CDN market scale was US$ 6.05 billion, with the compound growth rate in the past five years exceeding 22%. (8) IntelliSense facilities boost the development of the IoT. In 2016, the global cellular mobile communication networks bore approx. 800 million devices' access to the IoT and they are expected to witness explosive growth in the future. Internet

Chinese Academy of Cyberspace Studies
Beijing, China

© Publishing House of Electronics Industry, Beijing and Springer-Verlag GmbH
Germany, part of Springer Nature 2019
Chinese Academy of Cyberspace Studies (ed.), *World Internet Development Report 2017*,
https://doi.org/10.1007/978-3-662-57524-6_2

exchange points are expanding rapidly, promoting the networks to evolve in a flattening way. At present, there are nearly 500 exchange points in the world, distributed in 147 countries. (9) Basic resources of the Internet are unevenly distributed. Information infrastructure in some developing countries is not complete. There is a widening digital divide. Among all the Internet users of the world, only 10% come from Africa, but the population of Africa accounts for 16.6% of the world's total. (10) The future broadband will develop into gigaband and 5G mobile communication standard commercialization will bring about the popularization of the IoT. Space-ground Integrated Network (SGIN) will be available across the globe, network facilities will be more intelligent and open, and universal telecommunication service will boost the sustainable development of economy and society.

2.1 Broadband Networks: Upgraded into Super-Speed Networks

2.1.1 *Fiber Broadband Access is the Mainstream and the Number of Its Users is Increasing*

1. **All countries have set higher goals for their broadband development and household access capacity is rising towards 100 Mb.**

With the rapid development of broadband network technology and its application, all countries have set higher goals for their broadband development, so that everyone can be covered by digital economy (see Fig. 2.1). In September 2016, EU adopted *Connectivity for a European Gigabit Society*, which proposes that all public places (including schools) should have access to 1 gigabit of data per second (1 Gbps) and that all European households should have access to networks offering a download speed of at least 100 Mbps by 2025. Some EU countries have also set the same goals. For instance, according to *Digital Strategy 2025* launched by Germany, the country will provide 1 Gbps fiber network by 2025. China has modified the goal for network capacity in *Plan for the Development of Information Communication* (2016–2020) released in 2017. The *Plan* proposes that by 2020, the household access capacity in mid and large-sized cities will exceed 100 Mbps and that half of rural households will have the access capacity of over 50 Mbps.

2. **Fixed broadband network technology is evolving fast and the fiber broadband network will be the main network.**

The global fixed broadband access technology is developing from copper cable access technology represented by ADSL and cable-based television network broadband access technology to fiber access technology represented by VDSL, FTTx and FTTH. All countries are making effort to improve their broadband access capacity, make fiber broadband available to the users and choose different access technologies and evolution routes at the household terminals in accordance with

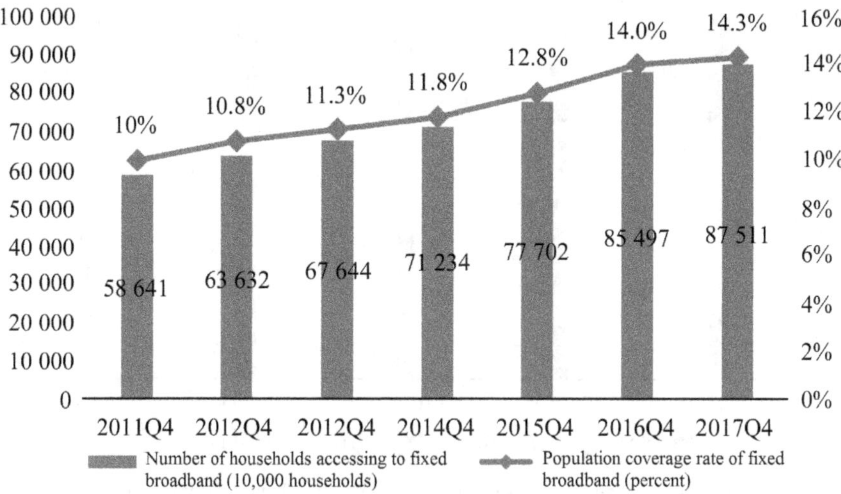

Fig. 2.1 Number of households accessing to fixed broadband and population coverage of fixed broadband in the world (*Source* Point topic)

their own situation. China, Japan and Republic of Korea are promoting household access to fiber broadband, the coverage rate of which is higher than that of other countries. The United States has more developed cable-based TV network broadband accesses. There, businesses like Comcast are introducing DOCSIS3.1 to provide gigabit service for the users, and traditional telecommunication operators like AT&T are developing fiber broadband in a large scale. In European countries, fiber and copper broadband are developing in parallelism. For example, the former develops fast in Sweden, Norway, Spain, and Portugal. Portugal ranks first with the fiber broadband covering 46% of its population, while in the United Kingdom and Germany it develops slowly. Copper-based G.fast broadband technology has been commercialized in large scale in many European countries.

3. **The number of fiber broadband users keeps increasing drastically.**

According to a Point Topic report (see Fig. 2.2), by the first quarter of 2017, the number of fixed broadband users in the world had amounted to 880 million, with its population coverage rate of 14.3% and the annual compound growth rate in the past five years of 7%; the number of fiber broadband users in the world had reached 500 million, of which 280 million were FTTH users and 220 million were FTTx users, with the percentage of fiber users in the past three years rising from 29.6 to 56. According to OECD statistics, at the end of 2016, the average number of fiber users in OECD countries accounted for 21.2% (see Fig. 2.3). China's fiber broadband develops fast. At the end of 2016, the number of FTTH households was 220 million, growing from 3.5% in 2011 to 76.6%.

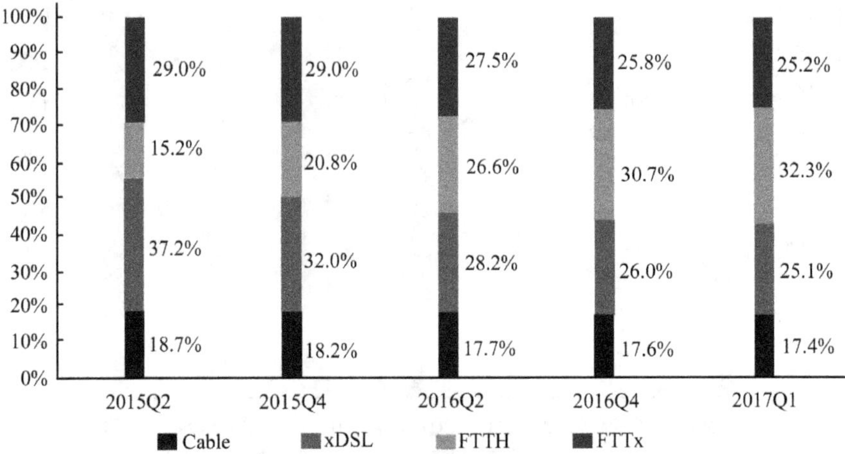

Fig. 2.2 Change in percentage of households using fixed broadband of different technologies in the world (*Source* Point topic)

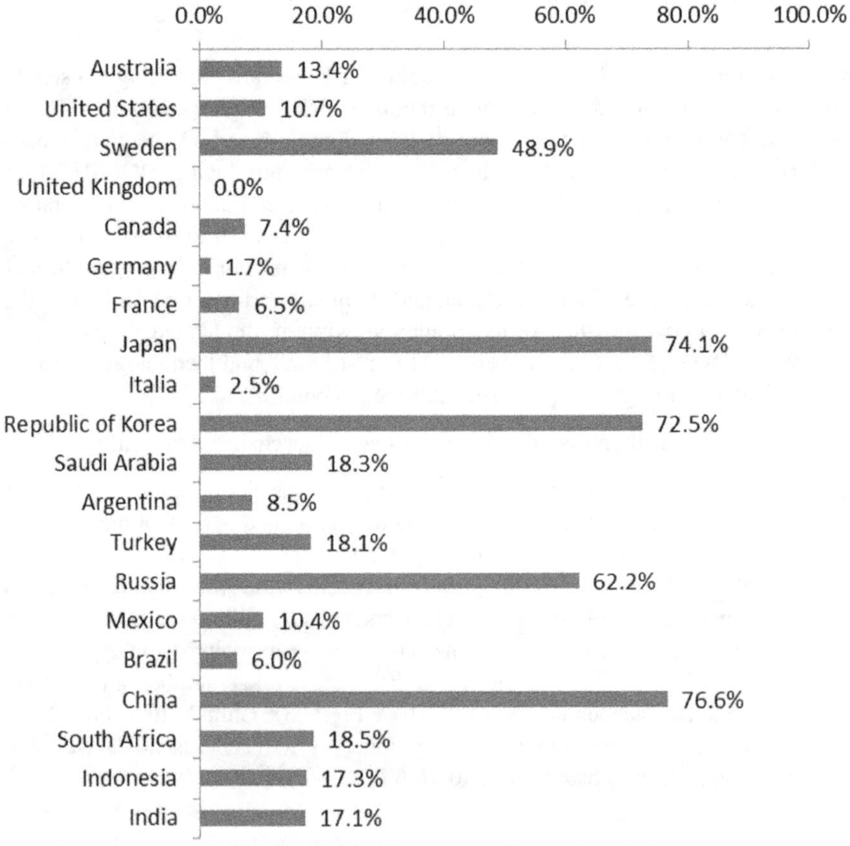

Fig. 2.3 Percentage of households using fiber in the world by the end of 2016 (*Source* Point topic)

2.1.2   4G Mobile Broadband is being Popularized and is developing Toward 4G+ and 4.5G

1. **4G networks are developing fast and the number of their users is growing rapidly.**

According to GSA statistics, by the end of June 2017, 192 countries and regions had opened 601 LTE (4G) commercial networks. Among them, 95 countries had laid out 195 LTE-A (4G+) or LTE-A Pro (4.5G) networks, 56 countries had laid out 98 TD-LTE networks (including 32 TDD/FDD mixed ones), and 57 countries had opened VoLTE in 109 networks. China now has the world's largest 4G network and 4G market, over three million 4G base stations, and 890 million users (households), with the penetration rate of 65%.

According to GSMA Intelligence statistics, by the end of June 2017, the number of mobile users across the world had reached 7.72 billion (including 5.04 billion independent mobile users), with the popularization rate of 103% and 4G penetration rate of 28%, an increase of 9% in comparison with that in the same period of 2016, but there exists regional imbalance in the development (see Fig. 2.4). GSMA Intelligence predicts that by 2020, the number of mobile users across the world will be 5.6 billion, exceeding the number of those enjoying power service at home (5.3 billion), bank account owners (4.5 billion) and those enjoying tap water service (3.5 billion).

The rapid development of 4G technology in the world has changed the user structure (see Fig. 2.5). 2G users have shifted to 3G/4G networks, so the percentage of 2G users is decreasing rapidly. 3G technology development witnessed its peak in 2015 and its users began to shift to 4G networks from 2016. The number of 4G users keeps increasing fast. By the end of June 2017, the 4G user percentage had reached 28, nearly equal to 3G user percentage.

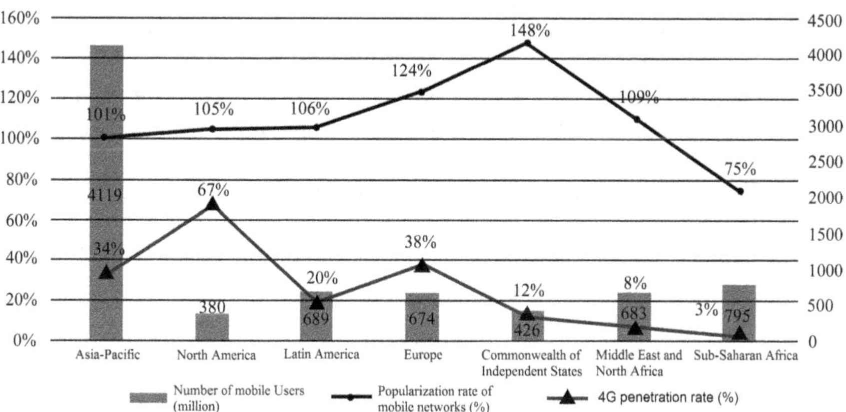

Fig. 2.4 Number of mobile users in different regions of the world (by the end of June 2017) (*Source* GSMA intelligence)

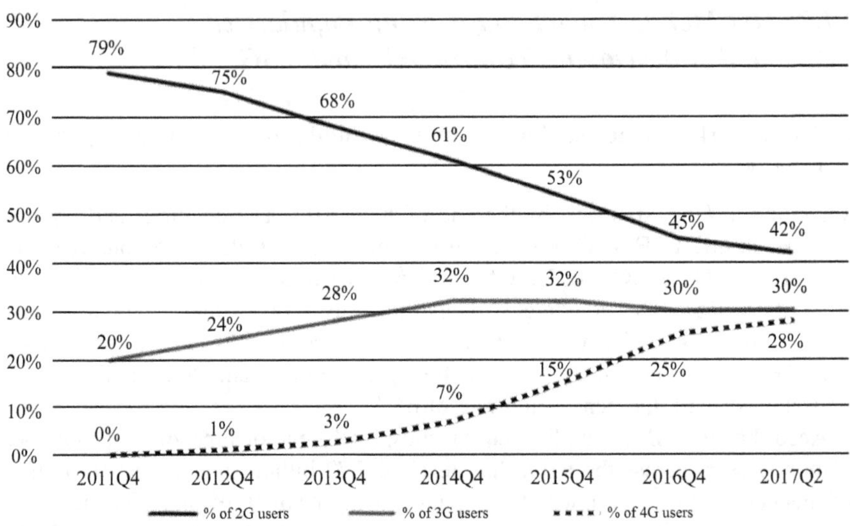

Fig. 2.5 Change of percentages of 2G, 3G and 4G users in the world (*Source* GSMA)

In terms of application, LTE-based special wireless broadband networks develop fast and the application scenario keeps expanding. Those special wireless broadband networks are especially developing fast in wireless government affairs, public security, social administration, emergency and disaster relief, and industrial application, playing an important role in ensuring information intercommunication and improving informatization and production efficiency. The United States released in 2010 the *National Broadband Program* supporting the layout of an inter-operable mobile broadband network of public security covering the whole country. In 2012, the country adopted an act to identify 700 MHz "D" frequency band as serving public security. Then, FirstNet, the wireless broadband network for public security of the country, was set up, taking LTE as the standard for the public security broadband network, and serving the US police, fire protection, first aid and other public security sectors. In Europe, special public security broadband networks are being studied. TCCA (TETRA + Critical Communication Association) and CCBG (Critical Communication Broadband Group) are seeking to formulate a strong LTE evolution route map for public security and other critical communication networks, taking digital service as the goal. In China, the governments (represented by the public security sector) of major municipalities and industries (such as communications and transportations, power and energy) are expressing their demand for the construction of special LTE broadband networks. In the latest frequency spectrum planning, 1.4 GHz (1447 ~ 1467 MHz) will be used for special governmental networks and 1.8 GHz (1785 ~ 1805 MHz) for special networks of power, railways and communications and transportation.

2. 4G is evolving towards 4G+ (LTE-A) and 4.5G (LTE-A Pro)

According to GSA statistics, by the end of June 2017, there were already 195 LTE-A or LTE-A Pro networks, and VoLTE had been opened in 109 LTE networks. SKTelecom of Republic of Korea upgraded its network to LTE-A network at the end of 2014, by adopting three-carrier aggregation, with the speed reaching 300 Mbps. In June 2016, it launched the LTE-A Pro network adopting five-carrier aggregation, with the speed reaching 500 Mbps, and it plans to provide LTE service with 1 Gbps throughout the country in 2019. To increase LTE network capacity, AT&T of the United States adopted three-carrier aggregation in 2016 and planned to adopt four-carrier aggregation in 2017. Japanese NTT DoCoMo improved in 2016 the speed "PREMIUM 4G" of LTE-Advanced service to downlink maximum 370 Mbps, and developed the TDD and FDD carrier aggregation technology. To improve LTE network capacity, China's telecommunication operators are speeding up the application of VoLTE and carrier aggregation technologies and promoting LTE networks to evolve towards LTE-A/LTE-A Pro. It is anticipated that the carrier aggregation will expand from hot-spot areas to more cities and towns of China. By the end of 2017, the VoLTE technology had been completely commercialized in the country. Mid and high-end mobile phones launched into the market in China since 2015 all support VoLTE and carrier aggregation technologies, laying a solid foundation for LTE-A/LTE-A Pro network development and application.

2.1.3 WLAN Promotes Universal Access to Internet and the Throughput is being Calculated in Gigabit

1. Demand for WLAN keeps increasing.

With the improvement of the convenience and security of WLAN, there is a tremendous demand for WLAN in households, public places, and businesses, which has promoted the popularization of the Internet. According to statistics of Cisco, by the end of 2016, there were 94 million public WLAN access points (APs) in the world, including about 85.1 million in households and 8.8 million commercial ones. It is anticipated that by 2021, the number of public WLAN APs in the world will increase by six times, to 541.6 million, including 526.2 million in households and 15.3 million commercial ones. The Asia-Pacific Region will see a rapid growth of the APs, which are expected to account for 45% of global total. According to the prediction of WBA, the number of terminals with WLAN or cellular function will be over 10 billion in 2018.

2. More market elements are introduced into governmental public WLANs.

The governmental public WLAN originated in the "Wireless Philadelphia" Initiative, which was aimed to provide free Internet access service based on WLAN. Due to the lack of sustainable commercial models, "wireless cities" initiatives like

"Wireless Philadelphia" across the world saw slow progress and some of the cities gave up or suspended their initiatives. At present, wireless cities are trying to promote their governmental public WLAN through the market. For instance, for i-Shanghai 2.0, third-party capacity has been introduced into the construction and operation and some commercial WLAN profit models have been developed through PPP. Governmental WLANs are aimed for public benefit, but market elements can bring about more vitality and commercial models for governmental public WLANs, which, in turn, will promote the sustainable development of governmental WLANs.

3. **WLAN is speeding up its evolution towards 802.11ax/ay.**

802.11a/b/g/n is the core standard of WLAN technology, 802.11n being the mainstream on the present market, with its throughput reaching 600 Mbps. 802.11ac and 802.11ad are the new-generation super-speed WLAN standards, working respectively at 5 and 60 GHz frequency band, with the throughput reaching gigabit. 802.11ah works at the free band below 1 GHz for IoT and coverage expansion application. 802.11af, through the adoption of TVWS frequency band, can expand the WLAN coverage greatly. Today, IEEE is accelerating the evolution of WLAN towards 802.11ax and 802.11ay to increase the transmission rate. 802.11ax is the upgraded version of 802.11ac, providing broadband connection for dense layout of scenarios, with its standard expected to be released in 2019. 802.11ay is the upgraded version of 802.11ad, providing wireless broadband connection at millimeter-wave band (60 GHz). It emerged in 2017. WAPI, the security standard for China's WLAN, has been applied in public networks, e-government affairs, customs, finance, power, medical care, and education. R&D as well as standard release is being promoted concerning interface specification, device testing and credential management. By August 2017, the relevant chips had passed the tests in protocol consistence and integrity as well as application interface, which has enhanced the interconnectivity and usability of WAPI products.

2.1.4 The Next-Generation Internet is developing In-Depth and IPv6 is being Laid Out at a High Speed

To solve the problem of the exhaustion of IPv4 addresses, IPv6 was officially released in early 1996, marking the beginning of the IPv6 era. However, since all kinds of alternative technology and IPv4 address transaction prolonged the use of IPv4 and IPv6 was faced with such problems as high investment, slow returns and shortage of killer applications, IPv6 was not applied in commerce in a large scale at that time. Today, with the acceleration of global informatization, and thriving of the mobile Internet, IoT and cloud computing, which has resulted in explosive growth of the demand for Internet addresses, IPv6 is witnessing new development opportunities. The next-generation Internet is developing fast, especially in the United States and Europe.

1. **The United States and Europe are promoting the development of IPv6.**

In May 2012, the U.S. Government issued *Planning Guide/Roadmap toward IPv6 Adoption*, in which IPv6 adoption is to be integrated into other governmental initiatives, such as cloud computing, data centers and trusted connection. Europe has taken the strategy of "mobile Internet priority" to build up its advantage in mobile communication. Japan puts its IPv6 focus on IoT and Ubiquitous Network. Driven by IPv6 strategies of different countries, the number of IPv6 users in the world keeps growing steadily. According to APNIC statistics, by July 2017, the number of IPv6 users had accounted for 11.25% on average (see Fig. 2.6).

2. **The support of large ISPs of the world for IPv6 is increasing.**

By the first half of 2017, over 240 ISPs had given support to IPv6. Among them, the support rate of Comcast, AT&T and Verizon in the United States had reached 46, 59 and 86% respectively. Japanese KDDI network's support rate had reached 27% while that of Indian RELIANCE JIO INFOCOMM, 77%.

3. **The number of mobile terminals supporting IPv6 keeps increasing rapidly.**

According to Cisco statistics, in 2016, about 3.5 billion mobile devices (43%) could support IPv6 while in 2015, the number was only 1.1 billion; about 2.6 billion smartphones and pads (68%) supported IPv6 while the number in 2015 was 2.1 billion.

4. **The layout of IPv6 in cloud infrastructure is accelerating.**

IPv6 can provide more IP addresses and higher performance and computing efficiency for cloud service, so large international cloud service providers are accelerating the IPv6 layout. GCP, Google's cloud platform, began to completely use IPv6 in 2008, providing dual services of IPv4 and IPv6. Microsoft's Azure had laid out services of IPv4 and IPv6 on Azure virtual machines in over 20 regions of the

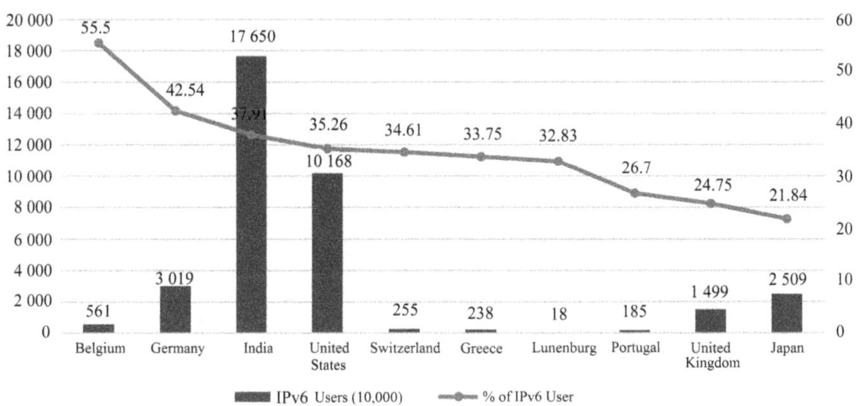

Fig. 2.6 Top 10 countries in terms of percentage of IPv6 users (*Source* http://labs.apnic.net)

world; Amazon's AWS has laid out virtual cloud in the east of the United States (Ohio State) to support dual service of IPv4 and IPv6.

5. IPv6 has started to be used in platforms of IoT and devices.

Since IPv6 has expanded IP addresses and complicated network management capacity, supporting IPv6 has become the foundation of the development of IoT. In 2013, IBM, a world's leading business in IoT, announced that it would jointly develop 6LoWPAN with Libelium, a wireless sensor business, to provide IPv6 support for sensors in the fields of IoT, such as smart cities, e-health and environment monitoring. In 2015, the gateway of AWS devices of IoT began to support both IPv4 and IPv6 at the same terminal, providing support for the users' seamless connection with IPv4 and IPv6 devices.

2.1.5 Submarine Fiber Cables Make Global Interconnection Faster, with Businesses as an Important Construction Force

1. Rapid growth of Internet bandwidth

Internet bandwidth has been growing rapidly and in the past five years its growth rate has been 30%. From 2013 to 2017, the bandwidth grew from 104 to 295 Tbps (see Fig. 2.7). With the advent of the era of big video and data, the traffic of the Internet will increase and the tremendous Internet bandwidth needs a new-generation ultra-large capacity submarine cables to bear it.

Seen from the intercontinental bandwidth change, there are closer ties between Europe and North America. By far, their intercontinental bandwidth has accounted for over 70% of the global total. With the continuous opening up of Asian, African and Latin American countries, the Internet traffic between these countries and

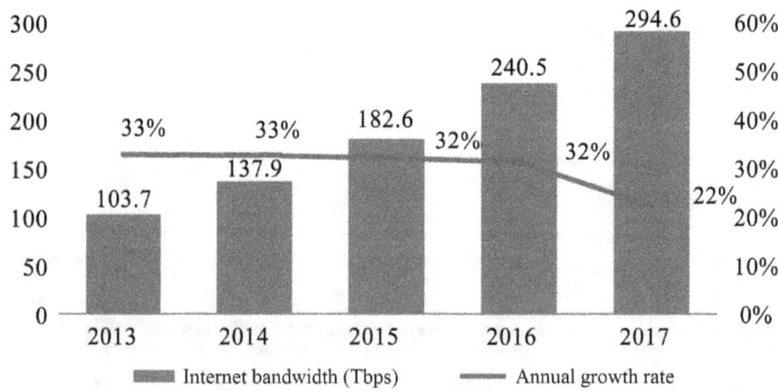

Fig. 2.7 Growth Rate of internet bandwidth (2013–2017) (*Source* Telegeography)

Europe and Americas keeps increasing, promoting Internet traffic to expand from Europe and America to the whole world. Since 2013, the percentage of intercontinental bandwidth between Asian, African and Latin American countries and European and American countries has increased by 8%. The worldwide intercontinental data transmission and information interaction require submarine cables to support the long-distance communication.

In terms of changes of Internet access bandwidth, all countries maintain high growth rate of Internet access bandwidth (see Fig. 2.8). From 2012 to 2016, the annual growth rate of Internet access bandwidth in 80% of countries and regions exceeded 30%, but that per capita remained low. In 2016, the average Internet access bandwidth per capita of the world was 0.49 Mbps per fixed broadband household, and the countries and regions whose Internet access bandwidth was above 1 Mbps per fixed broadband household only accounted for less than 30% of all the countries. It is an important measure of improving the Internet and supporting global economic cooperation to speed up the construction of Internet access and improve the Internet access bandwidth per capita. The layout of submarine cables will provide transmission network support for all countries' expanding the Internet and improving Internet accessibility.

2. Global interconnection relies on submarine cables.

At present, international communication and Internet connection rely on international undersea cables, cross-border land cables and international communication satellites. The submarine cable is the most important network interconnecting way in the world and the major transmission carrier and medium of cross-national communication because of its high communication quality, big system capacity, low unit cost and high security. According to Telegeography statistics, 98% of

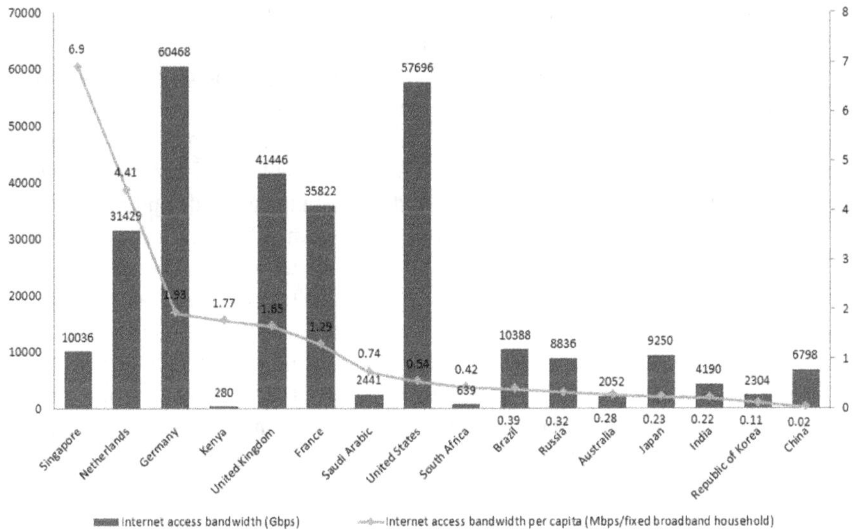

Fig. 2.8 Internet access bandwidth of some countries by the end of 2016 (*Source* Telegeography)

international data are transmitted through submarine cables. Since the world's first transoceanic submarine optical cable was laid and opened in 1988, 270 such cables have been launched into use and 10 are being laid, which connect all continents except Antarctica. North America—Europe, North America—the Asia and Pacific Region, North America—South America, and the Asia-Pacific Region—Europe are important connections by submarine cables. Backbone networks of international submarine cables have been formed with the United States (68 cables), the United Kingdom (51 cables), Japan (21 cables) and Singapore (22 cables) as transit centers. Optical cable hubs are New York and New Jersey of the United States, Cornwall of the United Kingdom, Hong Kong of China, Singapore, and Tokyo of Japan.

With the improvement of optical communication, submarine cable technology has developed greatly. The number of transoceanic submarine cable cores have increased from 3 pairs to 8 pairs and the transmission rate has increased from 280 Mbps every pair of cable cores to 100 Gbps (single wavelength rate) with the capacity of every pair of cable cores at 8 Tbps. The 400G large-capacity, super-speed WDM system is planned to be adopted in the new submarine cable system, which will improve the transmission rate and support the rapid flow of global data traffic.

3. Submarine cables are witnessing a new construction period.

The designed lifetime of submarine cables is generally 25 years, but some of them cannot be used as long as 25 years due to the limit of operation and maintenance cost, so before the end of the designed lifetime, it is necessary to consider the construction of new optical cables. According to statistics, 40% of global submarine cables were built before 2000, and they are now in the last period of their designed lifetime, so it is necessary to prepare for the reconstruction. In the coming years, global submarine cables will see the replacement of the old by the new.

The past 30 years have seen three periods of submarine cable construction. (1) From 1999 to 2003. From 1995, when the Internet emerged, relevant control, demand, technical progress and large capital input promoted the rapid development of the submarine cable industry and a large number of the cables were put into use. (2) From 2009 to 2013. In this period data centers were the major drive for the construction of international submarine cables, which witnessed another construction wave. (3) From 2016 till now. This period has seen the upgrade and replacement of cross-Pacific, cross-Atlantic and Asian and European submarine cables, increase of Internet traffic between Latin America and the United States, and between Europe and Africa, and effort of Southeast Asia and South Asia to build regional information centers. Influenced by these factors, there was a great growth of the number of new submarine cables in 2016, especially those around the Indian Ocean and those between Europe and Africa. With the interconnection of data centers and the growth of the demand for Internet bandwidth, it is estimated that in the coming years there will be a new wave of construction of submarine cables (see Fig. 2.9).

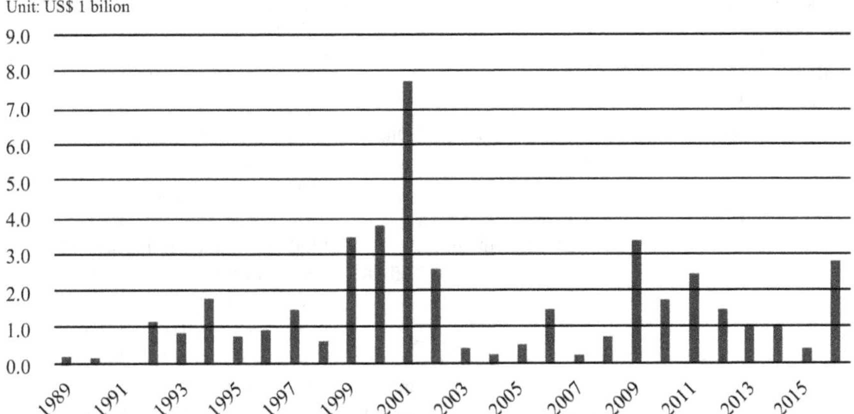

Fig. 2.9 Investment in submarine cable systems (1999–2016) (*Source* STF analytics)

4. **Internet giants are the major force in submarine cable construction.**

The thriving of the Internet urges Internet giants to participate in the worldwide submarine cable construction. In recent years, to meet the demand of their own data center traffic and business flexibility, Internet businesses no longer only rent the bandwidth of telecommunication operators. They have joined in the submarine cable construction and operation. In 2009, Google participated in the construction of UNITY, the cross-Pacific international submarine cable, marking the beginning of Internet businesses' involvement in international submarine cable construction. In the following eight years, Internet giants such as Google, Microsoft and Facebook participated in the investment and construction of more than 10 such cables worldwide, becoming one of the major forces in international submarine cable construction, which covers Asia, Europe, and Americas, including UNITY and FASTER connecting the United States and Japan, AMX-1 the United States and Brazil and other South American countries, APG connecting Japan, Taiwan of China, Vietnam and Singapore and other Asian countries and regions, and AEConnec connecting the United States and Europe. Internet giants' involvement in international submarine cable construction, operation and management has enhanced their power of discourse and management right in the international submarine cable construction (see Table 2.1).

Table 2.1 Major international submarine cables in whose construction internet businesses have participated

Internet business	Investment time	Partners	Undersea optical cables
Google	2009	Bharti, KDDI, Global Transit, Pacnet and SingTel	UNITY connecting the United States and Japan, 6200 miles in length, put into operation in 2010
	2010	Global Crossing, KDDI, Network i2i, Reliance, Telemedia Pacific and Telkom	SJC connecting Singapore, Indonesia, the Philippines, Hong Kong of China and Japan, 8300 km in length, with the capacity of about 20 Tbps
	August 2014	China Mobile, China Telecom, Global Transit, KDDI and Singapore Telecommunications Limited	FASTER connecting Japan and West Coast of the United States, with the capacity of 60 Tbps
	October 2014	Algar, Angola Cables and Antel	AMX-1 connecting the United States, Brazil, Mexico, Columbia and the Dominican Republic, with the capacity of 64 Tbps
Internet business	Investment time	Partners	Submarine cables
Google	November 2015	Antel Uruguay	Tannat connecting Uruguay and Brazil, 2000 km in length, with the capacity of 90 Tbps
Microsoft	2016	Aqua Comms	AEConnec connecting the New York State and the west coast of Ireland, 5400 km in length, with the planned capacity of 100 Gbps
	2015	GTT	Express, connecting Canada, Ireland and the United Kingdom, with the planned capacity of 10 Tbps
	Being planned	China Mobile, China Telecom, China Unicom, Korea Telecom and Softbank	NCP connecting the United States, the Chinese Mainland, Republic of Korea, Japan and Taiwan of China
Facebook	2016	NTT, China Mobile, China Telecom, China Unicom, VNTT, LG Uplus, Global Transit, Starhub	APG connecting the Chinese Mainland, Republic of Korea, Japan, Hong Kong of China, Taiwan of China, Vietnam, Singapore, Malaysia and Thailand, 10,400 km in length, with the capacity of 54.8 Tbps

Source Telegeography, Wikipidia and relevant Internet businesses

Table 2.2 Programs of low and medium-orbit constellations

Business	Orbit position	Program	Partners
O3b	Medium-orbit MEO	12 satellites already launched, and orders of 8 satellites signed in 2016	SES, Google and other partners
OneWeb	Low-orbit LEO	Planning to send the first 10 satellites in early 2018, to launch the overall project after 2019, including 720 near-earth orbit satellites, and to completely solve the broadband Internet access problems	Airbus, Coca Cola, Qualcomm, Virgin Group and other partners
SpaceX	Low-orbit LEO	Planning to lay out in the space from 2019 to 2024 4425 orbital plane satellites with its own Falcon 9 reusable rocket	

2.1.6 Space-Based Networks Are Emerging, Promoting Broad Band's Overall Coverage

1. **High-throughput broadband satellites are developing rapidly.**

Technical progress is promoting the development of high-orbit broadband satellites. Satellite businesses such as Hughes, Viasat and Eutelsat are active in developing emerging technologies, and laying out new large-capacity communication satellites' launching programs. The single capacity of high-orbit broadband communication satellites has increased dramatically and the designed capacity of Viasat-3 developed by Viasat will reach 1 Tbps. By the end of 2016, there were 59 high-orbit broadband satellites in the world, including 13 totally Ka-band satellites, and the Internet access capacity of global high-orbit broadband satellite system had reached 600 Gbps, able to provide service for 20 million users. By 2020, the world will launch another 30 high-orbit broadband satellites, including 10 total Ka-band satellites.[1] The service provided by the broadband satellite communication system is developing from low-speed service and voice service to high-speed Internet access and multi-media service. The Ka-band satellite communication system has been widely used in high-speed network access, HDTV, SNG, DTH and personal satellite communication in the countries and regions like the United States, Canada and Europe.

2. **The global satellite Internet system is thriving.**

In recent years, commercial satellite businesses represented by O3b, Oneweb and SpaceX are trying to construct the global satellite Internet system (see Table 2.2) based on the constellations of low and medium-orbit broadband satellites. At present, O3b has 12 medium earth-orbiting satellites, with the initial cluster capacity of 192 Gbps, serving more than 40 clients, of whom active clients are from

---

[1]Source: ITU.

31 countries, covering a large area of the world, including tropical areas, a number of island chains on the Pacific Ocean, marine drilling platforms and big cruises. OneWeb has finished application and coordination for the frequency orbit position resources, planning to launch 720 near-earth orbit satellites. It has completed the Internet coverage of ground areas and its designed capacity of the whole system reaches 8 Tbps, with the downlink speed of 750 Mbps and uplink speed of 250 Mbps.

3. **Floating-platform communication technology is progressing in exploration.**

In recent years, ground communication technology represented by optical fiber broadband and 4G technology is thriving, but there still exist blind spots of communication. For instance, ground communication networks are not available everywhere while it is an international vision for such networks to be available everywhere. About 60% of population in the world cannot access to the Internet through ground communication facilities. Facebook and Google are seeking to develop floating-platform communication technology to provide universal network access service. For example, Google is experimenting with the hot balloon project to provide cheap Internet access service and Facebook uses high-flying aircrafts to provide universal network access service. Europe, the United States and Japan are also doing research on and testing low platform communication systems like drones and hot balloons to meet the emergency communication demand, for instance, in disaster reliefs and rescues. FCC of the United States has issued a white paper entitled "The Role of Deployable Aerial Communications Architecture in Emergency Communications and Recommended Next Steps" and the FirstNet of NTIA has promoted the "deployable aerial communications architecture in emergency". The U.S. public security broadband network is considering using drone base stations in the areas subject to threats but not covered by the regional or ground facilities of the FirstNet. Two weeks after Hurricane Maria happened in 2017, 82% of the islands in Puerto Rico were still unable to reuse the mobile communication service. Then FCC issued the experiment permit to Google, permitting the company to provide Internet and cellular services with the large floating balloon to restore the communication service of Puerto Rico's basic facilities destroyed by Hurricane Maria.

2.1.7 *Basic Internet Resources Are Distributed Unevenly and the Network Strength of Different Countries Needs to be Balanced*

1. **The global IPv4 addresses have been distributed and the US ranks first in the number.**

The global IPv4 addresses were distributed completely in 2011, totally 3652 million, with the present broadcast rate of 78%. The United States has 1.61 billion such addresses (accounting for 44.08% of the total),[2] ranking first in the world. Then come China and Japan.

2. **The United States is the leader in the number of IPv6 addresses but the global broadcast rate remains low.**

In the first half of 2017, the number of distributed IPv6 addresses in the world was 215,970 blocks (/32), including about 43,970 blocks (/32) in the United States, accounting for 10.17% of the total. China and Germany rank second and third in the number of distributed IPv6 addresses. The use rate of the distributed IPv6 addresses remains low, 23.7% in the first half of 2017. Only in seven countries the broadcast rate was over 1%, with that of the United States as the highest, as high as 32.3% (see Fig. 2.10).

3. **The United States ranks first in AS Code number but the notice rate worldwide remains low.**

The total number of AS codes applied for in the world is 78,962 and the AS notice rate is around 60% (see Fig. 2.11). The United States has 24,944 AS codes, ranking first in the world.

4. **There are more domain name resources, and layout of resolution facilities is accelerating.**

By March 2017, there were over 330 million domain names in the world, an increase of 6.8%[3] in comparison with the number in 2015. Among them, there were 120 million registered with COM, ranking first in the world. With the increase of the number of registered domain names, the domain name facilities keep being improved. By August 2017, there are 788 global root servers and their mirror servers, covering 138 countries and regions and providing nearby root resolution service (see Fig. 2.12). Among them, 146 root servers and their mirror servers were in the United States,[4] accounting for nearly one fifth, so the country ranks first in this respect.

2.1.8 *Universal Telecommunication Service is being Carried Out and More People Can Share the Development Results of Broadband*

In recent years, broadband is playing an increasingly strategic, basic and public role in enhancing the national competitiveness and economic development, with obvious spillover effects. Most countries, based on their existing universal

---

[2]Source: www.resources.potaroo.net.

[3]Source: verisign.

[4]Source: www.root-servers.org.

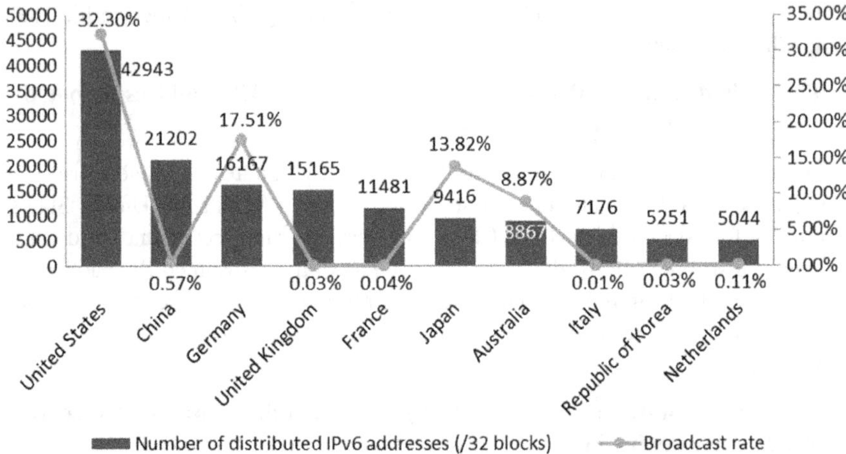

Fig. 2.10 Top 10 countries of distributed IPv6 addresses and their notice rate (*Source* www. resources.potaroo.net)

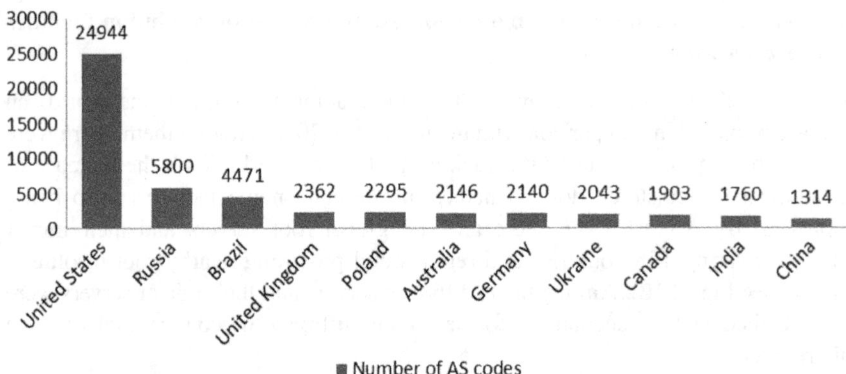

Fig. 2.11 Global application for AS codes (*Source* www.resources.potaroo.net)

telecommunication service, have started to innovate and implement mechanisms in accordance with their own situation to promote the popularization of broadband service in multiple ways.

1. **Universal telecommunication service is shifting to broadband service.**

Universal telecommunication service in all countries has been shifting from initial telephone service to broadband service. 71 countries have listed dial-up networking (DUN) into universal service, and 52 have channeled broadband service into universal service, so that more communities can be involved in the public service system based on broadband networks. The UK Government has adopted a law to recognize the users' right to use broadband, stipulating that every household or business in the country has the right to enjoy the broadband service at the speed

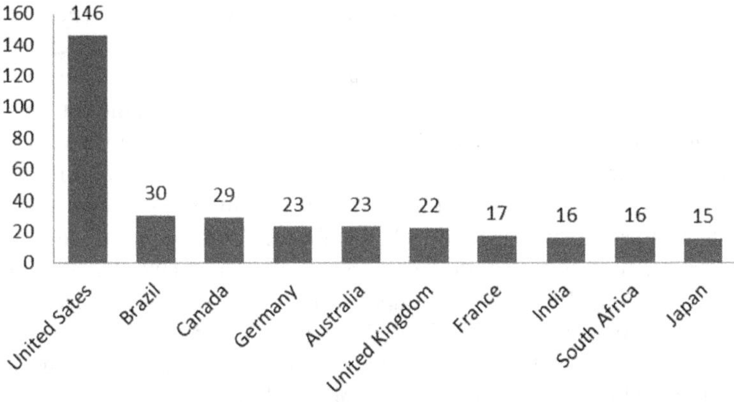

Fig. 2.12 Top 10 countries in the number of root servers and their mirror servers (*Source* rootserver)

of at least 10 Mbps. Though varying in universal telecommunication service content, all countries manage to extend their telecommunication service into rural and remote areas with high input cost and low return, where the market mechanism cannot work well; and increase the rate of broadband in urban communities and rural areas where the network service is backward, providing every household with certain broadband transmission and encouraging the coverage of fixed broadband and mobile broadband in remote areas.

2. **Universal telecommunication service ways keep being innovated.**

In the era of broadband service, all countries are innovating their mechanism. Many countries' governments adopt frequency spectrum auction and appropriate special fund to promote universal broadband service. The German Government plans to set up a gigabit fiber network construction foundation for rural areas through frequency spectrum auction to solve the shortage of fund for fiber network construction in rural areas. Ministry of Information and Communication Technology of Thailand has completed a bidding document for the national broadband project of THB15 billion (about US$430 million) to solve the broadband access in 39,000 villages. Besides, telecommunication operators are active in providing the universal telecommunication service. AT&T has adopted the "AT&T Connect" Program, providing the broadband service of 10 US dollars every month for low-income families that meet the service standard in Illinois State. British Telecommunication announces that by 2020, 10 Mbps broadband access will be provided nationwide in accordance with universal service obligation (USO).

There are altogether five models of promoting universal broadband service. (1) The universal telecommunication service foundation plays the dominant role while the short-term fund of the government acts as the complementary mechanism. This model has been adopted by the United States, India and Turkey. For example, in the United States, the Federal Communications Commission (FCC) set up Connect America Fund (CAF) in October 2011 and appropriated US$1.5 billion of fund in 2015 to expand the broadband connection in rural areas. (2) Public–private Partnership (PPP) promotes the broadband construction in high-cost areas. EU countries have adopted such a model. Take France as an example. The PPP model has reduced the areas uncovered by broadband by half since 2005, with the total investment of 3.4 billion euros, half of which come from local governments and the other half from private businesses. (3) At the "Digital Dividend" frequency auction, mobile broadband's prior coverage of rural population is taken as an additional premise. Germany and Sweden have adopted such a model. For instance, Germany was the first to auction 800 MHz "Digital Dividend" frequency band in Europe in May 2010, requiring the bid-winning operators to cover 90% of the population in priority in 13 white regions before they can use 800 MHz networking in urban areas. (4) National broadband network businesses are set up with the government playing the dominant role to undertake the national network construction. Australia and Singapore have adopted such a model. The former proposed in 2009 an investment of 43 billion Austrian dollars in setting up the NBN Company to construct a super-speed broadband network covering the whole country by making use of fixed, satellite and mobile broadband to expand the broadband coverage in rural areas. (5) Businesses participate in wireless network coverage in poor areas. Facebook has launched Express Wi-Fi to make the Internet accessible to everyone. It has launched interent.org to developing countries to help low-income and rural people from emerging markets to access to the Internet.

2.2 Application Infrastructure: Rapidly Developing into Large-Scale and Intelligent Infrastructure

2.2.1 Data Centers and Cloud Computing Platforms Are Centralized and the Collaboration Between Government and Market Is Successful

Because of the requirement for energy saving and consumption reduction, cost lowering and the best service, all countries' governments, public cloud service providers and third-party data center hosting service providers are promoting the centralized construction of data centers and cloud computing platforms. According to the data released by IDC, after the construction of data centers of the world thrived from 2001, now the number of newly-built data centers worldwide is decreasing, but the single center construction scale is increasing and some data centers with high energy consumption have been closed. By 2021, the number may decrease by about 15% in comparison with that in 2015.

1. **Global data centers are mainly located in the United States, China, Japan and the United Kingdom.**

In 2016, Synergy Research Group made a survey on the construction of big data centers of 24 leading cloud computing and Internet service businesses of the world. The result shows (see Fig. 2.13) that all countries are promoting the construction of data centers, but 45% of large ones are in the United States, where the number of ultra-large data centers ranks first. The number of large data centers in China and Japan is respectively 8 and 7%, ranking second and third in the world. Then comes the number of such centers in the United Kingdom, Australia, Canada, Singapore, Germany and India, with the percentage at 3 to 5.

2. **The government is the major force in promoting the integrated optimization and centralization of data centers.**

From 2010 to 2016, the United States launched a series of policies, including Federal Data Center Consolidation Initiative (FDCCI), Federal IT Acquisition Reform Act (FITARA) and Data Center Optimization Initiative (DCOI) to guide the integrated optimization and centralization of data centers, reducing the number of governmental data centers through cloud computing application and promoting the large-scale and green development of data centers. DCOI launched in 2016 proposes that the U.S. Government should at least close within three years 25% of Tier-level data centers (large data center facilities) and 60% of non-Tier-level data centers, which means that 52% of data centers will be closed by the Federal Government. EU has launched the EU *Code of Conduct of Data Centers*, requiring

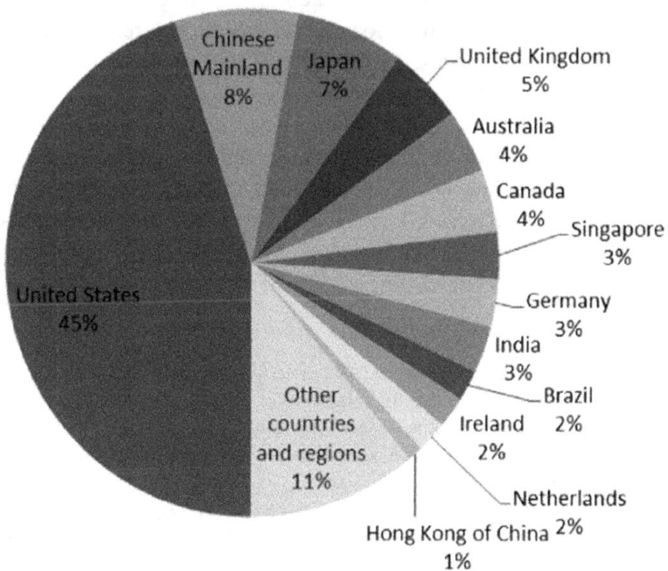

Fig. 2.13 Distribution of data centers of 24 leading internet businesses (till December 2016) (*Source* Synergy Research Group)

that the providers of all data center devices should develop and use energy-efficient and low-energy CPU and encouraging to adopt virtualized ways like software to manage energy consumption and to improve the efficiency of servers. China released in 2013 *Guiding Opinions on Data Center Construction and Layout*, giving suggestions on layout of data centers of different scales, requiring to integrate old data centers, encouraging administrative authorities to use cloud services provided by professional institutions and to reduce the number of governmental data centers, and urging enterprises and public institutions to use cloud service.

3. **Data centers and cloud computing platforms are mostly located in central cities.**

Economically developed areas attract commercialized data centers and cloud computing platforms because of the intensive resources like talents, industries, capital, and business headquarters, as well as high-level information technology application and high demand for IT services there. Globally, central cities, such as London, Singapore City, Frankfurt, Tokyo and Sydney, have high demand for data centers and cloud computing platforms. In the United States, the demand is centered in North Virginia (Washington), North California (San Francisco), Chicago and Seattle, which are all coastal cities or areas. In China, the demand is centered in Beijing, Guangdong, Shanghai, Zhejiang and Jiangsu, which are economically developed regions. According to the survey by Synergy Research Group, by December 2016, 24 businesses had constructed more than 300 ultra-large data centers, with Amazon, Microsoft and IBM each having over 40 (see Table 2.3). Data centers of Google and Oracle are widely distributed, while those of Apple, Twitter, Facebook and eBay are mostly located in the United States. Data centers of Alibaba and Tencent were mainly located in China before, but now they are expanding into the United States, Europe, Japan and so on through independent construction or co-construction.

Data centers of leading third-party data center hosting service providers like Equinix and Digital Realty Trust are also located in developed cities. Equinix is the leader on the global retailing hosting data center market. According to the information on its websites, it has 176 data centers worldwide, distributed in Americas, the Asia-Pacific Region and EMEA (Europe, the Middle East and Africa). Digital

Table 2.3 Global Cloud Infrastructure Layout of Amazon and IBM

Amazon AWS	Operating 44 available zones (AZ) in 16 geographical areas, all AZs distributed in developed cities with a high demand for cloud computing. Planning to expand to five areas, namely, the Chinese Mainland, France, Hong Kong of China, Sweden and the second AWS GovCloud zone in the United States; and to set up 14 new AZs
IBM BlueMix	Operating 39 data centers, distributed in North America, South America, Europe, Oceania and Asia, covering 16 countries, with over 450,000 servers, 17 centers in the United States, 3 in the United Kingdom, 3 in Australia, 2 in Holland, 2 in Canada, 1 in China, 1 in Germany and 1 in Italy

Source Amazon and IBM

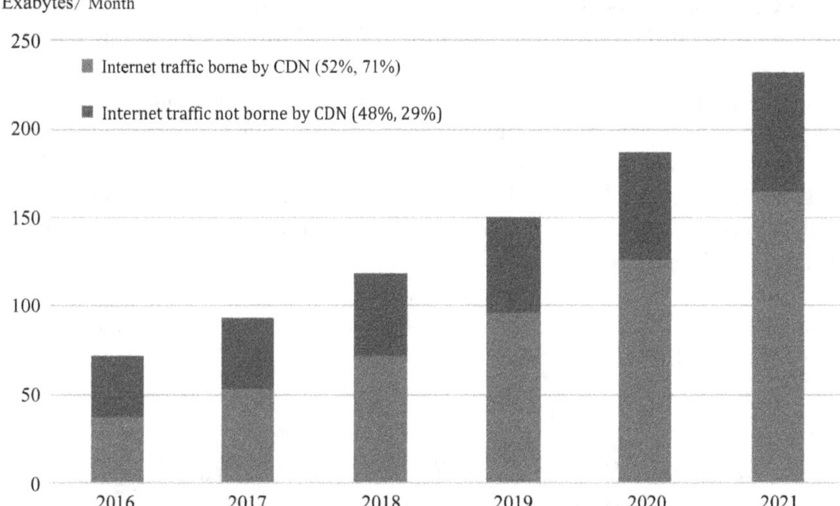

Fig. 2.14 Internet traffic borne by CDN in the world (2016–2021) (*Source* www.cisco.com)

Realty Trust is the provider with the largest share on the global wholesale hosting data center market, having 157 data centers in the world, 119 of which are in North America, almost monopolizing the US data center market. It has 31 data centers in Europe and 7 in the Asia-Pacific Region.

2.2.2   *Content Delivery Network is developing Fast and Its Service Availability Keeps Improving*

Content Delivery Network (CDN) is a delivery network made up of node server clusters distributed in different zones. Contents like Internet websites, online videos and online games are delivered to the edge of the websites close to the users, who can get the needed content from nearby, so that the cyber jam can be relieved, the response speed of Internet services be increased and the users' experience be improved. As an indispensable basic facility of the Internet, CDN produces revolutionary impacts on traffic control. It becomes a commercial service in 1998 and since then has witnessed rapid development. By 2016, the Internet traffic borne by CDN had accounted for 52% of the total.[5] According to Cisco, over 70% of Internet traffic will be carried by CDN by 2021 (see Fig. 2.14).

Akamai, the largest CDN service provider in the world, has laid out 230,000 servers in over 1600 operators' network of over 130 countries and regions. There is only one hop between 85% of Internet users and Akamai's CDN servers,

---

[5]Source: www.cisco.com.

whose CDN capacity peak can reach 30 Tbps. CDN's integration with cloud and expansion toward the edge enable all kinds of terminal users to get all kinds of content and service at the highest rate and the lowest cost and can complete smart processing of data. By sinking towards the user terminal and combining with P2P and edge computing technology, CDN can control the Internet traffic within local networks. According to MarketsandMarkets data, the CDN market scale in 2016 reached US\$ 6.05 billion, with the 5 year compound annual growth rate of 22%. North America is the leading market of CDN. 70% of the CND service vendors of Alex Top 1000 businesses are from the United States. With the steady increase of global Internet traffic, it is anticipated that by the year of 2022, the global CDN market scale will amount to US\$30.89 billion, with the Asia-Pacific market scale exceeding that of North America.

2.2.3 Intelligent Sensor Facilities Witness Integrated and Innovative Development and Internet of Everything Is Possible

1. **Low-Power Wide-Area Network (LPWAN) technology brings about new ways of wireless connection of IoT.**

IoT promotes the information society into the era of Internet of Everything. Especially, LPWAN, the technology specially designed for IoT, brings about new ways of wireless IoT connection, promoting the rapid growth of the IoT connection number. LPWAN is categorized into two kinds. The first kind includes NB-IoT and eMTC working at authorized frequency spectrum, based on cellular mobile communication technology. The second kind includes LoRa and SigFox, working at unauthorized frequency spectrum. According to Cisco statistics, the global cellular mobile communication networks bore 800 million IoT connections in 2016, and those borne by 2G, 3G, 4G and LPWAN respectively accounted for 29, 40, 23 and 8% and they are expected to account for 6, 16, 46, and 31% respectively by 2021. 4G and the subsequent technology will become of major IoT bearers. As predicted by Ericsson, the number of globally connected IoT devices will reach 16 billion, including 1.5 billion of IoT devices connected by cellular mobile networks.

2. **NB-IoT promotes the mobile communication industry to expand into the area of IoT.**

Standard organization like 3GPP and the mobile communication industry have shifted their attention to low-power big-connection scenarios represented by IoT and low-delay high-reliability scenarios represented by Internet of Vehicles and other wireless communication areas. Thus NB-IoT based on LTE technology emerged and promoted the mobile communication industry to expand into the area of IoT. At present, a number of operators in the world keep up with the NB-IoT development, carrying out verification, testing and layout of existing networks as

Table 2.4 Opened NB-IoT and eMTC network operators

Model	Number of networks for commercial use	Operators
NB-IoT	8	Telus in Canada, T-Mobile in Holland, Telia in Norway, Vodafone in Spain, German Telecom, Vodacom of South Africa, China Telecom and China Unicom
eMTC	2	Verizon and AT&T of the United States

Source GSA

soon as the standard protocol is finished. According to GSA statistics, by June 2017, 8 NB-IoT and 2 LTE-M (eMTC) networks had been opened for commercial use, and 55 NB-IoT and 16 LTE-M (eMTC) networks had witnessed construction of test networks (see Tables 2.4 and 2.5).

3. **IoT technology not based on cellular mobile communication is developing fast and has occupied the LPWAN market.**

LoRa and SigFox are typical LPWAN technologies working at unauthorized frequency spectrum. Compared with NB-IoT and eMTC based on cellular mobile technology, they were launched into market earlier and they have commercial networks laid out worldwide, with simple architecture and low operation cost. On the other hand, since they work at unauthorized spectrum, there may be potential disturbance and there is no clear evolution route for non-standard technology. Therefore, they are oriented to business applications with no high demand for reliability and security but for quick customization. By August 2017, there were 47 LoRa operating networks in the world, more than 350 cities were testing or laying out the technology and there were over 500 LoRa Alliance members. LoRa

Table 2.5 Layout of partly-opened NB-IoT operators

Operators	Layout
Vodafone	Spain: 1000 sites, covering Madrid, Valencia, Barcelona, Sevilla, Malaga and Bilbao Besides, Vodafone is constructing NB-IoT networks covering the whole Holland and Ireland
German Telecom	Germany: the first networks are laid out in Cologne, Bonn, Dusseldorf, Ruhr, Stuttgart and Berlin. German Telecom is laying out NB-IoT networks in other regions
T-Mobile	Holland: NB-IoT was first laid out in Amsterdam, Rotterdam, Hague, Eindhoven and the surrounding areas of the Schiphol Airport and was planned to cover the whole country in 2017.the United States: In July 2017, collaborating with Ericsson and Qualcomm, T-Mobile completed NB-IoT technology testing at a number of sites of Las Vegas by using AWS frequency spectrum of 200 kHz
Telia of Norway	At the end of 2016, Telia and Huawei launched the first NV-IoT network of North Europe in Oslo of Norway and the world's first smart agricultural service based on NB-IoT

Source GSA

supports 0.3–50 Kbps data transmission, applicable to smart terminals in need of long-time small-quantity data sharing. LoRa Alliance is promoting standard LRaWAN protocols worldwide, so that all devices conforming to LoRaWAN standards can be connected with each other. SigFox is a French company. For SigFox networks, UNB technology is adopted to provide a low-power and wide-coverage solution. SigFox has networks in 32 countries, including France, Spain, Holland and the United Kingdom, covering a population of 589 million and bearing 8 million devices.

2.2.4 Internet Exchange Points Are Witnessing Obvious Polymerization Effect

Internationally-used physical interconnection ways are Direct Circuit Interconnection and exchange center interconnection. Internet exchange points work as basic platforms for interconnection among different parties, so they have many access members and varieties of services, with high flexibility and scalability. They have become important interconnection facilities and keep developing worldwide.

1. **Internet exchange points are expanding worldwide fast into Latin America, Africa and emerging regions.**

Internet exchange points can promote traffic localization and improve source access quality, so they have witnessed rapid development in the world with the popularization of the Internet. By far, there are nearly 500 such centers in 147 countries[6] (see Fig. 2.15). The hot markets of Internet exchange points are shifting from Europe and America to Africa and Latin America. In 2016, there were 43 such centers in Africa, 1.26 times of the number in 2010.

2. **Internet exchange points are witnessing obvious polymerization effect, and there have emerged super-large ones.**

Internet exchange points attract the access of a large number of networks. According to PCH statistics, there are six large Internet exchange points with the access of over 500 networks (see Fig. 2.16). PTT Metro-São Paulo Exchange Point located in Brazil of Latin America has 1355 accessing networks, ranking first in the world. As for polymerization effect, the annual growth rate of the traffic of global Internet exchange points is 50–100%, and the traffic borne by large ones can be calculated by Tbps. PCH statistics show that there are seven Internet exchange points in the world whose peak traffic is calculated by Tbps (see Fig. 2.17). Six of them are located in Europe except PTT Metro-São Paulo in Latin America.

---

[6]PCH statistics, September 2017.

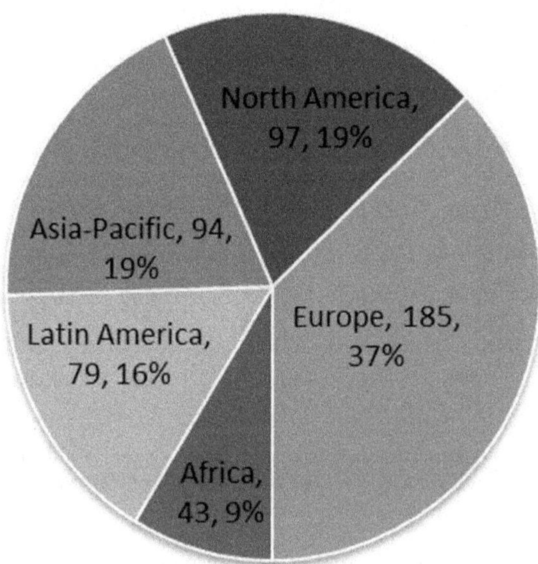

Fig. 2.15 Number of internet exchange points in different regions of the world (*Source* PCH statistics, and websites of different Internet exchange points, September 2017)

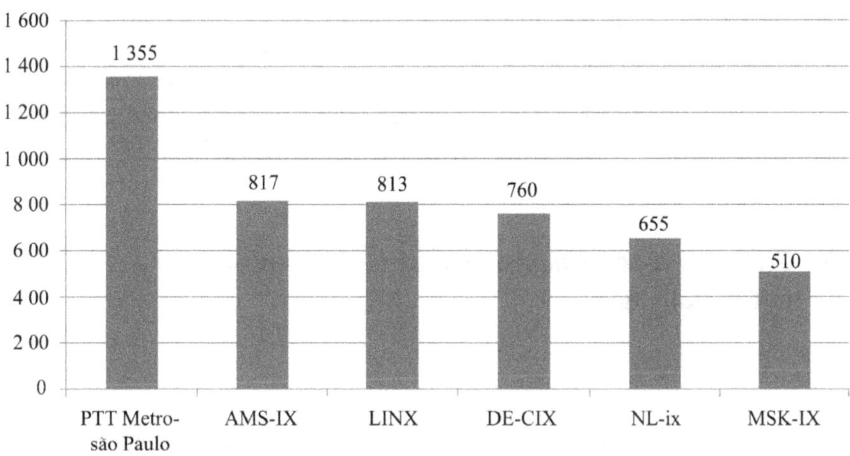

Fig. 2.16 Number of networks accessing to large internet exchange points in the world (*Source* PCH statistics, and websites of different Internet exchange points, September 2017)

3. **Internet exchange points promote global networks to evolve toward flattening architecture.**

With the rapid development of the Internet and the explosive growth of the Internet content, Internet service providers (ISPs) are not the only body in the network

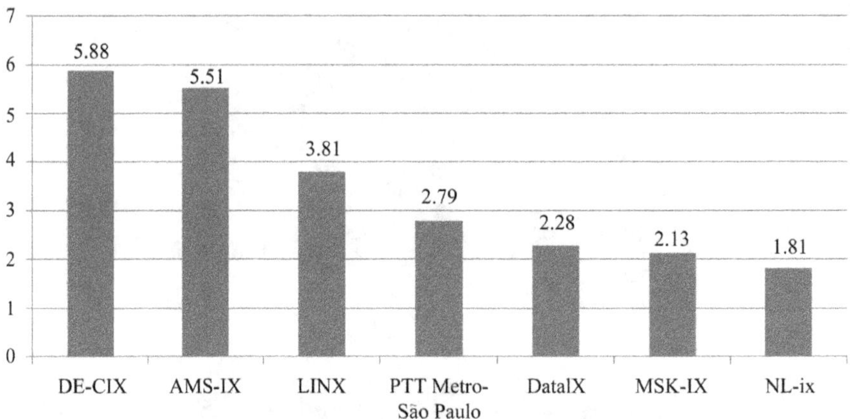

Fig. 2.17 Peak Traffic of Large Internet Exchange Points in the World (Tbps) (*Source* PCH statistics, and websites of different Internet exchange points, September 2017)

ecosystem. A large number of Internet content providers, data centers, CDNs, cloud service providers and domain name/Internet monitor facilities have independent networks, so they are in urgent need of interconnection. The emergence of Internet exchange points can meet the emerging networks' demand for low cost and wide coverage, so these exchange points enjoy high popularity among emerging networks, especially small and medium-sized ones. Euro-IX statistics show that Internet businesses have become the major force accessing to exchange points. With the diversification of accessing forces and increase of the number of exchange points, global Internet architecture has shifted from strict hierarchical division to flattening.

2.3 Future Trend: Coexistence of Opportunities and Challenges

2.3.1 Challenges

1. Rapid iteration of network technologies has widened the digital divide.

With the rapid development of information society, the international digital divide is becoming wider in the global development. Only 31.2% of the population in Africa uses the Internet and only 10% of Internet users in the world are from Africa, far lower than the percentage of African population (16.6%) in the world's total (see Table 2.6). In recent years, despite the improvement of technical infrastructure in some backward countries, with the accelerated application of cloud computing, big data, IoT, 4G/5G, and optical fiber broadband, the digital divide has been shifted from inequality in "number" and "access" into inequality in the function upgrade

Table 2.6 Global internet use and population statistics

Region	Percentage of population of the world's total (%)	Penetration rate of the Internet (%)	Percentage of Internet users of the world's total (%)
Africa	16.6	31.2	10.0
Asia	55.2	46.7	49.7
Europe	10.9	80.2	17.0
Latin America	8.6	62.4	10.4
The Middle East	3.3	58.7	3.8
North America	4.8	88.1	8.2
Oceania	0.5	69.6	0.7
Global total	100.0	51.7	100.0

Source www.internetworldstats.com

of the new-generation information infrastructure and the change of the users' experience and capability. Meanwhile, due to their technology, fund and personnel quality, digital divide among developing countries is widening. Those with backward information infrastructure will inevitably fall behind in the new-round economic development and the increasing information difference between countries will inevitably be a great barrier for the global social governance.

2. **Mobile broadband is attached importance to while fixed broadband is neglected.**

In comparison with fixed broadband (see Figs. 2.18 and 2.19), mobile broadband access can make the Internet visit available in any place at any time, and provide individual access with personalized applications, so it is the focus of development of all countries, who are promoting 4G network coverage and 5G technology R&D. Developing countries rely more on mobile broadband than on fixed broadband. In Africa, the popularization rate of household fixed broadband is less than 6% and most people "get in touch with the Internet" through mobile terminals. In most developing countries, backbone fiber optical networks and MAN networks are not improved, optical fiber access networks are universally backward and have little coverage, and the existing Internet access relies on wireless networks, which limits the rapid Internet application popularization and informatization in these countries and is unfavorable to the layout of 4G and 5G networks.

3. **Inland countries are lack of opportunities of accessing to the Internet.**

At present, the content sources of the Internet and data centers are located in the developed countries such as the United States, the United Kingdom, France, Germany, Sweden and Singapore. Many developing countries have to access to these sources and information through the Internet. According to an ESCAP report, the countries owning submarine cable landing stations can access to the Internet at a low price and a high speed and in a high quality, but those lacking submarine cable land stations can hardly benefit from the Internet. Due to the lack of feasible

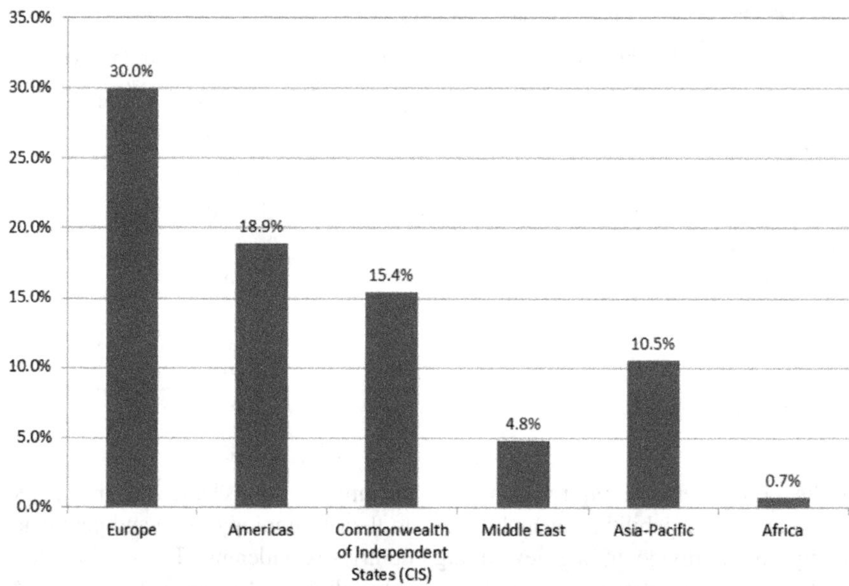

Fig. 2.18 Fixed broadband popularization rate in all continents and regions at the end of 2016

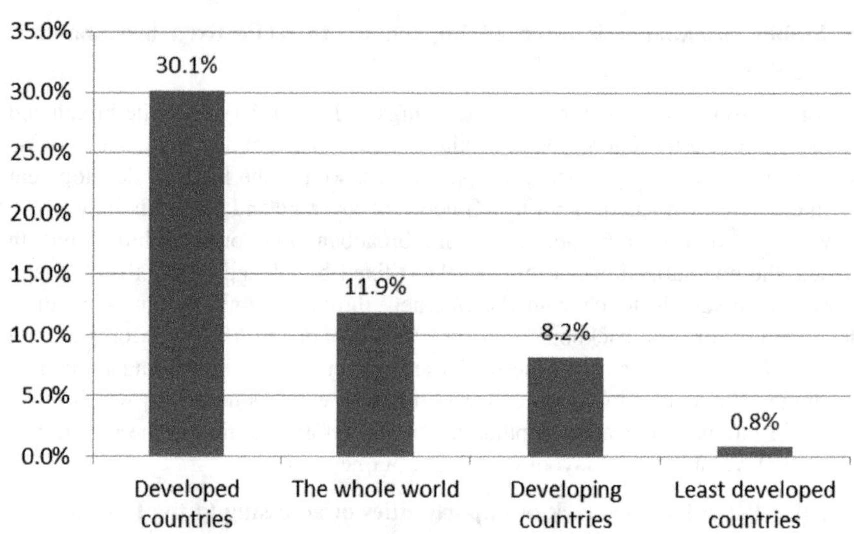

Fig. 2.19 Fixed broadband popularization rate in developed and undeveloped countries at the end of 2016

cross-border land cable operation models, the latter countries cannot access to submarine cable networks but rely more on international communication satellites to access to the Internet at a high cost and lower speed, which hinders the popularization of Internet services and applications in these countries.

2.3.2 *Opportunities*

1. **Development demand: digital economy**

With the development of information technology and digital economy, it is the new focus of global economic development to speed up the integration between digital technology and industry and to stimulate the new drive for all industries' development. In the past two or three years, the international community has been promoting the National Digital Initiative, taking digital economy as the key direction of tapping new sources for growth and improving the mid and long-term growth potential. Countries like Germany, the United Kingdom, France, Poland, India and Australia have all launched their digital strategies to comprehensively promote the in-depth integration between information technology and real economy, speed up the digitalization and intelligentization of traditional industries, and create new advantages in digital economy development to realize the prosperity and sustainable development of their economy.

The *Digital Economy Strategy* of the United States is oriented towards "economic transformation", with the aim to realize the transformation growth of economy through the use of the advantages of ICT. The measures include "enhancing infrastructure construction, giving full play to the basic role of the IT industry, strengthening the cultivation of IT talents and completing the digitalized transformation of economy and society". The *Advanced Manufacturing and Industrial Internet Strategy* of the United States, Industry 4.0 of Germany and New Industrial France share the core that the superimposed advantages and comprehensive advantages can be made the best use of through the integration of the IT and manufacturing, new materials and new energy to win the initiative of the new industrial revolution.

2. **Development foundation: upgrade of the new-generation ICT**

At present, the gigabit networks are being popularized rapidly and breakthroughs have been made in new mobile technologies of 5G and NB-IoT. Cloud computing platforms, CDN and IoT are important new types of application facilities and SDN/NFN technology has been introduced and laid out. Network infrastructure is developing rapidly in the direction of the integration of perception, transmission, storage, computing and processing. Upgrade is being launched with intelligence, sensitivity and openness as the goal. To be short, the upgrade of the new-generation ICT provides an important opportunity of the global network evolution and upgrade as well as the construction of advanced information infrastructure.

3. **Development guarantee: more governmental attention**

Since the financial crisis, all countries have listed the development of information infrastructure as priority. By the end of 2016, over 140 countries had implemented broadband strategies or initiatives, strengthening the support from governments and markets with the common goal of giving full play to the role of information infrastructure in the building of digital society. They are accelerating social digitalization by improving the coverage and popularization rate of broadband networks.

In the programs of digital economy development of all countries, information and communication infrastructure construction is the priority. In September 2016, EU adopted *Connectivity for a European Gigabit Society*, which proposes that all public places (including schools) should have access to 1 Gbps transmission by 2025 and that all European households should have access to networks offering a download speed of at least 100 Mbps. At the same time, EU launched the *5G Action Plan*, in the hope that the first pilot will be realized by 2018 and the technology will be commercialized by 2020. 5G technology will cover all urban areas and railways and roads by 2025. In March 2017, the United Kingdom issued *Next Generation Mobile Technologies: An Update to the 5G Strategy for the UK*, which proposes that from 2017 to 2018, the UK Government will collaborate with research institutions in carrying out 5G technology tests and providing end-to-end 5G tests from 2018 to 2019. The United States is the first country to have allocated large-scale high-frequency spectrum resources for 5G technology, which will be put into commercial use in 2018.

2.3.3 Prospect of Development

1. The evolution and upgrade toward gigabit networks will be accelerated.

The gigabit network will be the development direction of fixed broadband access, which has been universally recognized. Gigabit networks have become the goal of China and Germany in their development of information technology. With the emergence of VR/AR service and development of 4K/8K HD videos, the service carried by gigabit networks is becoming clear, and HD videos, VR, home security, smart home and tele-health will stimulate the demand for broadband. Telecom equipment manufacturers like Huawei and ALE have launched gigabit broadband solutions while China Telecom, AT&T, BT and NTT have begun to lay out gigabit broadband networks. Optical access, copper cable access and cable access can all leap into gigabit broadband access, with the evolution of wired access technology. On the other hand, fiber optic facilities are important to the future development of mobile and fixed broadband, so optical access will be the dominant access, which will promote the high-speed development of future broadband.

2. The R&D and testing of 5G devices will be accelerated.

5G technology is the strategic focus of the development of mobile communication. EU, the United States, Republic of Korea, Japan and China are leaders in 5G technology development, having proposed development plans to promote the R&D, testing and verification and commercial layout of 5G technology through major projects. 3GPP accelerates the formulation of complete 5G standards and the framework of the standard technology will be formed. It is anticipated that by the end of 2019, the complete 5G standard meeting the ITU demand will have been completed and will support three scenarios, namely, eMBB, mMTC and URLLC. eMBB will be the prior 5G scenario to be laid out by telecommunication operators while vertical industrial application will be centered in URLLC and mMTC. For instance, the layout model of Internet of Vehicles and industrial Internet will be clear and the popularization and application of IoT will become one of the important drives for technical progress.

3. **Space-air-ground networks will promote the popularization of broadband.**

The broadband network has developed from fixed one into mobile and it will develop into a space-air-ground one. Now, American and European powers and large Internet businesses are strengthening R&D of space technologies and trying to build new space Internet platforms. Multiple models have been primarily formed, including low-orbit satellite constellations, floating platforms like drones and hot balloons, and high-throughput broadband satellites. In the future, the progress of space-based Internet technology will help to accelerate the layout of space-air-ground networks, which will in turn speed up the global coverage of broadband networks, so that they can provide more reliable broadband services for rural and remote areas and some suburban areas not so accessible to the Internet, and improve the popularization of the broadband Internet.

4. **Network facilities will be more intelligent and open.**

Development of technologies like NFV, SDN and cloud computing will promote network facilities into integrated information infrastructure involving perception, transmission, storage, computing and processing. Leading basic telecommunication operators of the world are seeking to use SDN/NFN to promote intelligent transformation, and large Internet businesses are seeking to promote network facilities to be more open through general-utility hardware and open source of software. With the introduction of new-generation technologies like 5G technology and the upgrade of traditional devices, the networks will be more intelligent, sensitive and open.

5. **Universal services of broadband will boost sustainable development.**

The rapid upgrade of network technologies has resulted in the widening rather than narrowing of the gap between developed and developing countries, and between developed urban areas and rural areas. An increasing number of countries have realized the strategic role of broadband networks as basic facilities. They have also realized that the coverage, upgrade and popularization of broadband networks will contribute to sustainable development of their economy and society. Developed countries should, through guidance of policies, appropriate more funds to rural and backward areas to improve the coverage of optical fiber networks while developing countries should strengthen national information infrastructure construction from a long-term view and give policy supports for open competition and fund provision.

Chapter 3
Development of the World's Network Information Technology

Chinese Academy of Cyberspace Studies

Abstract Today, network information technology is developing fast, having been integrated into every area of economy, society and life. It produces great impact on the global economy, interests and security. The world's economy is more and more focused on the industry of network information technology, which has become a general tool integrated into all economic and social areas. The comprehensive competition between countries is becoming technical competition with the innovation of network information technology as the core. Major countries are promoting original and integrated innovation as well as popularized application, taking the development of network information technology as the foundation for transforming economic development and discovering new advantages for competition. To sum up, the present network information technology development is manifested in the trends as follows: (1) The iterative evolution of network information technology is being accelerated. In recent years, the innovative development and iterative evolution of all kinds of network information technology are being accelerated, including networks, software, IC, computing, storage and sensing. ICT has also witnessed accelerated development and optical communication technology is evolving toward super-speed and super-capacity technology, with its transmission capacity exceeding 560 Tbps. R&D and standardization of 5G network is progressing fast and it will be put into commercial use soon. Software technology tends to develop in the direction of mobile, networking and cloud technology. Cloud service software is becoming the direction of the development of all kinds of software. Massive production has begun for the 7 nm techniques of IC technology and great breakthroughs will be made. Especially, new materials, structures and techniques of IC technology will break the physical limits of Moore's Law. (2) Integrated innovation and AI technology are thriving. Network, hardware and software technologies are being integrated with each other fast, which has promoted the rapid maturity of mobile smart terminal technology. Smartphones, pads and wearable devices keep emerging. In 2016, the shipments of smartphones in the world reached over 1.47 billion. Storage, computing and analysis are shifting from the edge of networks to the center. The capacity of large-scale storage, computing

Chinese Academy of Cyberspace Studies
Beijing, China

© Publishing House of Electronics Industry, Beijing and Springer-Verlag GmbH Germany, part of Springer Nature 2019
Chinese Academy of Cyberspace Studies (ed.), *World Internet Development Report 2017*, https://doi.org/10.1007/978-3-662-57524-6_3

and data analysis keeps improving. AI technologies such as speech and natural language processing, computer vision and smart chip are becoming more mature and have been put into commercial use. AlphaGo has stirred a new wave of AI development. (3) Cross-border innovation and integrated development are being sped up. Network information technology is being applied in areas like biology, material, energy and equipment industries and thus it pushes them to develop in the direction of digitalization, networking and intelligence. The integration between network information technology and biological technologies promotes the in-depth and segmented development of biological technologies such as gene sequencing, brain science and organ repair. Network information technology is changing the model of R&D, design, manufacturing and application service of material, and it is reshaping the industry of energy and boosting digitalized, networked and intelligent production, transportation, transaction and utilization of energy. (4) IT is being used in all areas of economy and society. Network information technology is an important drive for the industrial innovation and development. It has boosted the comprehensive innovation of the model of corporate structure, service and business. New corporate structure model embracing networks, platforms and crowdsourcing stimulates the wisdom and potential of people. Service innovations including mobile service, targeted marketing, nearby supply, personalized customization and online and offline integration are stimulating and creating new consumption demands. Corporate development models including customization, everyone's involvement, experience in manufacturing, production and marketing integration, self-organizing collaboration and adaptive management are being formed rapidly. (5) Network information technology has become the strategic commanding point for all countries to create scientific and technical advantages. To reshape the national competitive strength, major countries take network information technology as the priority in their development strategies, trying to solve problems in development, cultivate new driving forces and speed up transformation to smart society through innovation and popularization of network information technology. Emerging developing countries and industrialized countries initiatively adjust their strategies for the development of network information technology, and less developed countries are speeding up the creation of new advantages for scientific and technical competition to conform to the development so that they will not fall behind but catch up with the new opportunities in the new-round international development.

3.1 Rapid Evolution of Network Information Technology

At present, innovation and development of global network information technology is being accelerated. All kinds of core technology like network communication, large-scale computing, IC, software, large-scale storage and perception technology of connectivity between things are being innovated and upgraded.

3.1.1 Network Communication Technology

The global network communication technology is in a new period of development. The innovation of mobile communication technology is being accelerated and the iteration cycle is being shortened. Optical communication is developing towards superspeed and super-capacity communication and becoming the catalyst of transformation and reform.

1. **Mobile communication technology**

Modern mobile communication technology has experienced five generations and become the research field of communication technology development with the greatest potential and market prospect. The first generation of mobile communication technology (1G) had analog signal technology as the focus, represented by Motorola mobile phones. In the second generation of mobile communication technology (2G) digital signal technology was adopted for the first time and model standards like GSM and TDMA were formed. The third generation (3G) saw the dramatic increase of transmission rate and TD-SCDMA of China, WCDMA of Europe and CDMA 2000 of the United States were the three major international standards. The fourth generation (4G) saw another dramatic increase of transmission rate, and two mainstream standards were formed: TD-LTE and FDD-LTE.

In 2015, the 5G mobile communication technology was put on agenda. Countries and regions like the United States, EU, China, Japan and Republic of Korea and international communication standard organizations like ITU and 3GPP have launched researches in the general network design, architecture, key wireless technology and component concerning 5G technology. So far, the 5G vision and demand have become clear. Key technical frameworks and routes have been identified, including new air interface technology, massive MIMO, UDN, full spectrum access, new multiple access, new multicarrier, advanced modulation coding, D2D technology, flexible duplex, FDX and frequency spectrum sharing. The R&D experiments of 5G technology are being done comprehensively. Qualcomm, Intel, Huawei, ZTE, Nokia and Ericsson and telecommunication operators from different countries are accelerating the layout and the industrial chain is developing fast. Breakthroughs have been made in network architecture, frequency spectrum use, air interface technology, millimetre wave and chips, communication devices and other key technologies and products. 5G standards formulators are ITU and 3GPP. The latter has completed the planning of 5G standards, and progress has been made in standardization of key specifications of 5G technology, such as Standalone of new air interface and core networks. Huawei and Qualcomm are speeding up the R&D of standards to take in the 5G "commanding heights". The planning schedule of 5G technology is shown in Fig. 3.1 and Table 3.1. According to them, 5G standards have entered the stage of output, and the mature standards are expected to be submitted in 2020.

5G will become the critical infrastructure of general technology and digital transformation of economy and society, producing impact on a number of sectors

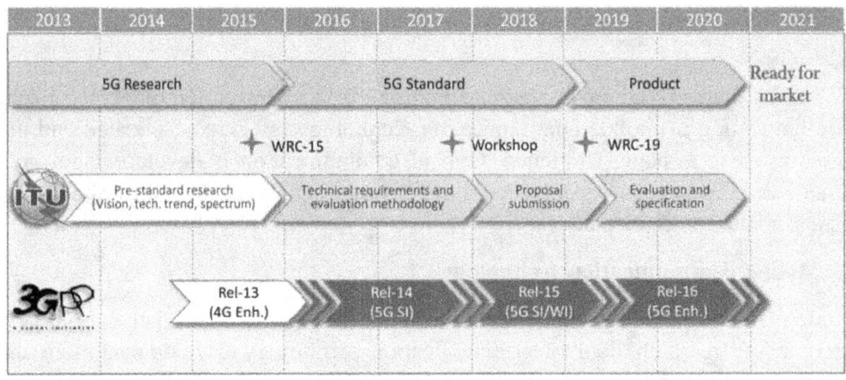

Fig. 3.1 R&D plan for 5G standards (*Source* IMT-2020 Group: *5G White Paper*, September 2015)

Table 3.1 3GPP R14/R15/R16 standard progress schedule

Stage	R14 (5G standard research)	R15 (First edition of 5G standards)	R16 (Complete 5G standards)
Stage 1	March 2016	June 2017	December 2018
Stage 2	September 2016	December 2017	June 2019
Stage 3	March 2017	June 2018	December 2019
Ratification	June 2017	September 2018	March 2020

Source 3GPP.org, June 2015

from both online and offline. ITU defined three scenarios of 5G at the 22nd ITU-RWP5D Conference in June 2015: eMBB, mMTC and URLLC, as shown in Fig. 3.2. eMBB corresponds to 3D and UHD video; mMTC, to massive IoT, and URLLC, to self-driving, industry automation and other ultra-reliable and low latency services.

According to an IHS Markit report, by the year 2035, the sales created by 5G technology will have reached US$12.3 trillion, accounting for 4.6% of the world's total output, including US$3.4 trillion in manufacturing (accounting for 28%) and US$1.4 trillion in information communication (accounting for 11.5%). The evolution route of mobile communication technology is shown in Fig. 3.3.

2. **Optical communication**

After experiencing the stages of PDH, SDH, ASON, MSTP, PTN, POTN, optical communication is witnessing its fast iterative upgrade and dramatic growth of optical network transmission rate and capacity. Breakthroughs have been made in 100G+ technology, single carrier 400G technology, Software Defined Optical Networks (SDON), WDM, SDM and silicon photonics. Bell Laboratory has successfully developed the MIMO-SDM technology and is expected to increase the

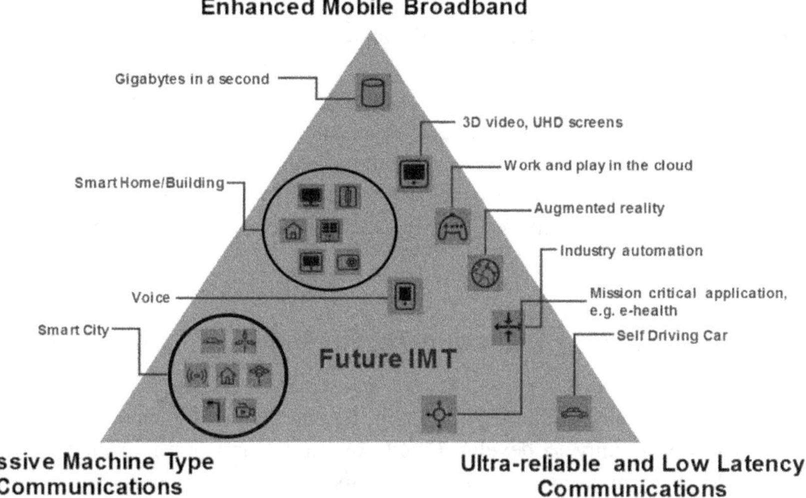

Fig. 3.2 5G Usage scenarios (*Source* ITU, September 2015)

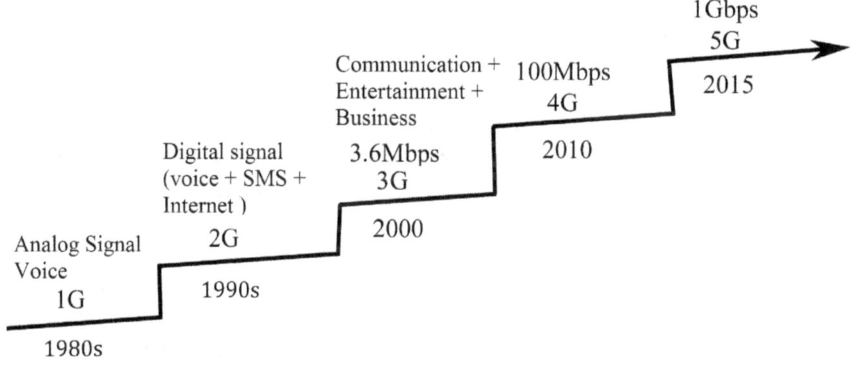

Fig. 3.3 Evolution route of mobile communication technology *Source* CCID

optical communication capacity by over 100 times. Wuhan Research Institute of Posts and Telecommunications has completed the 560 Tbps WDM and SDM transmission system experiments, enabling 6.75 billion pairs of people (13.5 billion individuals) to make phone calls on single fiber. In the future, with the rapid growth of online data traffic and development of 5G technology, the optical transmission technology application and revolution will be accelerated. There will be a rigid demand for super-speed, super-capacity, and ULH optical transmission of super-high frequency efficiency. 100 Gbps+ technology will be a dominant trend of Pbps transmission, and intelligent technology and software defined will become the trend of optical communication development. As the broadband strategy is carried out in all countries and gigabit fiber networks are laid out, the dominant global optical communication markets will be China, the United States, Japan, Republic of Korea and EU, among which Japan and Republic of Korea will be the leaders, since Mitsubishi, Sumitomo, Hitachi, Samsung, LG and SK are influential businesses with their advanced optical communication technology and equipment. China has great momentum in developing optical communication, which is being accelerated. YOFC and Hengtong Group and some other excellent businesses are developing fast. Chinese businesses are less reliant on the import of optical fiber preform. Domestic manufacturers' market share in that respect keeps increasing.

3.1.2 Large-Scale Computing Technology

Large-scale computing technology develops fast and network technology giants are seizing the opportunity to develop cloud computing and promote intelligent and virtual cloud computing development. Super-computing technology is being innovated and developed and the computing capacity keeps improving and becomes the major area of competition among countries.

1. **Cloud computing technology**

With the development of distributed computing, parallel computing, grid computing, the Internet and big data, Google launched Google 101 and officially put forward the concept and theory of "cloud" in 2006. Then Amazon, Microsoft, HP, Yahoo!, Intel and IBM all announced their "cloud program", which marked the beginning of cloud technology development. Breakthroughs were consequently made in the system structure, basic technology and commercial models of cloud computing. Service models like IaaS, PaaS and SaaS were soon popularized and then they became the mainstream computing technologies. So far, cloud computing has become the new engine for the development of software and information technology service, and Amazon, Google, Microsoft, Alibaba and Tencent have expanded their business into the cloud computing market to gain their new strengths for global competition. According to Gartner statistics, Amazon, Microsoft and Alibaba Cloud were the top three businesses on the public cloud market of the world in 2016, with their total market share exceeding 50%, and then came Google, Rackspace and IBM.[1] By 2021, the world's top 10 cloud service providers will occupy 70% of the public cloud market. Amazon, Microsoft and Alibaba will be the leaders in the field of IaaS, while Oracle, SAP and Microsoft will be the leaders in the fields of SaaS and PaaS.[2] In the future, with the accelerated innovation and application of core technologies of cloud computing, including virtualization technology, distributed computing, program mode and large-scale data processing, and with the integration of IaaS, PaaS and SaaS, the capacity of

---

[1] Gartner: *Public Cloud Services Market Share Report (2016).*
[2] Gartner: *Public Cloud Services Revenue Prediction Report (2017–2020).*

integration of cloud computing with AI, big data and edge technology will be improved rapidly and cloud computing will be extended to intelligent technology and network edge.

2. **Super-computing technology**

High-performance computing is a comprehensive manifestation of the economic and scientific and technical strength of a country, so it has become a strategic commanding height that all countries are contending for. After experiencing the development of vector machines and massively parallel processing (MPP), high-performance computing capacity has been growing rapidly into the Pflops era, with the annual growth rate reaching 50–100%. In recent years, the rapid development of technologies like the heterogeneous multi-core system, the fat/thin node cluster system, the high-efficiency system, processors, high-efficiency blade servers, 3D-Torus networks and high-efficiency network switches has boosted the computing capacity of high-performance computers, with the peak value of processing speed exceeding 125 PFlops. Sunway TaihuLight developed independently by China in 2016 is the world's first supercomputer whose peak speed exceeds 125 PFlops percent. With the development of cloud computing, supercomputing technology is becoming faster and more efficient, usable and reliable. The E-level high-efficiency computer at exascale is the hotspot of R&D worldwide. At present, the global pattern of super-computing technology remains stable. On the list of the world's top 500 supercomputers released in November 2017, the first two are from China, the third to the fifth ones are respectively from Switzerland, Japan and the United States. Altogether, among the top 500 computers, 202 belong to China, 143 to the United States, 35 to Japan and 20 to Germany. Seen from the computing performance, the total computing speed of the whole system has reached 749 petaflop/s, an increase of 30%[3]. Sunway TaihuLight and Tianhe 2 have been listed as the first two on the list for four times in a row, with their speed reaching respectively 93petaflop/s and 33.9 petaflop/s.

3.1.3 IC Technology

Integrated Circuit (IC) is the foundation of informatization and intelligent development of all industries. At present, IC technology development includes four steps, namely, design, manufacturing, packaging and test (see Fig. 3.4).

1. **IC design**

Today, SoC is the trend of IC design worldwide. Comprehensive solution design and coordinated design of software and hardware are the key to improving the

---

[3]*Source* The 50th top 500 list of supercomputers released by top 500 organization in November 2017, https://www.top500.org/lists/2017/11/

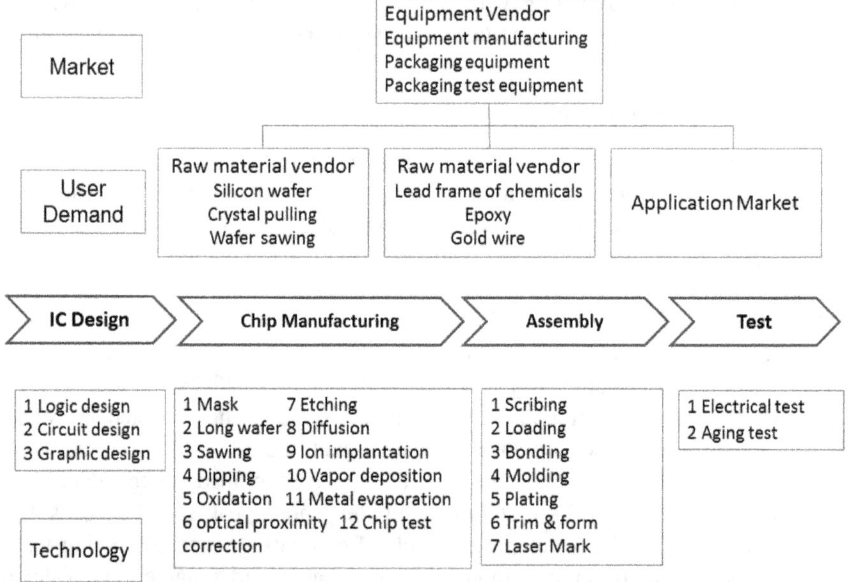

Fig. 3.4 Technical route map of IC technology *Source* CCID

product's competitiveness. To meet the demand of the mobile Internet, lower energy consumption and improve the reliability is becoming increasingly important. IC design is an intelligence-intensive industry, accounting for 25% of the industrial structure of the global IC technology. The total revenue of Qualcomm, Broadcom and MTK (Media Tek) accounts for 65% of total of the world's top 10 businesses and their leading position will not be changed in a short time.[4]

After nearly three decades of development and accumulation, China's strength in IC technology has been improved and the chip design capacity of the country has increased from one million gates to one billion gates and breakthroughs have been made in the development of high-end general chips. In China, the research results of many-core processors have been used in major national defense projects, the Sunway BlueLight supercomputer hitting one petaflops based on the Sunway 1600 processor was on the top of the Top 500 supercomputers, and the shipment volume of 8GBDDRII storage chips has exceeded 4.3 million, unprecedented in history. The overall IC design scale has exceeded that of packaging and test industry, accounting for nearly 40% of the total share of the country's IC industry, and the number and sales of IC design businesses in China have increased dramatically.[5] Especially in recent couple of years, thanks to the macro-policies of the country and the support from local governments, the number of IC design businesses has

---

[4]Source: WSTS statistics, 2017.
[5]Source: CSIA statistics, 2017.

reached 1500. On the other hand, there exist some problems, for instance, weak innovation capacity, low and mid-end products, fierce price competition among homogenized products, low profits despite high sales, and incapability of core technologies R&D or upgrade of products. To enable IC design technology to witness leapfrog development, China's IC design businesses are seeking to boost IC innovation through complete machine and application innovation. Innovative businesses of drones, robots, self-driving vehicles and smartphones are cooperating with domestic IC businesses in innovation to develop revolutionary products. For example, Huawei Hisilicon has launched Kirin 970 chip which has built-in Cambricon 16 nm AI chips designed by Institute of Computing Technology of Chinese Academy of Sciences to process mass AI data. The Kirin 970 chip is now carried by Huawei Mate 10 smartphones. China is developing the "reconfigurable computing chip technology" in which the chip is defined by certain software and RCP and X86 processors are integrated, trying to develop new server CPU oriented for smart processing.

2. **Chip manufacturing**

IC manufacturing technology has contributed to Moore's Law, which has been supposed to have revealed the rules of development of information technology. In 1965, Gordon Moore, one of the founders of Intel, proposed that when price remains unchanged, there will be a doubling every 18–24 months in the number and performance of components per integrated circuit. Since 1970s, the IC memory has seen a dramatic increase of efficiency, speed and capacity. Massive production of 10 nm chips has been launched by Samsung, TSMC (Taiwan Semiconductor Manufacturing Company) and Intel. 7 nm chips of TSMC were put into trial production in April 2017 and its 5 nm is expected to be put into trial production in the first half of 2019. Since the size of components is shrinking, Moore's Law is faced with challenges in the development of IC technology. Components' thermal noise and RC delay will reach the limitation, so the probability becomes higher that they cannot work by normal parameters, which will result in the unreliability of components, which, in turn, will result in the decrease of yield of IC products. Besides, the shrinking of the size gives rise to the increase of cost, which then increases the cost of the single product.

To solve the problems, the IC industry has begun to carry out R&D of new materials, structures and technologies of IC products. For example, high K metals, stacked device structures, systems and three-dimensional packaging can overcome the physical limitations proposed in Moore's Law, so they can promote the development of IC technology and industry. In general, IC manufacturing technology is developing in the direction beyond Moore's Law. New materials, structures and technologies are the important drive for the sustaining of Moore's Law. IC technology will be more widely used in IoT, automobile electronics, data centers and AI.

It can be foreseen that in the IC manufacturing technology, 7 nm will become a turning point following 28 and 14 nm. Its technical route is basically fixed and FixFET transistor technology will continue to play an important role in scaling. The technical route of the nodal point below 7 nm is not clear yet. As for 3 nm or below, the alternative solution is to integrate nanowire into hybrid GAA FinFET and vertical structure. For the nodal point below 2 nm (including 2 nm), new techniques, structures and principles beyond the present CMOS technology will be introduced, and the relevant 3D logic device integration, logic operation devices imitating the biological nerve system, quantum computing, single atomic layer 2D material devices are being studied and developed.

3. Packaging and test

Packaging and test are indispensable stages of the IC industrial chain. With the technical progress, 3D stacking technology and TSV have subverted the packaging test techniques in general sense. At present, in China, test is mainly done in packaging businesses. So the packaging and test industry has got its name. Promoted by the IC industry market and technology, IC packaging and test technology has developed in three stages as follows.

The first stage refers to the Through-hole devices (THD) age before 1980. At this stage, the jack was assembled to PCB and the representative technologies were TO and DIP, but the packaging density and the number of pins were hard to increase, so they could hardly meet the demand for efficient automatic production. The second stage refers to the SMT age starting from 1980s. Typical technologies at this stage were SOT, SOP and QFP, which made the package products light, thin and small and thus improved the performance of the circuit. Driven by the demand for small but multifunctional electronic products, there occurred techniques like replacement of leads by solder balls and surface mounting of distribution by area array at the end of the 20th century. At the same time, technologies like BGA, CSP, WLP and MCP emerged one after another. The third stage refers to the high-density packaging age starting from the early 21st century. As electronic products tend to be small and multi-functional, to increase the integration level by scaling became increasingly difficult since the size became smaller and smaller, so 3D packaging technologies represented by 3D stacking and through-silicon via (TSV) became the best choice to sustain Moore's Law. With 3D stacking, chips or structures of different functions are integrated and connected by signals in three dimensions in Z-direction. 3D packaging includes wafer level packaging, chip scale packaging and silicon cap packaging. 3D stacking is used for the integration of micro-systems. TSV is a cut-edging technology for interconnection between chips by making vertical electrical connection (via) that passes completely through a silicon wafer or chip.

The IC packaging technology now is mainly at the second age, since 3D stacking and TSV are being studied and developed, with only a few businesses taking the lead. For example, Samsung and CET have applied the new technologies in a few special areas. The packaging technologies are developing towards high end ones, such as TSV, which will become the important technical solutions to the limitation of Moore's Law. The 3D packaging technology will be an inevitable technology in the future.

The global IC industry scale from 2010 to 2016 is shown in Fig. 3.5.

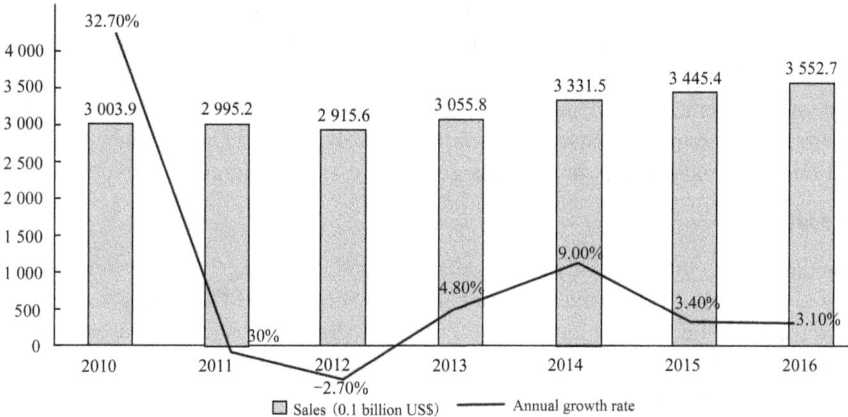

Fig. 3.5 Global IC industry scale from 2010 to 2016 *Source Survey Report on the Semiconductor Market* by American SIA, 2016

3.1.4 Software Technology

Software technology is the connector and fusion agent between information and the physical world. Software technology has experienced three stages of development. The first stage was symbolized by programmers' direct programming hardware and by software's relying on hardware. At the second stage, major businesses developed application software technologies and products for industries, and software development led the hardware development. At the third stage, with the emergence of the Internet, mobile Internet and cloud computing, software tended to be service and network software. Today, software technology is integrated into traditional industries, which will speed up the digitalization and intelligent development of the industries and thus usher in the industrial Internet era of Internet of Everything.

1. **Operating system (OS)**

OS is the fundamental system software of computers. System/360 series of computers developed by IBM used general OS for the first time. In 1970s, UNIX became one of the most influential stand-aloneOSs. In 1990s, Linux derived from UNIX became the most successful OS in the area of open- source code. In the same period, Microsoft OS was dominant in the market. In the early 21st century, Android, an open mobile OS based on the kernel of Linux, was launched, and was acquired by Google in 2005. In recent years, the mobile OS has been popular. For example, iOS of Apple and Android of Google are the representative products of the mobile OS. Meanwhile, with the maturing of cloud computing technology, the OS is accelerating to be the Cloud OS. AWS of Amazon, Cloud OS of Microsoft, PureSystems of IBM, Oracle Solaris of Oracle, OS of Inspur InCloud, Cloudview of Sugon, Apsara of Alibaba Cloud and FusionSphere of Huawei Cloud have all been launched into the market. Cloud OS needs to be constructed in a framework, through second development, based on virtualized technology, with the integration of other relevant technologies and services. It has the functions of resource access and abstraction, resource allocation and dispatch, application lifecycle management, system management maintenance and human-computer interaction. Common cloud OS frameworks are OpenStack, CloudStack, Eucalyptus and OpenNebula, with the last one surpassing others in business acceptance and community development.

2. Database

The database is the core software of data organization, storage and management of the computer system. Its development can also be divided into three stages, namely, network and hierarchical databases, relational databases and database systems characterized by object-oriented data models. Today, database technology is integrated with network communication technology, AI technology, object-oriented programming techniques and paralleled computing technology, which has promoted data mining technology. The unstructured data is the direction of the database development. Such technology supports the repeating fields, subfields and variable length fields, process variable length data and repeating fields, and store and manage in length the data items. It has advantages, compared with traditional databases, in processing continuous information (including full-text information) and unstructured information (repeating information and variable length data). It is a helpful complement for the relational database. Meanwhile, the combination of databases with disciplinary technologies has produced a series of new databases, for instance, Distributed Databases, Parallel Database, knowledge repository and Multimedia database, which all represent the direction of database technology development. The introduction of multimedia technology and visualization technology into Multimedia database will be the hot spot and difficult point of the future development of database technology. Open and open-source cloud technology is also an important trend. Big data processing will be used in businesses' application as the cloud service system, and cloud data gathering and cloud databases will be corporate data processing and storage means. At present, Oracle DBMS, Oracle MySQL and Microsoft SQL Server are representatives of relational database products. PostgreSQL, an open-source database, and MongoDB, a document database, perform well. Redis, a cache line database, Elasticsearch and Solr, the data storage and search service index, and Neo4j, a graph database, play a leading role in segmented areas respectively. With the rise of public cloud services, AWS, Azure and Alibaba Cloud all perform well. Oracle and Google's cloud services cannot be neglected, either. In the future, there will be fiercer competition in global database technologies and products. The hybrid cloud solutions will be the main battlefield. Alibaba Cloud, Tencent Cloud, Baidu Cloud and Huawei Cloud will continue to play an important role in data services.

3. Browsers

The browser is the first window for human beings to access to the Internet. The period from mid and late 1990s to the early 21st century witnessed the rapid growth of browser technologies and commercial models. They are able to support protocols like JavaScript, Java applets, CSS and SSL. IE of Microsoft, through the model of its bundling sale with OS, monopolized the browser market, totally defeating the Netscape browser. After 2003, with the launch of browser software like Firefox, Safari and Chrome, the monopolization by IE was broken. The technical progress of browsers is centered in the users' experience and access guarantee. All browser software providers have adopted JavaScript technology and asynchronous access models to improve the smoothness of videos and pictures. The development of sandbox access model, SSL and https has improved the security of browsers. In recent years, with the development of mobile Internet, browser technologies tend to pay attention to mobility, cross-platform technology, openness, componentization, personalization and socialization. By 2017, Google's chrome had occupied 60% of the market share, but Microsoft's IE's market share had dropped to 14% and kept dropping.[6] IE has been challenged by Firefox and Safari on the market. In China, after the monopolization of IE was broken, UC, Qihoo360, Sogou, Tencent and Baidu all developed their characteristic browsers based on their own strengths.

4. **Industrial software**

Industrial software is the core of industrial digitalization, networking and intelligent development, categorized into R&D and design software, production control software, information management software, and built-in industrial software. Its single application is shifting to wide coverage and deep penetration in manufacturers' R&D, production, operation, logistics and products. Then it is developing in the direction of integration, breaking the bounds between businesses, witnessing the reform oriented to the market and customers' business flow and production and operation modes, and supporting the formation of new industrial capacity. Model-Based Design (MBD) has broken the bottleneck that the product drawing cannot be reused and the machine cannot be read while Model-based Enterprise (MBE) creates the digital line running through the product's lifecycle. In the future, industrial software will, based on end-to-end data flow and supported by inter-connectivity of networks, create a series of new industrial models conforming to smart manufacturing which will incorporate personalized customization, collaborative R&D, collaborative manufacturing, targeted marketing through big data, and long distance operation and maintenance, by relying on cloud computing, big data, virtual reality and IoT. Autodesk takes a leading position in the market of R&D and design software, Dassault Systems, Siemens and PTC are prominent in Product lifecycle management (PLM) software and Synopsys, and Cadence and Mentor Graphics are representatives of Electronic Design Automation (EDA) software manufacturers. In application, industrial software will be more valuable in collaborative supply chain optimization, product lifecycle management based on MBE and digital ecosystem construction based on platforms as well as the innovation and application of new technologies using cloud, big data and virtual reality.

---

[6]Source: NetMarketShare, 2017.

3.1.5 Massive Storage Technology

Massive storage technology is evolving fast. Technologies like the memory and network storage are increasingly mature and the innovative capacity is being enhanced. Therefore, massive storage has become an important part supporting the development of economy and society.

1. Memory technology

Memory technology emerged in 1950s, when IBM launched the magnetic tape memory and the computer disk storage system, which laid the foundation for the development of magnetic tape storage. In 1970s, hard disk technology witnessed a great revolution. After 1970, when Intel developed 1 K DRAM, memories began to be widely applied in the internal storage of large computers. The memory capacity and reading speed are growing fast, with the storage capacity of mechanical hard disks (magnetic disks) and solid hard disks having reached the TB level. The size of storage devices of mainstream memory chips like DRAM, SDRAM and FLASH has been reduced to 1in and their reading speed to DDR4. With the rapid development of 5G technology, IoT and big data, there is an increasing demand for memories and for their storage capacity, speed, power consumption, reliability and service life. Emerging storage technologies of large capacity, low power consumption, high-speed reading and writing, ultra-long storage cycle and data security are developing fast. 3D NAND, 3D Xpoint, PCRAM, MRAM, RRAM and PCM are becoming the hot spot of development. The memory industry is highly centralized in some businesses. Samsung, SK Hynix, Micron and Toshiba occupy 95% of the market.[7] Samsung and SK Hynix, two manufacturers from Republic of Korea, occupy 70% of the DRAM market. The NAND Flash market is monopolized by Samsung, SK Hynix, Toshiba, SanDisk and Micron.[8] China is catching up in memory technology, having made breakthroughs in technologies and products of 130 nm PCM and 3D NAND flash memory of 32-level stacks.

2. Network storage technology

Before the emerging of the Internet, computer storage had been the stand-alone storage in accordance with the program storing idea put forward by Von Neumann. With the popularization of the Internet, e-commerce and AI have been developing fast and all kinds of massive data have occupied the storage space of systems. DAS, NAS, SAN, object storage, distributed storage and virtual storage have all witnessed their development. In particular, with the maturity of storage virtualization, RAID technology, software definition distributed storage and cloud storage system structure, virtual storage represented by cloud storage has become the mainstream. With the development of software definition, network storage technology tends to be virtualized and intelligent and hyper-integrated architecture and software

---

[7]Source: Statistics from TrendForce and IC Insights, 2017.
[8]Source: Global DRAM market data of the third quarter of 2017 from IHS Markit.

definition storage have become the trend. At present, cloud storage is mainly provided by large Internet businesses and cloud service providers, and there exists increasingly fierce competition. According to Gartner Magic Quadrant of global cloud computing and storage in 2017, the top businesses of cloud storage are Amazon AWS, Microsoft, Google, Alibaba, IBM, Rackspace, Tencent, Oracle and Virtustream.[9] In China, cloud storage is increasingly centered in large Internet businesses, such as Alibaba Cloud, Tencent Cloud, CTclouds and Wo Cloud, which are leaders in the cloud storage market.

3.1.6 Perception Technologies of IoT

With its popularization and development, IoT is entering the era of Internet of Everything. Millions of billions of devices have access to the networks and data are witnessing an explosive growth, which promotes the perception technologies of IoT to evolve fast in the direction of low power consumption, wide coverage and intelligence.

1. Sensor

The sensor is an important part of the perception layer of the IoT and the important entrance for data collection of the sensing layer of the IoT. Sensor technology, communication technology and computer technology are three major pillars of the information industry. Traditional structure sensors and solid sensors are being replaced by microsensors, networked sensors and multifunctional sensors. In 1999, University of California Berkeley released WeC microsensor nodes, opening up the new era of micro and intelligent wireless sensor nodes. As MEMS technology, laser technology, hi-tech material technology, microelectronic technology, IC and processing techniques are quite developed, R&D of sensors tends to cover high-end, diversified sensors, including new wearable sensors, optoelectronic sensors, molecular sensors and biosensors. With the development of the IoT technology, sensors tend to adopt new technologies, including MEMS, wireless data transmission, IR, new material, nanometer, ceramics, thin film, optical fiber, laser, compound sensor and multi-disciplinary integration. The United States, Japan and Germany are leading countries on the sensor market, with their sensor market scale in 2016 accounting respectively for 27.0, 20.5 and 14.0%. Leading businesses had the most market share. In 2016, top 30 sensor businesses' total operation revenue reached US$9.35 billion. The very top five businesses are Bosch, Broadcom, Texas Instruments, STM and Qorvo.[10] China's independent innovation capacity has improved and it relies less on imported sensors. In 2016, its import of sensors decreased to 60% and their core chips import decreased to 80%.

---

[9]Gartner: *Magic Quadrant of Global Cloud Computing and Storage in 2017.*
[10]Yole, *Status of the MEMS Industry*, 2017 edition.

2. RFID

RFID technology originated from the development of radar technology and it was commercialized in 1980s, with single-chip integrated RFID and the first ETC adopting RFID launched. In 1995, ISO/IEC JTCI set up SC31 to be responsible for the research of standard formulation of RFID. SC31 then formulated technical standards, data structure standards, performance standards and application standards. After 2000, RFID standards were formulated, which promoted the rapid development of active RFID, passive RFID and half passive RFID. In recent years, with the rapid development of RFID manufacturing and packaging technologies, there have emerged high-sensitivity, high-performance, flexible and low-power-consumption RFID products. In the future, with the application of new technologies like wireless communication, sensors, and xArray reader system, RFID technology will be integrated with Wi-Fi technology, sensors, GPS and biometric identification to promote sensors to have multifunctional identification capacity and to improve sensors' positioning coverage and precision and their application area and depth. At present, the global RFID chip manufacturing pattern keeps stable, but core and key technologies and products are in the hand of giant businesses of the world. Qualcomm, Philips, Siemens, ST and TI monopolize the RFID chip market while IBM, SAP, Sybase and Sun monopolize the RFID middleware and SI. Alien, Intermec, Symbol, Transcore, Matrics, and Impinj control the product and device market of RFID tags, antennas and readers.

3. LPWAN

LPWAN has not been implemented and applied in a large scale yet. There are two development paths of LPWAN. The first covers LoRa and SigFox under the unlicensed spectrum. LoRa is one of the widely-used LPWAN technologies. In August 2013, Semtech released a new U2H LoRa chip below 1 GHz, marking the birth of LoRa, the industrial chain of which is mature today. LoRa has become the commercialized LPWAN technology and a number of operators have started to lay out or are laying out LoRa networks. The second path of LPWAN includes NB-IoT and E-LTE supported by 3GPP under the licensed spectrum. NB-IoT is being developed by all countries. In 2013, Huawei collaborated with some manufacturers and operators in carrying out the cellular NB-IoT research named LTE-M. There occurred two proposed solutions, one based on the existing GSM evolution and the other on the new air interface technology of NB-M2 M. In May 2015, Huawei, Qualcomm and Neul jointly proposed the NB-CIoT new air interface technology. In June 2016, NB-IoT core protocol was approved, marking the completion of all researches on standard NB-IoT core protocols and the establishment of the NB-IoT standards. Huawei, Qualcomm, Media Tek and Intel as well as telecommunication operators from Europe and the United States have begun to accelerate the layout and have made breakthroughs in GM, endpoint operation, MMW, Bluetooth LE, high-frequency channel model and NB-IoT. The NB-IoT commercial layout is being accelerated. For example, a group of communication operators are strengthening the planning and layout of NB-IoT testing networks, 10 of which are already commercialized and some of which are used in smart city and smart home.

3.1.7 Blockchain

Blockchain technology is, in essence, an electronic cash system first proposed by Satoshi Nakamoto in the white paper *Bitcoin: A Peer-to-Peer Electronic Cash System* published in 2008. Blockchain refers to the establishment of a distributed ledger based on a list structure by using technologies like encryption, peer-to-peer networks and distributed ledgers. The payment can be made online by one party to the other without any financial institution in-between. To ensure the continuous work of blockchain, the mechanism for workload proof was put forward, which is called "mining". Every time the "miner" produces a new block, the first transaction of this block will be processed in a special way and the transaction will produce new electronic currency owned by the block creator. A method is provided for distributing electronic currency to the circulation field without any centralized authority issuing any currency. The difficulty of "mining" varies with the computing capability of the network. Coins used for "mining" award, according to the algorithm, will decrease step by step, so the blockchain theory can effectively control inflation. Blockchains are divided into public ones and union ones. Public blockchains are blockchain applications that the public participate in. Since the occurrence of ICO, public blockchain application projects have been launched in the fields of finance, insurance, logistics, medical care, fixed asset transaction and media. Union blockchains are used by an organization or union. The most mature union blockchains are Hyperledger Fabric and R3 Corda.

As for the application prospect, blockchain is faced with challenges in ID verification, legal compliance, transaction security, anti-money laundering and data privacy since it is an open cash system. There are not many blockchain technology businesses at present and in many of them the blockchain is in the stage of R&D and testing. It is a long way to go for it to be used in large scale. Decentralization of blockchain is not easy to be accepted by the government or supervision authority, so it is first used where there is no high centralization. According to the blockchain efficiency road map released by Mckinsey, the period from 2014 to 2016 was the period of blockchain technology assessment, when all kinds of technical and financial organizations assessed the value of blockchain technology; the period from 2016 to 2018 will be the period of concept verification, when whether the technology is feasible and expansible is verified, especially whether the performance, cost, speed and scale of blockchain can surpass the traditional financial system. In this period, some problems have been found in scalability and transaction speed. The period from 2018 to 2020 will be the period of formation of blockchain infrastructure. After 2021, the technology will be officially commercialized.

3.2 Accelerated Integration of Different Spheres of Network Information Technology

Different spheres of network information technology are being integrated. In particular, technologies such as the network, hardware, software, service and data are being integrated and innovated fast. The integration between hardware and software technologies has boosted the occurrence of new smart terminals like smartphones, pads and wearable devices. AI helps to release the energy accumulated by the previous technical and industrial revolutions and thus boost the occurrence of new products, services and activities. Quantum communication technology is developing fast and all countries are making related layout.

3.2.1 Smartphone Technology

In recent years, with the rapid development of mobile Internet, the number of APP users keeps growing rapidly, and mobile phones become the major means of Internet access. So the APP will enjoy development in broader space. The competition between Google and Apple in the area of software and hardware will be extended to more terminal areas.

The OS of mobile intelligent terminals is dominated by Android and iOS. Android mobile OS, launched by Google in 2008, is an open-source OS based on Linux. iOS, developed by Apple, has constructed CocoaTouch, Media, Core Service and Core OS. Thanks to its open source, Android has won support from a number of mobile phone manufacturers and has become the world's leading smartphone OS. Apple, based on iOS, has constructed a closed ecosystem to control the third-party application from beginning to end. The company has adopted the APP Store as the closed sales channel for applications. In recent years, with the thriving of mobile intelligent terminal devices, Google and Apple have been competing on the central control system of all kinds of intelligent terminal devices. The Android Platform has been extended to a number of intelligent terminal devices, for instance, Android Auto and Android Wear, while Apple has launched CarPlay, through which iOS devices can be interconnected with cars.

APP is witnessing an expansion of its use. The APP Store of Apple has led to the thriving of third-party application software. In recent years, with the rapid popularization of mobile Internet, the smartphone has become the major means of Internet access and the number of APP users is growing rapidly. In 2014, the number of the APP download mobile terminals exceeded that of PC terminals. APP has been expanded into the areas of news and information, socializing, shopping and entertainment. With the development of smart wearable devices, APP has been extended from smartphones to smart watches. In recent years, with the rapid development of big data and cloud computing, application programs driven by cloud is rising and mobile application software tends to be based on cloud and

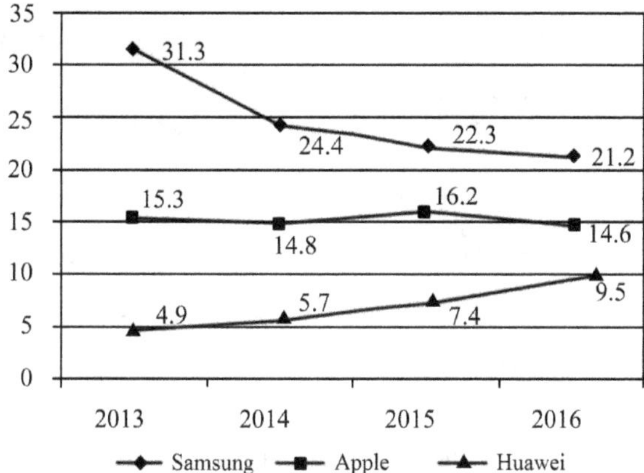

Fig. 3.6 Global market share of three major mobile phone manufacturers (2013–2016) (percent) *Source IDC Survey Report on the Global Mobile Phone Market in Q2 2017,* August 2017

intelligence. With the upgrade of the users' experience, interactive design is paid more attention to for mobile application software.

The global mobile phone market is expected to be led by three major companies, namely Apple, Samsung and Huawei. In 2007, Apple launched iPhone and became a famous mobile phone producer in the world. From 2012 to 2015, there were two leading companies (Apple and Samsung). Samsung took the leading market share while Apple influenced the trend of smartphones with its advantage in innovation. In recent years, their market share has accounted for 40% respectively. Nowadays, Huawei's market share keeps increasing. According to an IDC report, in the second half of 2017, Huawei's sales accounted for 11.3% of the world's total, only 0.7% less than that of Apple (12%).[11] Therefore, it is expected that the global mobile phone market will be dominated by the above three companies (see Fig. 3.6). Besides, OPPO, vivo and Mi smartphones of China are developing fast.

3.2.2 AI Technology

AI technology is experiencing the third development wave, following the first in 1960s and the second from 1980s to 1990s. A meeting held in Dartmouth College of the United States marked the birth of AI in 1956. At that time, AI was mainly witnessing its theoretical research. The mathematical proof system and knowledge inference system were the most mature research achievements. In 1980s, the technology based on the statistic model emerged and the artificial neural network began to be used in AI research. At the same time, speech identification and machine translation witnessed some progress, but with little practical value. In 2016, deep learning was put forward and developed, which promoted AI to leap

---

[11]Source: *IDC Survey Report on the Global Mobile Phone Market in Q2 2017,* August 2017.

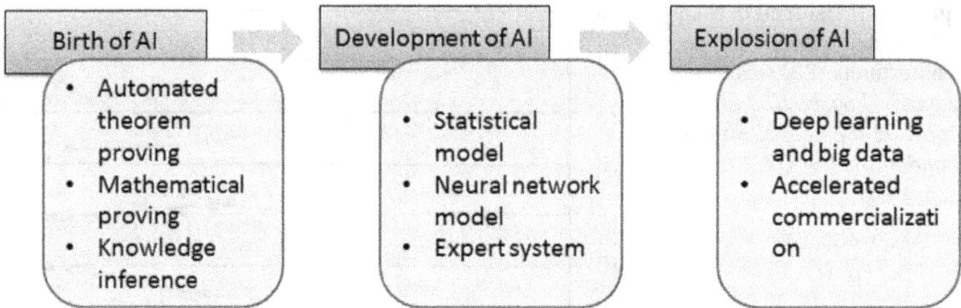

Fig. 3.7 Development roadmap of AI

forward. With the improvement of computing technology and the rapid development of big data and the Internet, AI began to witness its third development wave. The number of AI businesses founded in recent two years is the sum of the number in the past 10 years and the funding scale keeps increasing. Against this backdrop, AI will inevitably lead to technical and industrial revolutions of all industries (see Fig. 3.7).

1. **Natural language processing**

Great progress has been made in natural language processing. Speech and semantic identification is an important aspect of natural language processing research. In recent years, thanks to the development of deep learning development and big data corpus, the speech identification accuracy rate keeps improving, relevant product application is becoming mature and speech interactive technology is being adopted in industries, robots, house appliances and cars. In 2011, using the deep neural network model, Microsoft reduced the speech identification error rate to 30%, marking a new period of speech identification development. From 2011 to 2016, the error rate kept decreasing. Today, the error rate of Microsoft and IBM has decreased to within 5%. Chinese businesses like Iflytek, Baidu, Sogou, Aispeech and Unisound are engaged in R&D of Chinese speech identification. Except that Baidu, Sogou and Iflytek have launched voice input methods, all businesses tend to develop applications of hardware carrying speech identification, for example, Unisound's smart home speech identification system and Aispeech's automotive voice identification system. On the other hand, speech identification technology is at the perception stage of AI. Semantic comprehension and cognition will be the target of AI.

2. **Computer vision**

Computer vision is the foundation of machines' perception of the world and one of the major AI technologies. In recent years, influenced by the development of deep learning, computer vision technology is developing fast. Progress has been made in image identification, object detection and video analysis. Computer vision has been applied in finance, security protection, medical care and self-driving. In 2012, in the ImageNet Large Scale Visual Recognition Challenge (ILSVRC), Alex and Hinton reduced the error rate of Top 5 to 15% by using deep learning algorithm, a milestone in computer vision. In 2015, Microsoft reduced the rate to 3.6%, surpassing capability of human beings. In recent years, contestants of computer vision keep breaking the record and the identification accuracy rate keeps increasing. Today, the static content identification of computer vision has reached a high level and the dynamic content identification will be the direction of the future development.

3. Machine learning

Machine learning has witnessed dramatic progress and has been applied in many areas. It is an important means of making the computer intelligent, so it has been applied in AI areas. In 2016, AlphaGo developed by DeepMind of Google beaten Lee Se-dol in a Go game, attracting the universal attention to AI and machine learning. In recent years, machine learning keeps improving. Amazon, Baidu, Google, IBM and Microsoft all have their commercialized machine learning platforms. Deep learning is the most vital direction of machine learning since computer vision and natural language comprehension cannot be separated from deep learning.

4. AI application

AI has been applied in a number of areas, such as finance, transportation, industry, medical care and self-driving. In finance, AI has been applied in intelligent investment consulting, intelligent client service and security protection monitoring. The leading businesses in those areas are IBM, Ant Financial and Inno TREE. In transportation, AI is becoming the important technology improving and optimizing transportation. Since Alibaba Cloud's ET city brain was launched into cities like Guangzhou in 2016, traffic speed has been increased. AI robots have been widely used in industry. They are laid out in Industry 4.0 of Germany, and they are to replace manpower in Zhejiang of China. Altogether, there are over 30,000 robots working in industry. In medical care, AI has been applied to monitoring, diagnosis and intelligent medical equipment. IBM, Baidu and BGI Genomics have all developed such robots. In self-driving, AI has been used in object detection, goal identification and smart prediction. Robots can plan the path to avoid bump between vehicles. By May 2017, the self-driving test mileage of Baidu had exceeded three million miles. Self-driving is a big market and a command height contended for by all countries. Google, Baidu and Tesla are all leading businesses in this field.

Global AI businesses are mostly in the United States, China and the United Kingdom. The number of AI businesses of the three countries accounts for 65.73%[12] of the world's total. Generally speaking, the AI technology of businesses in other countries is more developed than that of China (see Table 3.2).

---

[12]Source: *Wuzhen Index: Global AI Development Report* (2016), October 2016. .

Table 3.2 AI development and application of representative businesses of china and other countries

Type of businesses	Businesses	AI layout
Foreign businesses	IBM	AI technologies: speech and semantics, and deep learning neural network (having acquired AlchemyAPI)
		Overall solution: providing a complete set of API, transfer from speech to text and vice versa, trade off analysis, Q&A, mood analyzer and visual identification
		Cloud platform: IBM Bluemix open cloud platform (PaaS+ seven IBM Watson services), and the machine learning platform SystemML
		Hardware: AI chip TrueNorth
		Industrial layout: Deep Blue Computer; intelligent robot cooperation (with Apple and SoftBank robot Pepper); IoT; medical care; and VR game
	Google	AI technologies: vision, speech, natural language, big data, neural network training + deep learning (having acquired Dark blue labs, Vision factory, Deepmind, Jetpac and DNNresearch)
		Cloud platform: the second-generation learning system TensorFlow (including all kinds of pre-training model, natural language processing, recommendation system, model identification and prediction function)
		Hardware: AI accelerator chip TPUs (Tensor processing units)
		Industrial layout: self-driving, APP applications and plug-ins, intelligent furniture (hardware watch, furniture pivot Google Home), VR ecosystem (Daydream platform, VR HMD and control, applicable mobile phones)
	Facebook	AI technologies: Visual Deep face (having acquired face.com), speech (having acquired Wit.AI), and neural network training + machine learning
		Cloud platform: developer platform Parse and open-source deep learning model in Torch
		Hardware: Big Sur (hardware system based on GPU and used for training neural network, open source)
		Industrial layout: speech assistant Moneypenny;VR ecosystem (having acquired the hardware Oclus Rift helmet, Sourroud360 panoramic camera promoting content development)
	Microsoft	AI technologies: speech, vision, natural language and distributed machine learning
		Cloud platform: Microsoft Azure (storage, computing, database, live, and media function); distributed machine learning tool kit DMTK (natural language processing, recommendation engine, model identification, computer vision and predictive modeling)
		Industrial layout: language assistant (Microsoft Xiaoice, Cortana and Tay), VR (Hololens)
	Apple	AI technologies: natural language (having acquired Vocal IQ), visualization map (having acquired Mapsense\GPS Coherent Navigation)
		Industrial layout: self-driving (automobile) and SIRI language
	Amazon	Cloud platforms: Amazon Web Services (storage, computing, model identification and prediction. Visual identification API acquiring Orbeus)

(continued)

Table 3.2 (continued)

Type of businesses	Businesses	AI layout
Chinese businesses	Baidu	R&D resources: three labs of Baidu Research Institutes, North America Valley AI Laboratory, Beijing Deep Learning Laboratory and Beijing Big Data laboratory
Type of businesses	Businesses	AI layout
		Research fields: having made progress in image identification, image search, speech identification, natural language processing, intelligent semantics, machine translation and precision advertising (over 500 international patents, including over 270 in neurolinguistics programming and over 120 in deep learning)
		Product forms: having opened its AI ecosystem and launched a number of application products in many fields
	Tencent	Research fields: soft and hardware service ecosystem based on the user system, especially face recognition and image identification
		Research sources: WeChat has established with Hong Kong University Science and Technology an AI laboratory, doing research in speech identification, image, identification and semantic comprehension
	Alibaba	Research fields: the first visualized AI platform DTPAI, having integrated Alibaba core arithmetic library including feature engineering, large-scale machine learning and deep learning
		Product forms: Xiaomi, an AI client service and the small smart program Ai working on the principle of neural network, social computing and emotion perception.
		Application fields: having invested in SBRH, having established the research cooperation and communication mechanism with SBRH in terms of machine vision, speech resolution, smart household control and smart network security to help the latter with the accumulation and expansion of basic technologies of smart home
	Iflytek	Research fields: single point breakthrough in speech identification and the general solution platform for it based on the speech system

China and the United States are different in AI industries. A report by Tencent Research Institute shows that the top three areas of U.S. AI businesses are natural language processing, machine learning application and computer vision and image while the top three in China are computer vision and image, intelligent robots and natural language processing. Chinese investors focus more on application, such as self-driving, driving assistance and drones while U.S. investors focus more on basic areas, with the fund for chips/processors accounting for 31%. The AI industrial chain has been primarily formed in China, but it has a long way to go to catch up with the United States. Data from Tencent Research Institute show that the number of Chinese businesses in basic, technical and application areas is respectively 42, 46 and 62% of that of businesses in the United States.[13] In basic algorithm and chip areas, the number of people engaged in basic area research (such as processor and chip research) in the United States is more than ten times that in China. In recent years, China's AI has developed fast. According to *Wuzhen Index: Global AI Development Report* (2017), from 2014 to 2016, the number of new AI businesses in China accounted for 55.38% of the accumulative number in the country,[14] with the fund reaching US$2.76 billion, only next to that of the United Sates. The annual number China's new AI businesses accounts for an increasing percentage in the world. The country has got 15,745 patents concerning AI,[15] ranking second in the world. In the past five years, the annual growth rate of AI patents in China is 43%.[16] As for chip research, Cambricon and Vimicro have certain R&D capacity. In terms of technology, Chinese businesses take the lead in computer vision and image, intelligent robots and natural language processing.

3.2.3 Quantum Communication

Quantum communication is new type of communication. It is a process of information transmission with the help of quantum entanglement, involving segmented areas of quantum cryptography communication, quantum teleportation and quantum dense coding. In recent years, with the breakthrough and development of single quantum state regulation and atomic laser cooling, human being's knowledge and control capability concerning microworld have been improved. Quantum teleportation is at the stage of testing, but it is the cutting-edge technology in quantum communication. The quantum private communication technology with key quantum distribution as the core can improve the security of existing communication technologies, so it has a promising application prospect in government affairs, finance and diplomacy. Now, it witnesses rapid development in technical research, pilot application and industrial popularization.

University of Science and Technology of China completed the MDOF quantum teleportation test in 2015, based on single photon polarization and orbital angular momentum for the first time, and in 2016 the quantum teleportation test of 30 km metropolitan area optical network for the first time. The two tests became the representative achievements in quantum communication. In 2015, Delft University

---

[13]Source: Tencent Research Institute: a comprehensive interpretation of the AI industry development of China and the United States, August 2017.
[14]Source: *Wuzhen Index: Global AI Development Report* (2017), July 2017.
[15]Source: *Wuzhen Index: Global AI Development Report* (2016), October 2016.
[16]Source: *Wuzhen Index: Global AI Development Report* (2017)—Industry, August 2017.

of Technology of Holland, overcoming the vulnerabilities in photon detection efficiency and transmission distance in traditional tests, fully verified the non-locality of quantum entanglement. In the same year, Battelle of the United States built the 650 km optical-fiber line of quantum secure communication between Ohio and Washington and released the plan of the construction of the quantum communication backbone network connecting the west with the east of the whole country. In 2016, UK Quantum Communications Hub began to construct the test network of quantum communication connecting Bristol, London and Cambridge, and Italy launched the quantum communication backbone network project, totally 1700 km long, running through the country. In 2017, University of Science and Technology of China published the test of quantum key distribution of the 404 km optical-fiber line based on an independent protocol of new-type measuring equipment, in which the transmission distance limit was increased. In the same year, Bristol University and Oxford University of the UK respectively released the news of chip quantum key transceiver based on photon IC and handheld quantum secret key terminals based on space light, laying a foundation for quantum key distribution equipment and application.

China's quantum communication has started from high and developed fast, with remarkable achievements. Both the research and application level takes the lead in the world. *The 13th Five-year Plan* issued by the Chinese Government in 2016 proposed the construction of quantum communication and ubiquitously secure IoT. The country listed quantum communication as a strategic emerging industry in a number of plans for scientific and technical industries and the information industry. In the same year, Ministry of Science and Technology set up the special project "Quantum Control and Quantum Information", laying out the strategic cutting-edge research of quantum communication and computing. As the most practical technology, quantum communication has witnessed a series of research achievements and pilot industrial applications in theoretical verification, practical application improvement and new application scenario development in China. Other countries have also begun to lay out quantum communication technology. In 2013, the United States adopted the "Quantum Information Science and Technology Development Program". In 2016, EU released *Quantum Manifesto* and in 2013, the United Kingdom released UK National Quantum Technology Programme. It can be anticipated that the research and application of quantum communication technology will witness a new period of explosive growth. In the future, the technology will be of strategic significance in the fields of scientific and technical competition, emerging industry cultivation, national defense and economic construction.

Network information technologies are developing fast and are being integrated into all areas of economy, society, production and life, changing the pattern of global economy, interests and security. The world's economy is shifting fast into the economy dominated by network information technologies, which will be a general tool industry integrated into all areas. The overall competition between countries will be the scientific and technical competition with the information technology innovation as the core. All countries are promoting the original innovation, integrated innovation and popularization of network information technologies, taking them as the important means of transforming the economy and seeking for new competitive advantages.

3.3 Network Information Technologies are being Integrated Fast into Other Areas

Network information technologies are being integrated fast into biology, material, energy and equipment, changing the modes of products, services and management in all those fields. Gene technology, brain science and organ repair are becoming mature and biological industry represented by precision medicine is being reformed fast. R&D of materials is being accelerated, resulting in new and intelligent materials. Energy production, transmission and utilization are totally reformed and the rise of Internet of energy has promoted the green development of economy and society.

3.3.1 Integration Between Network Information Technologies and Biological Technologies

1. **Network information technologies promote gene sequencing.**

Optimization of software algorithm is promoting the technical revolution of gene sequencing. High-throughput algorithm and sequence analysis technology are promoting the gene sequencing efficiency of American start-ups in biotech to grow exponentially. Since 2012, China BGI Genomics has detected millions of gene samples above the magnitude by laying out the computing and storage environment on Tianhe-1, the national supercomputer. In 2014, the company succeeded in designing and developing human's whole genome resequencing software pipeline on Tianhe-2. Alibaba has been cooperating since 2015 with Berry Genomics Corporation in building massive data cloud of Chinese genome so that precise interpreting can be done for individual genome.

2. **Network information technologies promote rapid development of brain science and technology.**

Brain function calculation and brain intelligence imitation are the cutting-edge areas of the integration between network information technologies and biological technologies. The United Sates, EU, China and Japan launched great brain research programs. The BRAIN launched by the United Sates during Obama's presidency was aimed to study the relationship between brain functions and human behavior. The Trump government will increase the input in brain science R&D in their budget for 2018, and lay out the plan for the drawing of Brain Activity Map. EU launched

Human Brain Project (2013–2023) to use the supercomputer to simulate human brain functions to strengthen AI. Japan has launched Brain/MINDS 2014–2024 to establish the nervous system model of primates' brain. China has identified the research on brain science and brain-like intelligence technology as an important part of the development of the new-generation AI.

3. Network information technologies promote progress in organ repair.

The exponential growth of the computing and storage capacity has promoted the integration between organ virtualization and big data analysis, thus bringing about some revolutionary tools for the repair of big organs. In 2016, the Wake Forest Institute for Regenerative Medicine of North Carolina, the United States, announced that they had made breakthroughs in 3D printers of human organs. Zhongshan Hospital of Fudan University of China developed in 2017 the new technology of taking gold as bio-printing ink. By far, bio-printing technology has been used in laboratory medicine testing. In the coming years, the first 3D printed organs will be used in clinical medical care.

4. Network information technologies promote clinical application of precision medicine.

Relying on deep learning algorithm, AI has its inborn advantages in improving the efficiency of health and medical services and disease diagnosis, so there have occurred all kinds of AI applications aimed to improve the medical service efficiency and experience. AI is used in precision medical diagnosis in two directions: to diagnose diseases with the aid of medical imaging based on computer vision; and to diagnose diseases through comparison of the contents in the disease databank and through deep learning. Precision medicine businesses like Editas Medicine have begun to make patient-tailored diagnosis and treatment schemes based on massive data and machine learning. AI will shift medical care to disease prevention and will be of significance to human health protection.

3.3.2 Integration of Network Information Technologies with Material Technologies

1. Network information technologies promote progress of material R&D technologies.

Development of software tools concerning material simulation, modeling and data analysis shortens the period between discovery and application of new materials. The United States promoted in 2011 and 2016 the material genomes program to accelerate the application of software tools in material preparation. The research achievements have been used in the manufacturing of the composite material for the fuselage of Boeing 787. EU launched Accelerated Metallurgy (ACCMET) and Metallurgy Europe: A renaissance programme for 2012–2022 respectively in 2011 and 2012, listing high-throughput technology software as the core for development and accelerating the discovery and application of high-performance alloy and other new materials. China has launched the material genomes program to accelerate the R&D and industrial application of new materials through data analysis models and computing software, and to promote the application of composite materials on the domestically manufactured commercial aircraft C919.

2. **Network information technologies improve material manufacturing capacity.**

3D printing technology is widely used in the industries of mold material manufacturing and industrial prototype design and has improved the production efficiency and precision of new material products. GM Group of the United States planned to apply in a large scale 3D printing technology in manufacturing fuel injection nozzles and sensor shells of the engines of jet planes within 2017. U.S. Navy and NASA use 3D printing technology in the manufacturing of components of naval vessels and rocket cooling systems to improve the production efficiency of complex industrial components. Harvard University has improved micro 3D printing technology, through which lithium battery units including conductive ink and micro electrode can be printed. Wuhan National Laboratory for Optoelectronics of China developed in 2016 a large 3D laser printer which can print high-precision metal components, turning the high-precision molding of complicated metal components into a reality.

3.3.3 *Integration of Network Information Technologies and Energy Technologies*

Technologies like IoT, through the integration of infrastructure of energy networks, have improved the energy information production and transmission efficiency, and thus promoted the development of Internet of energy. The United Sates has raised the new-type energy network construction to a national strategy. IBM, Intel, Microsoft, Google, Apple and Salesforce are all developing the Internet of energy in accordance with their strengths. By 2017, power supply technology based on cloud computing and IoT had become the mainstream of the energy supply system of the United States. The standard system of Internet of Energy is being established and the roadmap will be finished in 2018. FREEDM, Stem, and Opower of the United States, BDI and Green Packet of Germany, Powershop of New Zealand, and Envision and Huawei of China are seeking to construct the standard interfaces of the Internet of Energy. As for development strategies, the United States emphasizes the dominant position of the centralized large power grid in Internet of Energy and EU puts more stress on decentralized operation of Internet of Energy, while China aims to improve the robustness, compatibility and intelligence of energy transmission networks.

3.3.4 Integration of Network Information Technologies with Equipment Technologies

1. **Network information technologies make the manufacturing equipment industry more intelligent.**

Technologies like AI, perceptual identification technology, communication, cloud computing and virtualized manufacturing have promoted the emergence and development of digitalized manufacturers, and the interconnectivity between equipment, systems and production lines within the equipment manufacturer. They help to make production, management and decision into an organic whole and promote the equipment industry to be customized, flexible, intelligent and service-oriented. To make manufacturing intelligent, the United States was the first to put forward the plan in 2012 for industrial IoT driven by emerging Internet technologies and in 2013 Germany put forward the Industry 4.0 Strategy with intelligent manufacturing as the core. The United Kingdom put forward the Future of Manufacturing:A New Era of Opportunity and Challenge for the UK (Industry 2050 Strategy) in 2013 to accelerate the transformation of the equipment industry.[17] China released Made in China 2025 in 2015 and *Intelligent Manufacturing Development Plan* (2016–2020) in 2016 to make the country's equipment manufacturing intelligent.

2. **Network information technologies promote development of intelligent military equipment.**

Technologies of AI and perceptual identification have promoted the application of intelligent military equipment. The US Army has launched military robots named Bigdog into the battlefields. These robots can run at a speed of 12 km/h and carry the maximum load of 200 kg. They are bullet-proof and silent, able to carry supplies and act as companions. The US Department of National Defense put forward the Avatar Plan in 2013 to manufacture intelligent equipment that can be controlled by human brain from a long distance. DARPA has listed the HULC of warriors as a perspective plan. BLLEEX developed by Berkeley University and XOS developed by Raytheon Company will be gradually launched into the battlefield. Russia has put forward 2045 Project (also known as Russian Avatar Project) to promote the application of AI technology in HULC as military equipment and will organize a robot equipping force. In the future, intelligent military equipment will play an important role in military fights, backup supplies, battle field exploration and disposal of harmful substances.

---

[17]Source: IBM Analytics, *Industry 4.0 and IoT White Paper*, 2017.

3.4   Future Development of Network Information Technologies

With the emergence of network information technologies and applications like 5G and AI technologies, the era of Internet of Everything, involving human beings, programs, data and things, is coming, and network information technologies will play an increasingly important role as a foundation, tool and drive in economic and social development.

3.4.1   Network Information Technologies Will Keep Upgrading Fast

1. Information networks are developing fast in a ubiquitous and integrated three-dimensional way.

The new round of global acceleration of broadband is coming and the high-speed broadband will be upgraded comprehensively. The 5G mobile communication technology is being industrialized and commercialized fast. The Internet is developing into the Internet of Everything. The cyberspace is expanding to physical space of information. All kinds of networks are being integrated, covering telecommunication, broadcast and TV, Internet, mobile Internet, narrow-band IoT, satellite communication, and industrial Internet. Network access service is available everywhere and the number of access terminals is growing exponentially. The IPv6 network will be applied in a large scale. All countries will accelerate their exploration in constructing space Internet, providing inclusive network access service by means of hot balloons, drones and low and middle-orbit satellite communications. Innovation in space Internet and interstellar communication is active and will become the new focus of the new space competition among developed countries. Global cloud computing infrastructure will be formed rapidly and application infrastructure including CDN and IDC will be developed in scale, which will change the network structure and service mode.

2. All kinds of network information technologies, including networking, computing, storage and software, will be integrated.

In the future, with the popularization and development of network applications, all services including computing, storage and software will shift towards the center of networks. Cloud computing, cloud storage and cloud service software are becoming the mainstream service mode in the field of computing, storage and software. To meet the demand for online processing capacity of computing, storage and information service, the organization mode, service mode and transmission capacity of the network will keep being upgraded. Broadband networking will be the trend. Networks will develop fast, and it will be ubiquitous in three dimensions. They will be increasingly secure. Cloud computing will be the important operation infrastructure of economy and society, and its scale and service capacity will be increased. Its application will penetrate more areas of intelligent society and provide more convenient service through ubiquitous high-speed networks. Cloud storage technology is moving towards virtualization and intelligence and its scale will keep increasing to support he explosive and exponential growth of network information resources. Cloud service software will accelerate the deep integration between software industry and information service industry and restructure the industrial ecosystem of software and information service.

3. **Intelligent terminal technology is developing towards comprehensive integration and integrated innovation.**

Intelligent terminal technology will comprehensively integrate the achievements in the mobile Internet, cloud computing, big data, IoT and AI. It will promote intelligent terminals to be mobile, more intelligent and cloud-oriented. Mobile Internet, cloud service and intelligent perception will be the common characteristics of the future intelligent terminal technology, which will promote the accelerated integration of software and hardware technologies. The development of big data, IoT and AI will promote the accelerated integration of chip technology, OS and application program. Intelligent terminal technology will be integrated into technologies of automobile, household and carriers to make these traditional terminals digitalized, networked and intelligent to incubate new intelligent information services and applications like self-driving vehicles, drones and smart home.

3.4.2 Network Information Technologies Will Witness Cross-Border Integrated Innovation

1. **Network information technologies will become general tool technologies.**

With the development of economy and smart society, there is an increasing demand for the networking, platform-oriented and intelligent development of all areas of economy and society. Mobile Internet, cloud computing, big data, IoT and AI will become general tool technologies supporting the development of intelligent economy and society, and the necessary means, supports and innovative factors of the development of all industries. They will promote all areas to develop in the direction of mobility, networking, platform orientation and intelligence. The spillover effect of network information technologies will be exerted for new reforms of the modes of organization, service and business.

2. **Network information technologies will promote innovation and reform in all areas.**

Network information technologies will become the promoter of innovation and reform in all areas of economy and society, facilitating the overall innovation and upgrading of modes of organization, service and business. First, they will promote the innovation of the organization mode in all areas. Platform organization, network collaboration, crowdsourcing and crowdfunding will be adopted in all areas, and networking and flattening will be the trend to liberate the organizations and release the bonus of organizational reform. Network information technologies will promote the innovation of service mode. They will, by making use of the strength of cyberspace and the zero marginal cost effect of platforms, promote the popularization of online, mobile, nearby and personalized services and precision marketing. They will facilitate the mode innovation and service upgrading in all areas to better conform to the new normal of economy and the supply-side reform. They will also promote business mode innovation in all areas. Customized production, integration of production and marketing and service transformation will be widely adopted, and production and service mode will be reshaped to emancipate information productivity.

3. **Network information technologies will promote cross-border integrated innovation.**

Network information technologies will accelerate integrated innovation. The border between hardware and software, software and service, and network and application will be obscure. The integration between software and hardware, between software and service and between network and application will be a trend. Network information technologies will be integrated into biology, energy, material and equipment industries to promote the mode reform of R&D and design, manufacturing and after-sale service in the areas of biology, energy, material and equipment. Thus they can make the products intelligent and networked and make service upgraded, thus bringing about new products and services like man-made biology, intelligent energy, intelligent material and intelligent equipment. They will promote the cross-border integration of the primary, secondary and tertiary industries, whose R&D and design, manufacturing, logistics and transportation, marketing and sales and after sales will be closely related to each other. "Production + service" and "manufacturing + service" will be the important forms of cross-border integration of the industries.

3.4.3 Network Information Technologies Will Be More Intelligent

1. **The development and progress of information technology will provide guarantee and technical support for AI technology.**

The development of information technologies will provide guarantee and technical support for AI technology. The rapid development of the new-generation information technologies like mobile Internet, cloud computing, big data and IoT will provide high-speed ubiquitous network access, faster computing, deeper data analysis and stronger IoT for AI development. Fast perception, computing and analysis will bring about the new development wave of AI. Meanwhile, mobile Internet, cloud computing, big data and IoT are becoming more intelligent. With AI support, networking, computing, data analysis and sensing will also be more intelligent. To be short, AI will be an important element in information technology development.

2. **"AI+" will be an important form of the integration between information technologies and economy and society.**

With their integration into all areas of economy and society, network information technologies, as a general tool, will be an important drive for the innovative development of all industries, promoting the innovation of the mode of organization, service and business. New demands for intelligent economy, society and government will emerge, AI and network information technologies will drive each other and witness integrated development, which will make AI, just like network information technologies, applied in all areas of economy and society and turn it an innovative factor in economic and social development. "AI+" will become a new drive in the future.

Chapter 4
Development of the World Cyber Security

Chinese Academy of Cyberspace Studies

Abstract At present, cyber security is increasingly threatened by cyber attacks, crime and privacy disclosure, which are penetrating areas like politics, economy, culture and society. All countries are strengthening their attention to cyber security and have taken a series of actions in accordance with their own situation and the world Internet development to improve their capacity of cyber security assurance. (1) Cyber attacks are becoming more hidden but more destructive. Technically peacocking represented by CIH virus, Melissa and Code Red virus are replaced by purposeful and organized attacks represented by Worm.WhBoy.cw, Stuxnet and Ukraine Power Grid Incident. Governmental intervention is being deepened. (2) Web spoofing, DDoS, Trojan and botnet, and cybercrime are the major threats to cyber security. In 2016, at least 255,100 phishing attacks happened worldwide, an increase of over 10% in comparison with the number (230, 300) in 2015. Actually, the number in 2016 was unprecedentedly high. Every year, about 500 million host computers are attacked by botnets, 18 computers per second, resulting in a loss as high as US$110 billion. (3) Advanced persistent threats are becoming increasingly serious with great power of destruction, imperceptibility, durability and complexity, posing great threat to critical infrastructure of finance, business, communication and transportation, and national defense. (4) The number of ransom software attacks is growing explosively. By the end of 2016, such software had evolved into a family-like development mode, with 44,300 new varieties having been discovered and 114 countries having been affected. RaaS is developed into a black industrial chain. From 2016 to 2017, the market scale of the underground market of ransom software saw an increase of 2502%. (5) Large-scale data breach happens frequently, threatening the security of data from governments, businesses and individuals. From 2008 to 2016, over seven billion pieces of online ID information worldwide was embezzled, which indicates that every person's information is embezzled once. The governmental data of Sweden, Mexico and the Philippines were disclosed, which should arouse the attention of all governmental sectors. (6) With the maturity of IoT and AI, the equipment of IoT and AI may be the major targets of cyber attacks. By the end of 2016, it had been discovered that

---

Chinese Academy of Cyberspace Studies
Beijing, China

© Publishing House of Electronics Industry, Beijing and Springer-Verlag GmbH
Germany, part of Springer Nature 2019
Chinese Academy of Cyberspace Studies (ed.), *World Internet Development Report 2017*,
https://doi.org/10.1007/978-3-662-57524-6_4

2526 control servers had controlled 1.254 million intelligent devices of IoT of the world. Therefore, we should attach equal importance to security and development and take actions in advance to prevent any attack. (7) The United States, Russia, China, Germany and Singapore have launched national cyber security strategies and established special organizations to strengthen the protection of critical infrastructure. They also enhance cyber content security supervision and promote the development of cyber security industries. (8) The position of cyber security has been elevated in the overall national security. Some countries have channeled cyber security into military security to enhance its importance. The United Sates has elevated Cyber Command to be the highest United Combatant Command. Germany and Israel have also founded their cyber commands to improve cyber military fighting capacity, which, at the same time, brings about new factors of instability of cyberspace.

4.1 Threats to Cyber Security

Since the emergence of cyber attacks, the purpose, technique and damage of the attacks have seen great changes. From 1980s to the early 21st century, the purpose was simple, only the showoff of personal techniques. Following C-Brain, the first computer virus which occurred in 1986, there emerged CIH, Melissa, Code Red, Blaster and other viruses. With the rapid development and popularization of the Internet, cyber attacks are no longer aimed to show off personal techniques but to seek for profits. They have shifted from unorganized attacks to organized ones, from purposeless to purposeful. They are increasingly imperceptible with increasing destructive power. The Worm.WhBoy.cw and Storm Scandal and other incidents indicated that the black organized industrial chain had been formed. Stuxnet, PRISM, Ukraine Power Grid Incident and the theft from the hacker organizations' kits all showed that cyber attacks with national support had emerged, that the concept of "cyber warfare" has been established and that cyber warfare already has some scale. Cyber security threats can be classified into the following.

1. Web spoofing

Web spoofing often directs users to click and get personal information at a fake webpage, which is nearly identical to the legitimate one. By web spoofing, the hacker can obtain millions of US dollars of profit every year. The year 2016 saw the highest number of phishing attacks and of phishing websites, since the phishing website is a common form of web spoofing. The United States had the most phishing websites, China's phishing incidents became the mostly complained incidents and Republic of Korea saw an increasing number of phishing websites. Besides, the attack quality and methods keep improving and traditional phishing attacks have seen a drop in number while other ways of phishing are on the rise. In 2016, the new trend was shown by the combination of phishing and social media. The latter was used to obtain information concerning work, habits and organizational structure. In general, web spoofing remains an attack with low cost, high benefit and a wide affected range.

2. **DDoS**

DDoS, not so frequently seen as phishing and malicious programming, may do greater damage to data or systems and even give rise to more serious results. In October 2016, Dyn, a DNS in the United States, was subject to a large DDoS attack, which resulted in the network paralysis of the whole country. In Australia, almost every DDoS attack has been successful. It can be seen that such a threat is hard to resist. For e-government, online business and e-commerce, the break caused by DDoS can be disastrous and can produce serious impact on the whole economy. Besides, DDoS embodies the obvious purpose of making profits and committing crime, and it is a commodity in the black industrial chain.

3. **Trojan and botnet**

Trojan and botnet's harm and threat is the key cause of privacy disclosure, secret divulging, spams and large-scale DDoS. Such attack is harmful but develops slowly. According to estimation, the loss caused by botnet to the world has been as high as US$110 billion, with 500 million host computers being attacked every year, 18 computers per second. The number of Trojan and botnet control servers in China remains stable in recent five years. Except that in Hong Kong the new botnet Mirai saw an increase in the fourth quarter of 2016, the number of botnet incidents keeps decreasing throughout China. The number of botnet incidents in Malaysia also saw a decrease in 2016 and the number of botnets traced by CERT-In in India remained stable in the same year. IoT is generally the target of botnet attack. Hackers are becoming more patient in launching attacks. The botnet is also used by some high-level threat agents to attack special groups of people.

4. **Cybercrime**

Cybercrime can be divided into two kinds: crime through the Internet as a tool and crime with the Internet as the target to damage. There is considerable economic loss caused by cybercrime. In the past five years, the United States has received 280,000 complaints about cybercrime every year, with 298,728 complaints in the single year of 2016, when the economic loss amounted to US$1.4507 billion. Middle-aged and senior citizens (30–39 and over 60 years old) are usually the victims and the latter suffer greater losses. In Australia, 56% of organizations discovered one or two cyber security incidents in the previous year and the remaining 44% did not discover any incident, which reflects that cyber security incidents are not all sensed and that different organizations have different definition of cyber security incidents. Every year, millions of Americans are hurt by cybercrime, but only 15% of the victims report to law enforcement agencies, which indicates that cybercrime sensibility and complaint initiative should be improved.

5. **APT**

APT refers to the threat to cyber security launched by hackers to certain target in a persistent and advanced way. More and more APT organizations and attacks are discovered (for example, Stuxnet and Black Energy). APT has its clear target, i.e.

critical information infrastructure. It is disastrous, long lasting and complicated. Now, APT may happen anywhere at any time.

6. Ransomware

Ransomware is a popular Trojan program. By disturbing, terrifying and even hijacking the user's file, it makes the user's data assets or computing assets unable to be used unless a ransom is paid. In recent years, the number of ransomware attacks has grown explosively and they have caused great losses and produced great impacts. Kaspersky found over 40,000 new ransomware varieties in 2016 and over 100 countries suffered attacks from ransomware, which is target-oriented, technical, destructive and easy to spread. Ransomware has grown internationally. The model of "ransomware as a service" has been developed into a black industrial chain. The research of Carbon Black shows that from 2016 to 2017, the underground market of ransomware increased by 2502%, with the value amounting to US$6.2 million.

7. Massive data breach

Recently, a series of massive data breach, covering as many as tens of millions of pieces of information, have been exposed. A Symantec report reveals that from 2008 to 2016, over seven billion pieces of online ID information was embezzled in data breach, which means that every one's information on the earth is embezzled once. Personal ID information is the common type of information stolen, then comes personal financial information. Among the causes of data breach, malicious acts account for the largest percentage and cause more serious economic losses. Besides commercial data, data of governmental sectors are faced with security risks, which should arouse the attention of all governments.

8. Security of IoT

With the rapid development and popularization of intelligent devices of IoT, the risks of the security of IoT are on the rise. More and more attacks on IoT happen every year. The attacker can obtain the authorization of device control through vulnerabilities and then steal the user's data, hijack the traffic and form large-scale botnets. In 2016, massive network outages on the east coast of the United States and exceptions of telecommunication users' access in Germany happened one after another. They were related to Mirai, a malicious program that can control large botnets of IoT. Android phones have been found to be used to construct massive botnets and security hazards are hidden in IoT using Bluetooth and wireless networks. Smart homes and smart cities are also faced with the threats. Notably, there is no sufficient protection for the security of IoT, which is also reflected by the recent attacks on it.

9. AI security

All countries have input a lot of manpower and material resources in AI. According to an IDC report, the income from cognition and AI worldwide in 2020 will be US $46 billion. In the past two years, the emergence of AlphaGo and AlphaGo Zero manifests the cognitive competence of AI, indicating that the technology will produce deeper and wider impact on human beings. Therefore, we should pay more attention to the security of AI to make better use of the technology. Innovative solutions should be found to tackle AI security problems so that AI technology can better serve human beings.

4.2 Evolution of Cyber Security Threats

The Internet now covers a population of more than three billion, so it has become an indispensable part of people's life and work, but there are more and more problems of cyber security. Due to the vulnerability of the network system, there are inevitably bugs in OS, application software and hardware. There are also security hazards in the protocol design of networks. All these provide potential opportunities to be exploited by hackers to attack the network. So far, changes have taken place in the purposes, techniques and harms of attacks. At the very beginning, the purpose might be simple, just being to show off personal techniques. The target of attacks was also simple, and the virus and traces could be easily analyzed and resolved. However, with the rapid development and polarization of the Internet, the attacks have evolved from showoff of personal techniques to seeking for financial benefits, from disorganized to organized, and from purposeless to purposeful. They have become increasingly imperceptible and destructive.

1. **Stage one: showoff of personal techniques**

From 1980s to the early 21st century, the attacker's purpose tended to be simple, just being to show off their techniques to have the sense of achievement. The first computer virus C-Brain was born in 1986. It already had the complete characteristics of a computer virus. The designer of C-Brain did not aim to attack a target but to prevent the copyrighted software from being embezzled. Later, viruses like CIH, Melissa, Code Red and Blaster were all aimed to show off personal techniques.

(1) CIH

In June 1998, the first report on CIH virus was released from Taiwan, China. It was a file virus programmed by an undergraduate. The early CIH virus was spread with piratic CDs in the United States and then in the whole world through the Internet. With the updating of its versions, CIH witnessed three major outbreaks, respectively from July to August of 1998, from April 2000 to April 2001, and in 2003.

(2) Melissa

In March 1999, a virus named "Melissa" began to spread worldwide. Through Outlook software of Microsoft, Melissa was disguised into an email with "important message" from a colleague or friend and then spread everywhere. When a user opened the email, the computer would be infected and send the first 50 recipients in the address book emails carrying the virus. Though it did not harm the system files in the computer, it could paralyze the server by overloading it. According to a report, Melissa infected about 20% of commercial computers in the world. Its fast spread made Intel and Microsoft unable to make full preparations to tackle it, so they had to shut down the whole email system.

(3) Code Red

Code Red was a computer worm observed on the Internet in July 2001. It is different from the previous file infecting viruses and boot viruses. The worm used the technique known as "buffer overflow". It existed in memory and attacked through the network of servers, which could cause slow access or access break. It was reported that after 20 days of self-spread of the worm, the number of infected hosts reached 400,000 and the number of infected computers, one million. Afterward, the worm launched DDoS attacks on some IP addresses, including the IP address of the White House.

(4) Blaster worm

Blaster worm started spreading in August 2003 by exploiting a buffer overflow in the DCOM/RPC released in July 2003. The infected computer system would have abnormal operation and repeated restarts until even breakdown. At the same time, the worm launched DDoS attacks on the upgrading servers of Microsoft, so that Microsoft was unable to upgrade the system for its users. Blaster worm would launch attacks initiatively, exploit vulnerabilities, block the networks and cause repeated damages. It spread itself by scanning, attacking and copying. Most anti-virus software could not block it. According to statistics, 130,000 computers worldwide were infected within one week, and half of them were from the United States.

2. **Stage two: organized threats**

Since the beginning of the 21st century, attackers were no longer satisfied with the showoff of their techniques, but intended to get benefit from their attacks. With the rapid development of the Internet, the threshold of network technologies has been lowered and so has the threshold of cyberattacks, which to some extent, has given rise to the unchecked spread of cyberattacks. There are more organized gang attacks. Black industrial chains like malicious program development, malicious program spread, cyberattack, cyber fraud and data or information selling have clear division of work. As cybercrime, especial cyberattack, becomes increasingly organized, the threats to the stable operation of the Internet are becoming more serious, causing greater harm to the global economic and social development.

(1) Worm.WhBoy.cw

From the end of 2006 to early 2007, Worm.WhBoy.cw spread itself in the network. It infected the computer through file downloading. It can spread itself and infect the hard disk, with great destructive power. The infected computer users might find their executable files changed into a panda holding three incenses. The designer of the virus stole the user name and password through the virus and sold the infor-mation by himself or via others. He also sold the virus to 120 people on the Internet, making an illegal profit of RMB100,000. Then the purchasers of the virus spread it further, causing serious damage to cyber security.

(2) Storm Scandal

One day in May 2009, the recursive resolution service of many provincial domain names witnessed breakdown because of the large numbers of DNS requests and the Internet domain name resolution service in some regions became abnormal, which affected the Internet operation in China on that day. That was a scandal caused by cybercrime. According to the investigation, the criminal had done business in online game servers for a long time and helped others to attack other game servers or other servers by renting servers to make profits. That night, the criminal launched an attack on the domain name resolution servers (provided by DNSPod for free) of other private game server websites, which resulted in the paralysis of DNSPod domain name servers (DNS). Since DNSPod provided Storm software with DNS resolution service, the attack caused the block of the DNS requests from Storm software. Since the software had a large number of terminals with advertising pop-ups and frequent upgrade requests, a huge number of requests were launched frequently, but the DNSPod servers were attacked and paralyzed, so many requests flew to the DNS recursive servers of the operators, which ended up as the Storm Scandal, a major incident of cyber security.

3. **Stage three: more and more government background**

In June 2010, Stuxnet worm was uncovered, arousing the unprecedented attention of all governments worldwide to cyber security. They realized that cyber security will affect the security of the physical world. From then on, cyber attacks with national background came into people's view. PRISM in June 2013 made the US global surveillance known to the world, and then all countries began to rethink the importance of cyber security and decided to enhance their cyber security. In recent years, APT incidents have been revealed one by one, and high-level organized network confrontations between countries are on the rise. The concept of "cyber warfare" has been established and cyber warfare has already gained some scale. The cyber security threats' destruction will be unimaginable.

(1) Stuxnet Worm

In July 2010, Stuxnet Worm began to spread. It was a worm sweeping the global industrial system, especially taking the industrial infrastructure as the special target. Exploiting at least four zero-day flaws (including three new ones) of the Microsoft Windows operating system, the worm used effective digital signature for the derivative programs. Through a series of invasion and spread flows, it broke the physical limit of the special networks of industry to launch the attack. According to statistics, the worm infected 45,000 networks within two months, with the largest number in Iran, accounting for 60% of the number of total infected hosts. Different from the worms in traditional sense, Stuxnet took important targets as the attack targets. Region-oriented, it has obvious purposes, so it is an elaborately designed cyber attack.

(2) PRISM Scandal

In June 2013, Edward Snowden, former employee of CIA, gave some confidential material to *The Guardian* of the United Kingdom and the *Washington Post* of the United States, revealing the secret program PRISM and other secret programs of the United States, which caused a sensation worldwide. The incident was called PRISM Scandal. According to the disclosed information, the United States inputs a large fund into monitoring and collection of information through the Internet, communication networks and businesses' servers. The monitoring covered metadata and communications on the Internet, social network information, telephones and messages. The monitored targets were leaders and high-ranking officials, diplomatic systems, media networks, and large business networks of many countries, and international organizations.

(3) Ukraine Power Grid Cyber Attack

The Ukraine Power Grid Cyber Attack took place in December 2015. Due to the attack on the power grid, some areas of the capital of Ukraine Kyiv and millions of residents in the west of the country were left without electricity, which led the people into panic. By cheating, the attacker made a worker in the Ukraine Power Company download the malware BlackEnergy, which disconnected the host computer of Ukraine Power Company from the transformer substations. The attacker then planted KillDisk, another kind of malware, in the system to make the computers unable to work. At the same time, the attacker disturbed the telephone communication of the power company, so that the affected people were unable to get in touch with it. The massive power cut-off aroused the world's attention to the security of critical information infrastructure.

(4) Theft from the hacker organizations' kits

In March 2017, news from online came that the CIA, to collect intelligence, had developed hacker kits to attack the mainstream operating system of Microsoft, Android, Apple and Linux and smart devices like Samsung smart telecommunication, SVOS and routers. The theft from the hacker kits shows that the United States had strong online arsenals and the online transaction of kits will bring about hazards to the Internet security. In April 2017, Shadow Brokers organized the distribution of equations to organize some tool files, including many high-risk vulnerabilities and tools targeting the Windows operating system and other office and email software. Such tools are highly intensive and some are quite efficient in supply and use. Within one month, the Wannacry ransomware attack took place worldwide. It was a network worm that caused great harms to the interests of the world's Internet users.

4.3 Major Threats to Cyber Security of the World

At present, the number of threats to cyber security keeps increasing, involving cyber attack, cybercrime and privacy protection. Cyber security is a matter closely related to the security of all countries and regions.

4.3.1 Web Spoofing is Rampant Across the Globe

Web spoofing intends to mislead users by creating a webpage which is highly similar with a targeted website. Phishing is a common form of web spoofing, usually transmitting information through email spam, live chat, mobile phone message or false advertising on the webpage. Usernames and passwords may be disclosed after the users visit the phishing website. The attackers take advantage of people's negligence and disguise themselves as a trustworthy entity to attract the click by the users, so that they can get private information. A web spoofing attacker can get illegal benefits of millions of US dollars every year. It is difficult to counter the attacks.

1. **Victims of web spoofing experience different attacks across the globe.**

According to *Global Phishing Survey: Trends and Domain Name Use* in 2016, the year saw the unprecedented number of phishing attacks and domain names of phishing. At least 255,065 phishing attacks took place, 10% more than in 2015, when there were 230,280 such attacks. 75% of the malicious domain names are.-com,.cc,.pw and.tk, among them 454 are TLDs.[1] The United States has the most phishing websites. According to FBI, at least 7000 American companies have been victims of phishing attacks since 2013, with a total loss of US$740 million.[2] According to *IT News Africa*, South Africa is the second targeted country of phishing attacks, having suffered a loss of tens of millions of US dollars, with US $320 million of loss in 2013, accounting for 5% of the global total.[3] *Report on China's Internet Security (2016)* shows that in 2016, CNCERT found 178,000 phishing websites targeting the websites within China, with a decrease of 3.6% in comparison with the number in 2015, and about 20,000 IP addresses had phishing websites, with overseas IP addresses accounting for 85.4%. At present, the number of phishing attacks has surpassed that of vulnerabilities, ranking first in all website complaints.[4] *APCERT Annual Report 2016* points out that HKCERT2016 (Hong Kong, China) tackled 1957 phishing attacks in 2016, accounting for 32% of all the attacks tackled. Among the cyber security incidents handled by JPCERT (Japan),

---

[1] APWG:*Global Phishing Survey: Trends and Domain Name Use in 2016*.

[2] FBI: *Business E-Mail Compromise-An Emerging Global Threat*.

[3] *IT News Africa*: South Africa is second most targeted for phishing attacks.

[4] CNCERT: *Report on China's Internet Security* (2016).

15.3% are phishing attacks, ranking third in all the incidents handled. According to statistics released by KrCERT (Republic of Korea), the number of phishing websites keeps increasing, causing leakage of private information and usernames. KrCERT received 749 reports on phishing in 2016, the number increasing by about 60% in comparison with that in 2015. Among the 134,375 cyber security incidents tackled by VNCERT (Vietnam) in 2016, 10,057 were phishing attacks, twice the number in 2015. In 2016, the number of phishing incidents that happened in Thailand accounted for 26.4% of the country's total number of cyber security incidents and 309 IP addresses were affected. MNCERT (Mongolia) jointed APWG in January 2016, the number of phishing attacks tackled that year decreasing by one half, which showed great progress.[5]

2. **Threats are becoming increasingly serious and the means are being diversified.**

According the *ENISA Threat Landscape Report 2016*, from November/December 2015 to November/December 2016, though the number did not increase, the quality and means of phishing attacks kept improving. The new trend in 2016 was the combination of phishing attacks with social media, which were used to obtain information about jobs, habits and organizations.[6] According to APWG's *Phishing Activity Trends Report: 4th Quarter 2016*, there happened 2000 fraud attacks on businesses and individuals, and 304 of them were from phishing websites and two were paid search engine phishing activities. The number of traditional phishing attacks is decreasing, while the number of other forms of such attacks is increasing. Most of the attacks are from the five countries, namely, the United States, Ireland, Brazil, France and Canada, for the reason that most servers carrying phishing websites are in the United States.[7]

3. **Web spoofing remains a frequently used low-cost but high-return attack.**

ACSC (Australia) Cyber Security Survey 2016 shows that among all the successfully attacked organizations, 42% were attacked by email phishing and cheating. In most cases, their brands are imitated or attacked in other phishing ways. Almost all organizations are faced with phishing and fraud attacks and 84% of them have experienced phishing attempts, most of the activities being concerned with remittance and a small number of the activities being attacks of malicious programs on data of systems.[8] According to *NCSC 2016–2017 Unclassified Cyber Threat Report*, phishing, as a cheating means, is a common cyber threat in New Zealand. Though people are more aware of the threat today, phishing remains a low-cost but high-return cyber attack means, easily used by attackers.[9] According to a cyber

---

[5]ENISA: A*PCERT Annual Report 2016*.

[6]ENISA: *ENISA Threat Landscape Report 2016*.

[7]APWG: *Phishing Activity Trends Report: 4th Quarter 2016*.

[8]ACSC: ACSC 2016 Cyber Security Survey.

[9]NCSC: *2016–2017 Unclassified Cyber Threat Report*.

security test done by KPMG among New Zealand businesses, one tenth of people from the country have suffered phishing attacks, the frequency of the attacks keeps increasing and the means is becoming more complicated.[10] LACNIC, an organization of Internet resource distribution and management in Latin America and the Caribbean, points out that the number of phishing attacks remains the highest among all the cyber crimes, accounting for one third. Among the incidents tackled in the previous year, 32.8% were related to the theft of private data and financial information.[11]

4.3.2 Huge Destructive Power of DDoS

The DDoS attack is an attack in which the targeted machine or resource is flooded with superfluous requests in an attempt to overload systems, or an attack on special targets so that the targets may stop providing service. DDoS attacks are not so frequently seen as phishing and malware attacks, but they may cause damage to data or systems, or even more serious results.

1. **Smaller number but greater impact of DDoS attacks**

According to *Global DDoS Threat Landscape Q2 2017* by Incapsula, the United States suffered the most DDoS threats, accounting for 79.7% the world's total.[12] In October 2016, Dyn, a US domain name server, was attacked by large-scale DDoS. The hacker used the Mirai virus to infect nearly one million devices of IoT, such as routers, web cameras and DVRs, leading to the inaccessibility of the websites of Twitter, Netflix, CNN, *The Wall Street Journal* and Amazon. The media reported that "The attack has paralyzed half of the U.S. networks." According to ACSC 2016 Cyber Security Survey, the DDoS attempts accounted for 23% of the total cyber attacks in 2016 and almost the same number of the attacks were successful, which shows that it is hard to prevent or resist such attacks. According to ACSC Threat Report, from July 2015 to June 2016, AusCERT made response to 14,804 cyber security incidents against businesses of Australia, in which banks and financial and communication industries suffered more. According to the official website of MyCERT (Malaysia), MyCRET received a report of 6274 cyber security incidents in the first three quarters of 2017, and among them 37 were DDoS attacks (accounting for 0.59%), more in January, July and August, with 11, 8 and 6 such incidents respectively, causing great impact on the operation of relevant business. DDoS attacks are especially destructive to e-government, online service and e-commerce and they can seriously affect the economy of the country.[13]

---

[10]KPMG: New Zealanders get hooked by phishing attacks.
[11]LACNIC: *Phishing Accounts for One Third of the Total Number of Incidents.*
[12]INCAPSULA: *Global DDoS Threat Landscape Q2 2017.*
[13]MyCERT: MyCERT Incident Statistics.

2. Clear intention to obtain benefits and commit crime

DDoS attacks have two aims: to obtain benefits and to commit crime in the form of service. According to *ENISA Threat Landscape Report 2016*, in that year, cyber criminals used DDoS as a major channel to launch many attacks, and the number increased dramatically. The common DDoS was the Web Browser Impersonator, accounting for 45. 36% of the attacks on applications bypassed the existing protection measures, the number of the attacks having increased by 6% in comparison with that in the previous year. The number of single-vector attacks keeps increasing, accounting for 50% of the total number of the attacks. The increase resulted from the increase of NTP reflections. In 2016, the number of multi-vector attacks kept increasing, accounting for 35–50%, with an increase of 10%. DDoS attacks discovered in early 2016 were used as blackmails, which shows that the motivation for DDoS attacks is shifting to obtaining benefits. DDoS is a commodity of transaction in the underground industrial chain. Its price in 2016 decreased from US$25–30 per hour to US$5, making it easily available. There is difference between the attacks at the network level and the application level. Three economies, namely, the Chinese Mainland (50%), the United States (17%) and Taiwan of China (5%), were subject to the most DDoS attacks on networks. Three other economies, namely, Brazil (25%), the United States (23%) and Germany (9%), were subject to the most DDoS attacks on the application level. Among the attacks on networks, games account for 55%; software and technology, 25%; and finance, 5%. Among those on applications, retailing accounts for 43%; hotel and tourism industry, 10–20%; and financial service stations, 12%. In general, 73% of organizations have fallen victim to DDoS attacks, and 85% of them have been attacked for many times. In most cases, DDoS attacks were used for other purposes, including virus infection (46%), malicious program activation (37%), network attack (25%), customer's trust loss (23%) and customer data theft (21%).

3. Pertinent measures should be taken to reduce damages of DDoS attacks.

According to *Report on China's Internet Security (2016)*, CNCERT, as the coordinator of telecommunication and security organizations, announced in 2016 the establishment of China Cyber Threat Governance Alliance (CCTGA), and made effort to counter DDoS attacks. Through collaboration, damages of DDoS attacks have been reduced. In 2016, CNCERT monitored 452 DDoS attacks of over 1Gpps every day, the number decreasing by 60% in comparison with that in 2015. The number of big-traffic attacks kept increasing in 2016 and that of daily attacks of over 10Gbps in the fourth quarter increased by 1.1 times in comparison with that in the first quarter, amounting to 133, accounting for 29.4% of the total daily DDoS attacks. Besides, there were over six DDoS attacks of over 100Gbps every day. Among all the DDoS attacks in 2016, over 60% came from overseas; 67% were related to the underground industrial chain; reflected attacks were dominant; and besides PC "broilers" and IDC servers, the attacking devices corresponding to the IP addresses of the attacker include smart devices gradually.

4.3.3 Threats from the Trojan and Botnet have been Reduced

A Trojan (or Trojan horse) is a malicious computer program installed and run in the victim's computer. It allows an attacker to access users' personal information and remote control their computers. The program is made up of controlling and controlled terminals. A botnet is a computer program used to construct massive attack platforms. It is a computer cluster controlled by hackers. The hacker can control a host computer infected by a Trojan or botnet program to execute the same malicious actions through one-to-more command and control channels. The damages and threats caused by a Trojan and botnet are the causes of private information leak-out, confidential information disclosure, email spam and DDoS attacks.

1. **The overall damage is serious but the growth tends to slow down.**

Intralinks points out that FBI expressed when providing testimony to Botnet Sub-committee of the Senate that the loss caused by the botnet to American victims had amounted to US$9 billion in contrast with the loss US$110 billion to the world. Every year, about 500 million host computers are attacked, 18 of them per second.[14] According to *Report on China's Internet Security (2016)* released by CNCERT, in 2016, about 97,000 Trojan and botnet control servers controlled 16.99 million hosts within the border of China, the number of the control servers decreasing by 8.0% in comparison with that in 2015 and the growth slowing down in the previous five years. About 4896 botnets controlled over 100 hosts and 52 botnets controlled over 100,000 hosts. According to APCERT Annual Report 2016, except that the number of botnets increased in Hong Kong of China in the fourth quarter of 2016 because of the breakout of attacks from the new botnet "Mirai", the number of botnet attacks decreased throughout the country. In Malaysia, the number of botnet attacks in 2016 decreased, with the lowest number in September and October. In the same year, VNCET removed the malicious programs of the botnet from thousands of computers of governmental institutions, and it will make a national plan for botnet monitoring and clearing. In 2016, CERT-IN (India) traced a stable number of botnets, 2,232,231.08 every month. MonCERT (Mongolia) tackled 468 cyber security incidents in 2016, the number decreasing by two thirds in comparison with that in 2015.

2. **Hacker' attack means and patience keep being enhanced.**

In July 2017, it was exposed that Twitter had suffered an unprecedentedly huge malicious attack. A botnet with the code name SIREN was discovered, including approximately 90,000 forged Twitter accounts, which released 8.5 million tweets containing malicious links, inducing the users to make 30 million malicious clicks within weeks. Most of the accounts controlled by SIREN had many active tweets

---

[14]Intralinks.Botnets.*What are They? and How can You Protect Your Computer?.*

records and most of them had been set up in the year before. The botnet successfully bypassed the monitoring system of Twitter, which showed that the hacker had become more patient. According to a Symantec report, with the development of IoT, it becomes common for botnets to use routers, modems, network connection or storage devices, closed-circuit TV systems and industrial control systems to launch attacks, which should arouse close attention.[15] According to *ENISA Threat Landscape Report 2016*, the botnet was one of the major attack tools in 2016 and the attacks were shifted towards IoT, especially DDoS attacks, driven by profit making. Major botnet types include spam, malicious program delivery, DDoS, cheating advertising and so on. According to statistics, 18% of DDoS botnets at the application level can resist cookie and Java Script, and imitate legal browsers to attack applications. Botnets have also been used by some advanced threat agents to attack special communities, with 19,000 of them having fallen victims to the attacks. China has the highest number of attacked IP addresses, then come the United States, Vietnam and India. Most botnet-controlled servers are located in the United States, Germany, Russia, Holland and France.

4.3.4 Other Malicious Threats Like Attacks on Technical Vulnerabilities have Emerged One After Another

Besides web spoofing, DDoS attacks, Trojans and botnets mentioned above, there are some other types of malicious threats to the Internet. For instance, vulnerabilities due to defects of improper configurations of software, hardware or communication protocols in the information system can inspire attackers to access to or damage the systems without any authorization, which results in security risks of the information systems. These risks and all kinds of ransomware, information theft and junk information sweeping the world these years all pose threats to cyber security.

1. **The number of recorded vulnerabilities is increasing while the reporting measures are in place.**

According to *ENISA Threat Landscape Report 2016*, browsers and plugin vulnerabilities are massively taken advantage of. IE has the most browser vulnerabilities, followed by Chrome, Safari and Mozilla. Adobe and Apple have the most plug-in vulnerabilities, while the number of plug-in vulnerabilities of Chrome and Active X is decreasing. According to endpoint protection businesses, 78% of websites have vulnerabilities, 15% of which are major ones. *APCERT Annual Report 2016* points out that according to the latest network vulnerability landscape traced by CERT-In, 325 vulnerability notices were issued in 2016. Japan provides vulnerability information and counter measures concerning software products

---

[15]Symantec: IoT Devices Increasingly Used to Carry out DDoS Attacks.

through the vulnerability notification platform JVN. In 2016, JPCERT/CC coordinated the tackling of 342 vulnerability incidents and launched the result on the platform JVN. When discovering major vulnerabilities, KrCERT/CC released notices on its homepage. The software vulnerabilities discovered by KrCERT/CC in 2016 covered Adobe Flash Player, Java and Microsoft IE. China's CNVD recorded in 2016 10,822 general software and hardware vulnerabilities, the number having increased by 33.9% in comparison with that in 2015. Among the 10,822 vulnerabilities, there are 4146 high-risk ones (accounting for 38.3%), the number having increased by 29.8% in comparison with that in 2015, and 2203 zero-day vulnerabilities, the number having increased by 82.5%. 24,246 vulnerability incidents were reported in 2016 concerning governmental institutions, key information system sectors and industries, the number having increased by 3.1% in comparison with that in 2015.

2. **The number of ransomware and information theft incidents is increasing.**

According to *ENISA Threat Landscape Report 2016*, the number of malicious program samples is on the rise, with 600 million incidents every quarter, mainly of ransomware and information stealing. In particular, the number of mobile malicious programs is growing fast and their interfaces are becoming more and more mature. In 2016, the amount of ransomware on mobile devices increased by three times. The steady growth of the number of malicious programs is due to the occurrence of the Malware-As-A-Service tools, with which the attacker can launch attacks only by paying a rent of several thousand US dollars every month. In terms of infection channels, the email attachment is on the top, followed by the Web drive-by and the email with malicious URL links. As for the transmitted number of malicious programs, the Trojan accounts for 60% and it is a major infection source. The top five infected economies are the Chinese Mainland, Turkey, Taiwan of China, Ecuador and Guatemala. Europe is the least infected. According to ACSC 2016 Cyber Security Survey, malware as another most commonly-used attack means, was used in 42% of the attacks on the successfully attacked organizations. Most of the tools are ransomware or other malware. According to ACSC Threat Report released in October 2016, among the attacks the center tackled, governmental institutions and private organizations were the targets, and energy and mining industries received the most malicious emails.[16] According to *NCSC 2015/2016 Cyber Threat Report*, malicious network attackers take the key network of New Zealand itself as the target, which has resulted in ID information disclosure, economic loss or theft of important governmental information.

3. **The number of spams is decreasing, but their threat becomes more serious.**

According to *ENISA Threat Landscape Report 2016*, the spam as one of the transmission means of malware and malicious URL has been on the decrease since 2013, from 85 to 55%. In June 2015, the percentage of spams dropped to below 50,

---

[16]ACSC: *ACSC 2016 Threat Report.*.

the lowest since 2003. Despite the decrease in number, the quality has improved, making people easily cheated. In 2016, the number of spam URLs dropped. The top countries as spam sources are the United States (11%), Vietnam (10%), India (10%), China (7%) and Mexico (4.5%). According to Spamhaus, an international non-profit organization, spam senders have begun to directly develop and send computer viruses, and launch DDoS attacks on Spamhaus through botnets. 60% of spams are sent through infected computers. In most Internet service providers and networks in North America, Europe and Oceania, 90% of received emails are spam and over half of them contain links targeting IP websites in SBL. Today, over 80 public DNSBL server networks of Spamhaus distributed in 18 countries and regions give feedbacks to hundreds of millions of public requests every day.[17]

4. **The number of attacks on network components and control systems keeps increasing.**

According to a survey by Trend Micro, in the past year, malware has been more and more used in attacks on SCADA, because the attackers have got more knowledge of SCADA technology. The attacks are launched in two ways: disguising malware as normal and effective SCADA applications, and using malware to scan and identify a SCADA protocol.[18] According to *ENISA Threat Landscape Report 2016*, the number of attacks on network components has risen to the second place in all attacks, only after DDoS attacks. Mal-operation (for instance, installation, configuration and maintenance) of CMS is the source of attacks on websites. The reason is that the plug-in of CMS has been out-of-date. The top three types of infections are backdoor, installation of malware and attack on search engine optimization.

4.3.5 Cybercrime

Cybercrime is crime in which the criminal, by using a computer and a network, launches attacks on and does damage to the system or information of the computer or network. Cybercrime can be classified into two kinds: offences in which the Internet is used and offences in which the Internet is the target of an attack or damage.

1. **Cybercrime has caused great losses and middle-age and senior citizens tend to be the victims.**

According to *2016 Internet Crime Report* released by Internet Crime Complaint Center (IC3) of the United States, they have received 280,000 complaints every year in the past five years. In 2016, they received a total of 298,728 complaints with reported losses in excess of US$1.4507 billion. Cybercrime topics for 2016 were

---

[17]Spamhaus: DDoS and Virus Attacks on Spamhaus.
[18]Trend Micro: *Report on Cybersecurity and Critical Infrastructure in the Americas.*

business email compromise (BEC), ransomware, tech support fraud and extortion. Those aged from 30 to 39 and over 60 were mostly affected. Their total count was respectively 54,670 and 55,043, with the latter group suffering the greatest loss, about US$340 million. The top countries by victim were Canada, India and the United Kingdom, the top crime type by victim count was non-payment and non-delivery, with the total number of victims at 81,029, three times the number of victims of personal data breach. BEC/EAC caused greatest losses, which amounted to US$360 million.[19]

2. **Perception capacity and complaint initiative need improving.**

According to *Cybercrime & Security Survey Report 2013* released by CERT Australia, 56% of organizations discovered one or more cyber security incidents in the previous year but 44% did not, which shows that a number of cyber attacks are not detected or that different organizations have different definition of the incidents. Among those who discovered incidents, 35% discovered 1–5, 12% discovered 6–10, and 9% discovered more than 10. In comparison with the number in the previous year, the number of the organizations discovering cyber security incidents increased, from 56 in 2012 to 76 in 2013, which shows that the perception capacity and investigation initiative need improving.[20] *2016 Internet Crime Report* released by IC3 shows that millions of people in the United States are victims of Internet crimes each year, but only an estimated 15% of the nation's fraud victims report to law enforcement authorities. According to a case study by Microsoft, the number of the cyber attacks on computers and networks increased by five times in the previous five years, and over 856,000 people in New Zealand (accounting for one fifth) were victims of cybercrimes every year. In 2016, cyber crimes caused a loss of US$257 million in the country.[21] The number of cyber security incidents received by NCSC from July 1st, 2015 to June 30th, 2016 was 338, less than twice the number (190) in the previous year.

4.4 Countermeasures Taken by Different Countries

In recent years, all counties have enhanced their attention to cyber security and have taken a series of measures in accordance with their situation and Internet development landscape, to ensure the security of the nation, economy, society and the public. These measures include: enhancing the cyberspace layout, launching strategies and laws and regulations, establishing the coordinating and leading institutions, protecting critical infrastructure, strengthening network information

---

[19]Internet Crime Complaint Center (IC3): *2016 Internet Crime Report.*
[20]CERT Australia: *Cyber Crime & Security Survey Report 2013.*
[21]Microsoft: *Case study: cybercrime escalating in New Zealand.*

security supervision, promoting the industrial development, improving the cyberspace military strength and reinforcing the cyber security cooperation with other countries.

4.4.1 Formulation of Cyber Security Strategies

Since the United States became the first to launch international cyberspace strategies and initiatives in 2011, over 67 countries have formulated cyber security strategies. In the United States, a powerful country of the Internet, Obama and Trump governments have launched *Cybersecurity National Action Plan (CNAP), Presidential Policy Directive—United States Cyber Incident Coordination, Strategic Principles for Securing the Internet of Things (IoT), Guide for Cybersecurity Event Recovery* and *A Framework for a Vulnerability Disclosure Program for Online Systems*. Russia has released a new *Info Security Doctrine*, an amendment of *Info Security Doctrine 2000*, to define the general goal of new policies, including promotion of armies, management of the Internet and defense against the attacks from overseas. China released *National Cyberspace Security Strategy* and *International Strategy of Cooperation on Cyberspace* respectively in December 2016 and March 2017. *Cybersecurity Law* was put into enforcement in June 2017. It is China's first basic law regulating cyber security administration. The German Government released in November 2016 the *Cyber Security Strategy 2016*, which points out that to resist cyber threats to governmental institutions and critical infrastructure, Germany will establish a quick responding army led by the federal cyber security office, and set up emergency responding groups in the federal police bureau and intelligence agency to tackle more and more threats to governmental institutions, critical infrastructure, businesses and citizens. Singapore released in October 2016 *Cyber Security Strategies* which contain four key points: (1) Building a Resilient Infrastructure; (2) Creating a Safer Cyberspace; (3) Developing a Vibrant Cybersecurity Ecosystem; and (4) Strengthening International Partnerships. Ukraine released in April 2016 the new *Cyber Security Strategy* to reduce hackers' attacks on its energy equipment. The strategy contains the cyber security standards drafted by the national bank for the financial system and expands Ukraine's cooperation in cyber security, for which Ukraine National Security and Defense Commission is responsible. Australia released in May 2017 the annual amendment of the *National Cyber Security Strategy*, which covers 33 cyber security plans to counter cybercrime, unite relevant communities to improve IoT equipment security, and reduce the supply chain risks of the governmental IT systems. The country has accelerated the launch of joint central plans for cyber security to improve the cyber security for small and medium-sized businesses. Ashraf Ghani, Afghan president, released the amendment of the *National Cyber Security Strategy*, covering 33 cyber security plans to counter cybercrime and unite relevant communities to improve IoT equipment security. The Turkish Government released in September 2017 the *National Cyber Security Strategy and Initiative*, which contains five strategic goals,

41 independent initiative themes and 167 practical steps, working as the comprehensive blueprint for cyber security of the country to counter the increasingly serious domestic and international threats to cyberspace.

4.4.2 Establishment of Special Agencies for Cyber Security

In recent two years, Japan, the United Kingdom and Australia have enhanced the construction of cyber security systems by setting up cyber security leading and coordinating organizations or special agencies. Ministry of Foreign Affairs of Japan set up the Office of Cyber Security Policy in July 2016 to tackle more and more cyber attacks on governmental sectors, promote the rule of cyberspace by law and help developing countries to build their cyber security capacity. The Australian Government founded in August 2016 the cyber intelligence monitoring department to counter online financial crimes. The department is affiliated to Australian Transaction Reports and Analysis Centre (AUSTRAC). It is responsible for online payment platforms and counters all sorts of online financial crimes. The British Government officially put National Cyber Security Centre (NCSC) in operation in February 2017. The center has four goals: (1) to understands cyber security, and distils this knowledge into practical guidance; (2) to respond to cyber security incidents to reduce the harm they cause to organisations and the wider UK; (3) to use industry and academic expertise to nurture the UK's cyber security capability; and (4) to reduce risks to the UK by securing public and private sector networks. Indian National Cyber Security Coordination Centre (NCSCC) was put into operation in June 2017. It is the highest cyberspace intelligence agency of the country. Besides, the Indian Government established in March 2017 the Cyber Swachhta Kendra (Botnet Cleaning and Malware Analysis Centre) and will set up CERT groups for special industries like power and communications. The Singaporean Government appropriated 8.4 million Singapore dollars (about 5.93 million US dollars) for the establishment of the cyber security laboratory in National University of Singapore in March 2017 to support cyber security research and testing by the academic and industrial staff. The laboratory can simulate over 1000 computers in executing simultaneously a number of tasks of creating cyber security incidents and has a large database of malware.

4.4.3 Enhancement of Critical Infrastructure Protection

To enhance the critical infrastructure protection, the United States, the United Kingdom, Russia and Singapore have formulated laws and regulations on cyber threats to critical infrastructure, established monitoring and protection systems, recruited and cultivated talents and launched security maneuvers to ensure the national security. First, they have enhanced legislation on cyber security. In May

2017, President Trump of the United States signed the executive order aimed at "strengthening the cybersecurity of federal networks and critical infrastructure", requiring to take a series of actions to strengthen the cyber security of federal networks and critical infrastructure. The executive order stipulates the measures on cyber security from three areas, namely, the Federal Networks, critical infrastructure and the Nation. The US House of Representatives has passed the *Department of Homeland and Security Appropriation Bill* to appropriate US$1.8 billion to National Protection and Programs Directorate (belonging to the United States Department of Homeland) in charge of cyber security and critical infrastructure administration. Of the US$1.8 billion, US$1.37 billion will be spent on cyber security and critical infrastructure protection. In August 2017, the UK Government launched a new security bill to ensure the security of critical national infrastructure, requiring energy and communication service suppliers to make security emergency plans to reduce the impact of power defaults or environment disasters. Russian Duma has adopted the *National Infrastructure Protection Law* to protect the critical information and communication infrastructure from hackers' attack, requiring critical information and communication infrastructure only to be owned by Russian citizens or legal entities, or foreign entities who cooperate with branches registered in Russia. Secondly, the countries have established security systems. The United States began in 2003 to implement the Einstein Program, which is a detection program that monitors the network gateways of governmental agencies in the United States. The software was developed by the US-CERT to detect the intrusion and protect the security of the governmental network system. In 2009, the United States launched CNCI, into which "Einstein Program" was integrated. CNCI was then renamed NCPS, which was still called Einstein Program. In 2017, the United States Department of Homeland proposed to list the national election system into the critical national infrastructure and planned to add resources to improve the cyber security defense capability of the national election system and to set up relevant authorities before the end of 2017. Germany has set up the General Election Firewall to prevent Russian hackers from attacking the German network during the general election. Christian Democratic Union (CDU), of which Merkel is a member, calls for the formulation of laws to allow the government to take "anti-hacker" actions and destroy servers launching cyber attacks. Thirdly, they recruit and cultivate talents in relevant fields and carry out manoeuvres. The National Cyber Security Center of Ireland are recruiting cyber security experts to protect critical infrastructure, including power, hydraulic engineering, port, airport and medical facilities, against cyber attacks. The Singaporean Government held NetStar Manoeuvre, covering 11 designated critical infrastructure sectors, to test the effect of the governmental cyber incident administration and emergency plans.

4.4.4 Enhancement of Cyber Security Supervision

The governments of all countries take initiative measures for cyber security supervision to reduce the risks and challenges resulting from cyber threats. First, measures have been taken to enhance content security supervision. Germany has passed an act targeting social network platforms like Facebook and Twitter, which will be fined 50 million euros at the highest if they fail to eliminate the "obviously illegal" expressions of hatreds and slanders. Russian Duma has passed an act, requiring to completely forbid VPN and other agent services throughout the country. Indonesia has announced that they will block the "telegram" application of social network, because the police of the country hold, through long-term intelligence analysis, that the "telegram" application has become the transmission platform for information threatening the national security due to its point-to-point encryption. The Thai Government has released the *Report and Option on Social Media Use and Reform*, which points out that social media prevalent in Thailand have caused many malicious incidents. The Government encourages to develop localized social media software to reduce cyber crimes. Secondly, cyber security censorship is enhanced. Swedish Transport Administration has outsourced the information technology maintenance to IBM in recent years, which has given rise to an unprecedentedly large-scale data breach in Swedish history, with personal data of millions of citizens and large amounts of confidential information having been disclosed. To prevent similar incidents from happening, Sweden is formulating laws with stricter provisions on outsourcing concerning national security. Russian Federal Security Bureau (FSB) requires Western technical giants like Cisco, IBM and McAfee of the United States and SAP of Germany to provide the source codes of security products for the convenience of censorship. In this way, they can avoid the codes being duplicated or changed, and thus guarantee no "backdoor" used for espionage. Most businesses allow Russian authorities to examine the source codes of their products used as security facilities.

4.4.5 Promotion of the Cyber Security Industry Development

All countries are promoting the development of the cyber security industry by cultivating relevant talents and increasing the fund for it.

1. Cyber security talents cultivation

At the conference of Cyber Security Strategy Headquarters held in March 2016, the Japanese Government formulated the plan for cultivating talents working in the key positions of cyber security protection. The plan proposes that they will produce nearly one thousand experts in the coming four years, especially for the Tokyo Olympics and Paralympic Games in 2020, to reinforce the countering against cyber attacks. In June 2017, the Japanese Government launched a new project which aims to cultivate young people in mastering highly sophisticated technologies. The youngest participant is only ten years old. The British Government recruits a large group of cyber security talents from civil society. In 2016, it endowed volunteers with expertise with some privileges to provide help and support for the police in countering cyber crimes. Ministry of Defense of Denmark began to recruit hackers nationwide in April 2017 to tackle and prevent foreign network attacks. After receiving some training, the competent hackers are selected into the Intelligence Agency of Ministry of Defense to resist the intrusions in the networks. China released in June 2016 *Opinions on Enhancing the Construction of Cyber Security Discipline and Relevant Talent Cultivation*, which proposes that the country should speed up the construction of the major and department of cyber security in higher education institutions, innovate the mechanism for cyber security and the team-building of teachers in that field, promote collaborated innovation by higher education institutions and businesses, enhance on-the-job training of cyber space staff, and improve the public awareness and skills of cyber security. Australia has held the Challenge Contest on Cyber Security of National Defense Industry to improve the country's information security technology and to cultivate cyber security talents to meet the rapidly growing demand for such talents.

2. Increase of the fund for the cyber security industry

To encourage the development of network defense tools, the United States Department of Energy appropriated US$4 million to support four cyber security businesses in developing new technologies to protect the country's power supply system from hackers' attacks. In July 2017, the United Kingdom launched the Cyber Schools Programme, encouraging the students to register for the online school program of 20 million pounds and thus guiding the school-age children in mastering some cyber security techniques and providing middle school students with high-end cyber security technique education. In the coming three years, the country will appropriate US$19 million for the establishment of the cyber security technology development and innovation center to enlarge the national cyber security talent team and provide cooperation opportunities, technical guidance and operation support for large businesses and startups of cyber security technologies. Australia announced in December 2016 that it would appropriate 4.5 million Australian dollars for the establishment of Academic Centres of Cyber Security Excellence to enhance the capacity training concerning cyber security for the students who are going to graduate and find jobs so that the country can enlarge the national cyber security team, lead the world's cyber security research and provide management and education training for industrial and governmental sectors.

4.4.6 Promoting Cybersecurity Education and Awareness

The United States, Europe and Japan hold Cyber Security Awareness Month every year while China and Australia hold National Cybersecurity Week to enhance the public awareness of cyber security, popularize cyber security knowledge, create healthy and civilized cyber environment and safeguard cyber security. The Cyber Security Awareness Month of the United States is held by the Department of Homeland Security in every October guide the public and businesses in paying attention to cyber security and improving their security awareness, and to provide the public and businesses with tools and resources needed for secure access to the Internet and to improve the country's capacity of restoration from cyber attacks. European Cyber Security Month is also held every October to arouse the universal attention to cyber security. It is aimed to enhance EU citizens' awareness of it and personal protection capacity in that regard and change their passive attitude towards cyber attacks through a variety of cyber security publicity and education activities. Japanese Cyber Security Month is held from February 1st to March 18th to help the Internet users to know more about how to construct a secure cyber environment, observe cyber security laws and regulations, protect themselves from cyber attacks, identify changeable new-type cyber hazards and make corresponding response. The publicity activities in Japan are famous for the vivid cartoons. In 2017, the promotion ambassador of the Cyber Security Month was Kirito, the hero of *Sword Art Online*, a popular cartoon movie. In 2017, China's Cybersecurity Week was held from September 16th to 24th nationwide, with the theme "cybersecurity for the people and by the people". During the week, a cybersecurity exposition and achievement exhibition, a cybersecurity technology summit, a Theme Day and the appraisal and election of first-class demonstrative cyber security colleges were organized and excellent cyber security workers were awarded.

4.4.7 Improvement of Cyber Force

Major countries have channeled cyber security into the military security and set up cyber armies to improve cyber force and deterrence. Cyber security is increasingly important in the overall security of a nation. First, many countries have set up the Cyber Army. American President Trump announced in August 2017 that the United States Cyber Command would be elevated into a full and independent United Combatant Command to be accountable for enhancing the task guarantee of the network of the Department of Defense, deferring and defeating threats to American interests or network infrastructure when necessary and providing necessary support for other commanders. Israel Defense Forces announced in July 2016 the establishment of the Cyber Command, which is the control center protecting military data and online communication and enabling the Defense Forces to exert its full strength even when confronted with serious cyber attacks. The German armed forces announced in April 2017 that they would found their Cyber Command, which is equal to Germany Army, Navy and Air Force in position. The Cyber Command aims to protect German information technology and weapon systems from being attacked. Secondly, different countries are recruiting more talents. Ministry of Defense of the Republic of Korea has released the *Mid-term Plan for National Defense 2018–2022*, which proposes that the country will appropriate 250 billion Koran won for cyber security building within five years to protect the defense intranet from being threatened by hackers from Democratic People's Republic of Korea. The Thai Army announced in January 2017 that they would recruit "cyber soldiers" from civil society to enhance the defense capacity of the IT systems and networks of the cybercrime security center.

4.4.8   *International Cooperation in Cyberspace*

Because of the openness, interconnectivity and boundlessness of the Internet, there are an increasing number of cross-regional and cross-border cyber threats and security incidents, which makes cyber security incident response and attribution more difficult. The international community has realized the necessity and imperativeness of international cooperation in cyber security, so communications among international organizations have been carried out in that regard and a series of cooperation agreements have been reached between countries to promote cyberspace sharing and joint governance. First, the cooperation in cyber security between international organizations has been enhanced. In February 2016, NATO declared that they had signed a technical agreement with EU to strengthen their cooperation in cyber security. Both parties will make arrangements for information communication and experience sharing between cyber emergence sectors. In July 2016, the plenary session of European Parliament adopted *EU Network and Information Security Directive* to enhance the cooperation in network and information security among the EU members and to improve EU's capability of tackling IT defaults and countering malicious attacks by hackers, especially cross-national cyber crimes. In April 2017, G7 Ministerial Conference released a declaration of national accountability for cyber security to encourage all countries to observe laws, show mutual respect and help each other and establish mutual trust in using ICT. In June 2017, 28 EU members reached an agreement on joint punishment of hackers. EU Council announced the joint framework of the "cyber diplomacy kit" to guide EU in centralized tackling of malicious cyber activities. The framework also suggests that for the responsible party the EU members will jointly implement economic sanction, forbid entry into all of them, freeze assets and issue comprehensive bans. Secondly, bilateral cyber security cooperation has been carried out. In December 2016, the White House of the United States and the Canadian Government released *U.S.-Canada Electric Grid Security and Resilience Strategy*, promising to strengthen the electric grid security in North America. In March 2017, Japan and Germany passed the *Hanover Principle* to promote the international standards and specifications formulation and development concerning high-end technologies like IoT and AI. In May 2017, Japan reached an agreement with the U.S. Department of Homeland Security. According to it, Japan officially joined the automated indicator sharing platform to deepen the cyber information sharing between the two governments. In June 2017, Australia and Thailand signed an agreement on cybercrime to strengthen their cooperation in tackling cybercrime in the Asia-Pacific Region and reinforcing the regional business security. Australia will provide support of "digital forensics of cybercrime development" for the Thai Royal Police and security and diplomatic affairs officials.

4.5 Cyber Security is faced with New Challenges

Today, new network technologies and applications develop fast. While bringing convenience for people in different countries, they give rise to changeable threats and challenges to cyber security. Therefore, all countries have to be alert and tackle the problems in joint effort.

4.5.1 Increasingly Serious APT

APT refers to the persisted threat to a special target launched by the hacker through advanced attack means. Today, more and more APT organizations and attacks are known by the public. First, APT has special targets, usually first critical information infrastructure of the government and diplomacy and then finance, business, military and defense sectors and basic facilities. It is very destructive.[22] Secondly, APT is pertaining and complicated. It cannot be detected in a short time and the attacks may occur repeatedly. Stuxnet can be traced back to 2005, but it was not discovered until 2010. BlackEnergy began to be used at least in 2007 for the creation of a botnet for DDoS attacks and reached its peak in 2011. Since 2014, it has occurred every year. Thirdly, APT tends to be seen everywhere in cyberspace. Information systems of traditional infrastructure and emerging critical information infrastructure will be the focus of attacks. With the expansion of attacks and the decrease of cost, the attackers will find it easy to launch APT, which, therefore, will be difficult to resist.[23]

---

[22]Qihoo 360: *Research Report on APT in China* (2016).
[23]Antiy: *Retrospect and Prospect of Cyber Security Threats* (2016).

4.5.2 Explosive Growth of the Number of Ransom Attacks

Ransomware is a popular Trojan program which prevents the users' data assets or computing resources from working normally by harassing, terrifying or even binding the users' files, until the ransom is paid. In recent two years, ransomware attacks have grown explosively and caused great impact and economic losses. According to Kaspersky research, by the end of 2016, ransomware had evolved into a family-like development model, 44,287 new varieties had been detected and at least 114 countries had been affected.[24] The spread of ransomware worldwide is due to its pertinence, technology, destructiveness and spreading capacity.

(1) Ransomware has its target. It cannot be traced when a ransom is paid. A large amount of ransomware tends to target the users' important data and the payment is done in Bitcoin through the encrypted Tor. The whole transaction is anonymous.

(2) The technology and function of ransomware keep being upgraded. It is hard to mine the encryption. A series of wars have been fought between CryptXXX Ransomware Creators and security researchers, who provide the victims with free tools to unlock the locked files so that the victims needn't pay a high ransom. But the ransomware creators usually adjust the code to frustrate the decryptor, so that the researchers had to look for new ways to destroy the enhanced ransomware. The process tends to be repeated for times.

(3) Ransomware is destructive and it spreads quickly. The ransomware Locky that occurred in February 2016 spread at the speed of 4000 computers per hour and the number of infected computers amounted to 100,000 every day, involving over 10 countries, such as Germany, Holland and the United States.

(4) RaaS (Ransomware as a service) model keeps being developed and a black industrial chain has been formed, which has accelerated the spread of ransomware and enabled more primary cyber criminals to use the relevant services and tools to launch attacks.

4.5.3 Frequent Large-Scale Data Breaches

Since 2016, a series of critical data breaches have been exposed, involving hundreds of millions of pieces of information. In September 2017, Equifax, a US consumer credit reporting agency, admitted that cybercriminals accessed approximately 143 million Equifax consumers' personal data, including their full names, social security numbers, birth dates and addresses. In October 2017, Verizon, the mother company of Yahoo!, claimed that personal data of three billion Yahoo! users had been leaked out, including the users' names, email addresses, telephone

---

[24]Kaspersky laboratory: Kaspersky Security Bulletin 2016.

numbers, birth dates and encrypted passwords. Based on the statistics of the data breaches that have happened, some features can be summarized. First, the scale of data breaches is shockingly large and each incident involves more and more data disclosed. According to a Symantec report, in the eight years before 2016, over 7.0 billion pieces of online ID information had been stolen, which indicated that the information of everyone on the earth had been stolen once.[25] Secondly, the most common data stolen are personal ID information and financial information. According to Symantec statistics, the stolen personal data accounted for 42.9% of the total stolen data in 2016, when nearly one million pieces of personal information were stolen in every data breach and over 1.1 billion pieces were embezzled, almost twice the number in 2015, which was 563 million. Personal financial information accounted for 39.2% of the stolen data and financial data theft was a serious problem in all data thefts, including the detailed information of credit cards or debit cards or bank transaction records, the abuse of which would bring financial risks directly. Thirdly, among the causes of data breaches, malicious actions account for the most and produce more serious economic impact. According to the 2017 reports of IBM Security and Ponemon Institute, most data breaches were done by hackers and internal staff and 47% of data breaches were caused by malicious or criminal attacks, at the cost of US$156 for every recorded breach while the cost for every data breach caused by system default is US$128 and that for the breach caused by mistake or negligence is US$126.[26] Fourthly, besides the theft of commercial data, data of governmental sectors are faced with all kinds of security risks. In July 2017, the Swedish Government admitted that large-scale data breaches happened during its Internet project outsourcing, involving the information of bridges, subways, roads and ports, and information of the vehicles of Swedish police and army. In April 2016, security researchers found that the MongoDB storing the data of 93.4 million Mexican voters was open for access on Amazon AWS, including the voters' names, addresses, birth dates, voting ID numbers, parents' names and occupations. In March 2016, security researchers found that the databank of COMELEC of the Philippines was attacked by hackers, and the data of 55 million registered voters of the country were leaked out, including data of passports and fingerprints. All these incidents deserve the attention of the governmental sectors of all countries.

4.5.4 Increasing Risks of IoT Smart Devices' Being Controlled

With the rapid development and popularization of IoT smart devices, there are an increasing number of attacks on IoT. The attackers can obtain the device control

---

[25]Symantec: *Internet Security Threat Report*, ISTR, Volume 22.
[26]Ponemon Institute: *2017 Cost of Data Breach Study: Global Overview*.

authority through the device vulnerabilities and then steal the users' data, hijack the network traffic and thus form large-scale botnets. IoT smart devices have become the key targets of attacks by malware and there are new features of the attacks. First, when the IoT smart devices are controlled, the botnets of such IoT are more destructive than traditional botnets. The latter are made up of traditional computer terminals and servers while the former of smart devices closely related to people's daily life. In 2016, large-scale network outrages on the east coast of the United States and abnormal accesses to telecommunication networks in Germany happened frequently because of Mirai, the malware that can control large IoT botnets. Secondly, IoT smart devices are online 24 h a day, so the infection of malware cannot be easily sensed, so malware has become the "fixed" attack source. According to China CNCERT/CC's monitoring of the Mirai botnet, by the end of 2016, 2526 control servers had controlled 1.254 million IoT smart devices, which was a big threat to the operation of the Internet.[27] Thirdly, it has been discovered that mobile phones with Android can be used to construct large botnets. In August 2017, the joint group investigation by Akamai, Cloudflare, Flashpoint, Google, Oracle Dyn, RiskIQ and Team Cymru showed that the attackers' over 300 pieces of malware spread through Google's application stores had constituted Wirex, the first Android platform botnet, which launched DDoS attacks by using several hundred thousand Android mobile phones. At the peak time, it controlled over 120,000 IP addresses in over 100 countries and regions, but fortunately, the joint group took actions to counter it.[28] Fourthly, there are more IoT smart devices with Bluetooth and Wi-Fi, but there are potential security hazards in them. In September 2017, Armis, a startup of IoT security, found the new IoT attack means "BlueBorne". The attacker, via 8 Bluetooth transmission protocol vulnerabilities, can remote control the IoT smart devices using Android, iOS, Windows and Linux, steal data and construct large-scale botnets, without any traditional network media or any interactive operation.[29] In October 2017, the Wi-Fi protection protocol standard WPA2's vulnerabilities were disclosed. These vulnerabilities allowed the attackers within the Wi-Fi range to monitor the Wi-Fi traffic between the computer and the AP, so all soft and hardware supporting WPA/WPA2 was affected. Fifthly, with the development of the Internet, the number of security hazards in wearable devices, smart homes and smart cities will increase. According to Gartner, there will be 26 billion IoT devices by 2020.[30] IDC predicts that wearables' shipments will reach 213 million worldwide in 2020.[31] and Statista predicts that the smart home market scale will amount to US$43 billion.[32] According to IHS statistics, smart city devices

---

[27]CNCERT: *General Description of China's Cyber Security in 2016.*

[28]Akamai: *Q3 2017 State of the Internet Security Report.*

[29]Armis: The Attack Vector "BlueBorne" Exposes Almost Every Connected Device.

[30]Gartne: The Internet of Things Installed Base Will Grow to 26 Billion Units By 2020.

[31]IDC: Wearables Shipments to Reach 213.6 Million Units Worldwide in 2020 with Watches and Wristbands Driving Volume While Clothing and Eyewear Gain Traction.

[32]Statista: Smart home-Statistics and Facts.

shipments will increase to 1.2 billion by 2025.[33] However, the problem is that many IoT devices are not protected sufficiently or are not protected at all in terms of security, which is reflected by the IoT attack incidents in recent years.

4.5.5 *Precautions for AI Security*

So far, all countries have input considerable manpower and material resources in AI technology. According to an IDC report, the world's total income from cognition and AI in 2017 amounted US$12.5 billion, with an increase of 59.3% in comparison with that in 2016, and in the coming five years, the global expenditure of businesses on cognition and AI solutions will increase, and see a compound annual growth rate of 54.4% in 2020, with the income amounting to over US$46 billion.[34] AI technology will influence human beings' future in depth and in a wide range, so we have to be aware of the energy variables and security hazards brought about by AI development. First, as high-end technical teams make progress in machine learning, AI has shown some cognitive capabilities. For example, AlphaGo developed by Google's DeepMind defeated the Go World Champion, which caused a sensation worldwide. DeepMind has released the latest research result AlphaGo Zero, which can start from zero to learn intensively by itself and will defeat AlphaGo without referring to any experience of human beings in playing Go. It is no longer limited by human being's cognition but can find new knowledge and develop new strategies. Secondly, there are different opinions on AI's impact on society. Elon Musk, CEO of Tesla, says that AI is the biggest threat to civilization because it will not only replace human being in work but also cause the Third World War. M.E. Zuckerberg, CEO of Facebook, is optimistic towards AI, opposing the idea of AI being a threat. He holds that in the coming five to ten years, AI will bring positive changes to people's life. Masayoshi Son, CEO of SoftBank, holds that in the coming 30 years, the IQ of computers will surpass that of human beings and that AI will be a crisis and the solution to it. Thirdly, AI, as an emerging technology of the Information Revolution, will witness more possibilities in path innovation and efficiency and will contribute more to cyber security. CSAIL of MIT and PatternEx have jointly developed a totally new AI system called AI^2, which can detect 85% of cyber attacks. Its false alarm rate will be 5 times lower than that of the present solution and its accuracy rate will be 2.92 times that of the automated cyber attack detecting system.[35] We are obliged and held accountable for using AI to create a safe network environment to benefit society.

---

[33]HIS: Smart City Devices to Top 1 Billion Units in 2025.

[34]IDC: A Trillion-Dollar Boost: The Economic Impact of AI on Customer Relationship Management.

[35]AI^2: Training a Big Data Machine to Defend.

Chapter 5
Development of the World's Digital Economy

Chinese Academy of Cyberspace Studies

Abstract Digital economy is the major economic form in the era of information revolution, representing a new technical model, new resource endowment and a new institution and system. It is the most promising new drive for the mid and long-term growth of the global economy, marking that human beings have entered a new stage with digital productivity as the monument. Digital economy has become the direction of all countries in guaranteeing their steady economic growth and rebuilding their competitive strength as well as one of the cooperative areas of international community. (1) In major countries, digital economy is developing fast, with its speed higher than that of GDP. In 2016, the global digital economic scale accounted for over one fourth of the global total GDP. (2) China and the United States are the "twin stars" in the field of digital economy. The latter ranks first in the scale of digital economy while the former is the largest e-commerce market of the world, with its e-business transaction volume accounting for 40% of the world's total. (3) Platform economies have new ecosystems of digital economy. Global Internet businesses have adopted the platform development mode. So far, the top 10 Internet businesses in market value have surpassed the top 10 traditional multinationals and become the leading force in the global digital economy development. (4) Digital economy has become the "wind gap" of global entrepreneurship and innovation, attracting a large proportion of technologies, capital and talents of the world. (5) Digital economy is ushering in a new round of globalization. In the global service trade, digital trade accounts for an increasing percentage. It is expected that the transaction volume of global cross-border B2C business will reach US\$994 billion by 2020. (6) Intelligent manufacturing has become the major direction of global manufacturing and will bring about revolutionary changes in traditional industrial manufacturing. (7) China's sharing economy has become the biggest highlight of the world. In 2016, the transaction volume amounted to RMB3.45 trillion *yuan*, with a year-on-year growth of 103%. It is taking the lead in global sharing economy.

Chinese Academy of Cyberspace Studies
Beijing, China

© Publishing House of Electronics Industry, Beijing and Springer-Verlag GmbH
Germany, part of Springer Nature 2019
Chinese Academy of Cyberspace Studies (ed.), *World Internet Development Report 2017*,
https://doi.org/10.1007/978-3-662-57524-6_5

5.1 Development of Digital Economy is the Global Consensus

Today, digital economy not only promotes the digitalization of labor tools, service orientation of labor objects and popularization of labor opportunities at the level of productivity, but also promotes resource sharing and organizational platformization at the level of production relations. Digital technologies have penetrated all areas of life and production, leading to innovations at the supply side, such as new industries, new industrial forms and new models. Digital transformation has become the dominant force of major countries and regions in improving their competitive strength in global industries and cultivating new economic drive.

5.1.1 Consensus has been reached on "What Digital Economy Is"

Despite the different stages, focuses and orientations of digital economy in different countries, major countries all have a clear understanding of digital economy. In September 2016, the G20 Summit held in Hangzhou adopted the *G20 Digital Economy Development and Cooperation Initiative*, which embodies the consensus on digital economy. Digital economy refers to a series of economic activities with digital knowledge and information as key production elements, modern information networks as the carrier, and efficient use of ICT as the drive for efficiency improvement and economic structure optimization. The rapid progress of digital technologies has promoted the information industry to develop into a strategic emerging industrial sector with active innovation and rapid growth. Meanwhile, digital technologies, as basic general technologies, have been widely applied to all areas of economy and society. Digital economy has gone beyond the information industry sector. With the emergence and development of cloud computing, big data, IoT, VR and so on, digital economy's connotation and denotation will be expanded.

Digital economy includes digital industrialization and industrial digitalization (see Fig. 5.1). Digital industrialization is centralized in information production and application, involving information technology innovation, and information product and service production and supply. Its corresponding industrial sectors are information manufacturing, information communication and software service as well as new industrial forms and models of information technology service based on Internet platforms. Industrial digitalization is centralized in information technology application in traditional industrial sectors. Digital input has improved the number of products and the efficiency of production in traditional industrial sectors, so industrial digitalization is an important part of digital economy. Specifically, industrial digitalization includes the contribution of digital input to agriculture, industry and service industry.

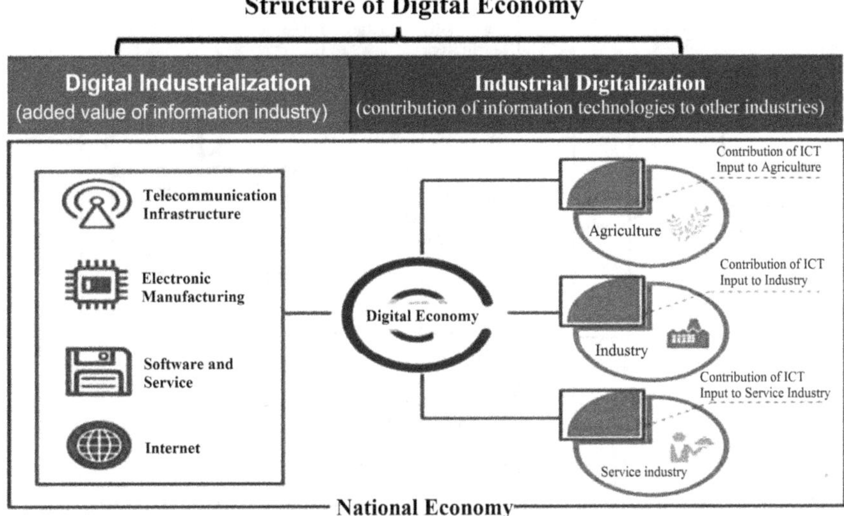

Fig. 5.1 Structure of digital economy (China Academy of Information and Communications Technology: *White Paper of China's Digital Economy Development 2017*)

5.1.2 Digital Economy is the New Drive for Global Economic Growth

1. **Digital economy represents a new technical form promoting the resurrection of global economy.**

Technical progress promotes industrial revolution. A series of network information technologies like mobile Internet, cloud computing, big data, AI and IoT are basic overflowing complementary technologies that can be widely used, so they can promote fine labor division and production tool intelligentization, improve production efficiency, reduce transaction cost, and boost efficient match between supply and demand, thus promoting the digital economy development of all countries and becoming the new drive for global economic resurrection. Especially at present, when the global economy is witnessing a sluggish growth, digital economy is taken as a new leverage for global economy. Research shows that over one fifth (22%) of global GDP is closely related to digital economy covering skills and capital. It is predicted that the application of digital skill and technology will increase the global economic output by US$2 trillion by 2020 and that by 2025 half of the global economic output will come from digital economy.[1]

---

[1] Accenture: *Digital Disruption: The Growth Multiplier.*

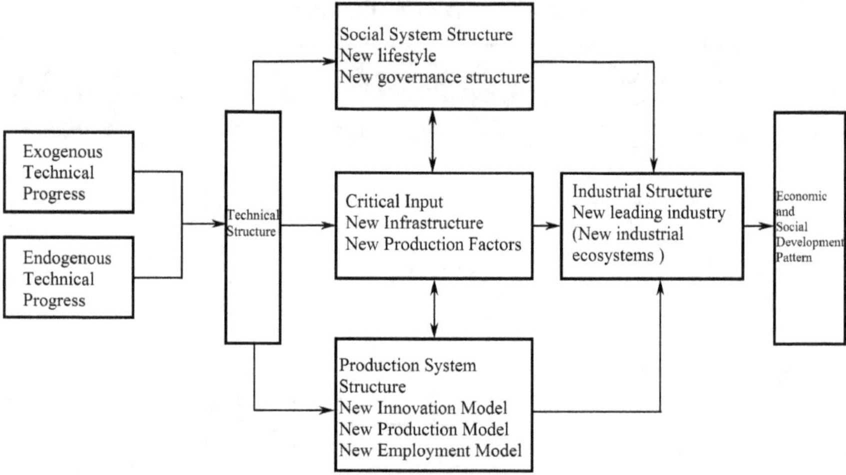

Fig. 5.2 Basic structure of the digital technology–economy model (ChinaInfo 100: Rise of Information Economy.)

2. **Digital economy represents new resource elements and consolidates the data endowment for sustainable global growth.**

Digital economy can effectively promote the networking sharing, intensive integration, collaborative development and efficient utilization of labor, capital, land, technology and management, and improve the economic operation efficiency and total factor productivity. Every increase of 10% in digitalization will lead to an increase of 0.5–0.62% in GDP per capita.[2] Data resource endowment like replicability, sharability and unlimited growth and supply breaks the limit of the limited supply of traditional factors to growth, laying foundation and providing possibility for sustainable growth, so it has become a new production factor. The richness and development of data resources determines the efficiency of resource allocation and produces impact on the worldwide production, circulation, distribution and consumption.

3. **Digital economy represents a new institutional system and facilitates the global economic system into a new round of reform.**

In terms of the production system structure, the open innovation model has emerged, so has the intelligent production means. The scale of flexible employment based on platforms keeps increasing and the ecosystem of platform industry is developing rapidly. In terms of the social system structure, digitalized lifestyle is becoming increasingly popular and the new governance structure and model are taking shape.

---

[2]WEF: *Global Information Technology Report 2015.*

Table 5.1 Accumulative impact of digitalization improvement on GDP of major countries by 2020 (Accenture: *Digital Disruption: The Growth Multiplier.*)

Country	Change in 2020 GDP (%)	Change in 2020 GDP (US$ billion, 2015 prices)
Australia	2.4	34
Brazil	6.6	120
China	3.7	527
France	3.1	80
Germany	2.5	90
Italy	4.2	81
Japan	3.3	146
Netherlands	1.6	13
Spain	3.2	43
United Kingdom	2.5	84
United States	2.1	421

The basic structure of the digital technology–economy model is seen in Fig. 5.2 and the impact of digitalization improvement on GDP of major countries by 2020 is shown in Table 5.1.

5.1.3 Major Countries Have Made Plans for Their Digital Economy Development

Digital economy is the commanding height of global industrial competition in the new era, so to plan for its development has become the common choice of countries (see Table 5.2). By 2015, among 34 OECD member countries, 27 have made national strategies for it and the remaining ones are considering the formulation of relevant strategies or have released some policies targeting some areas and sectors.[3] Traditional developed countries and regions, such as the United States, EU, Germany, the United Kingdom and Australia, have launched systematic strategies and measures for their digital economy development; developing countries like China, India and Mexico are taking opportunities for digital economy development and have formulated "Made in China 2025", "Digital India" and "National Digitalization Strategy" respectively.

---

[3]Tencent Research Institute: *Rapid Rise of Digital Economy Has Become a Strong Drive for Economic Development.*

Table 5.2 Digital economy strategies and plans of major countries

Country/Region	Overall planning strategies	Specific planning policies
EU	Digital Agenda for Europe, The Digital Single Market Strategy, and Digitising European Industry: Reaping the full benefits of a Digital Single Market	"Horizon 2020", Framework Programme 7, "Towards Data-driven Prosperous Economy, A Project on Digital Competence (DIGCOMP), Data Value Chain Strategy, Strategic Research and Innovation Agenda on Big Data Value, and A European Agenda for the Collaborative Economy
Japan	Strategy of Developing the Country through IT	ICT Comprehensive Strategy toward 2020, Declaration of Building the Most State-of-the-art IT Country, New Strategy of Robots, e-Japan Strategy, u-Japan Strategy and i-Japan Strategy
Republic of Korea	Manufacturing Innovation 3.0 Strategy	IoT Basic Planning, and IT Integrated Development Strategy
United States	Digital Economy Agenda	AMP, Federal Big Data Research and Development Strategic Plan, Federal Cloud Computing Strategy, Data Innovation 101, and Report on Securing and Growing the Digital Economy
China	Made in China 2025	Action Outline for Promoting the Development of Big Data, Guiding Opinions on Vigorously Advancing the "Internet Plus" Action, and Guiding Opinions on Enhancing the Integrated Development of the Manufacturing Industry and the Internet
Germany	Digital Agenda 2014–2017, Digital Strategy 2025	Industry 4.0, and Digital Germany 2015
United Kingdom	UK Digital Strategy, Digital Economy Strategy 2015–2018, and the Digital Economy Act	SPARC, High Value Manufacturing, A Strategy for UK Data Capability, and Digital Britain
Australia	Australia Digital Economy Update, and Digital Economy Strategy	Australian Public Service Big Data Strategy, and the National Cloud Computing Strategy

5.2 Global Digital Economy is Growing Dramatically

At present, global digital economy is witnessing its overall popularization, deep integration, accelerated innovation and transformation leading, producing fundamental and overall impact on the growth drive, economic and social operation, production mode and lifestyle. It has become an important drive for the resurrection of national and regional economy.

5.2.1 The Scale of Digital Economy in Some Countries Has Reached Several Trillion US Dollars

The global digital economy is growing exponentially at an unexpected speed.[4] It is becoming increasingly obvious that major countries are trying to occupy the commanding height of the global competition (see Fig. 5.3). First, the scale of digital economy keeps increasing. In 2016, the total volume of the U.S. digital economy amounted to US$11 trillion; China's total volume, US$3.8 trillion; Japan's US$2.3 trillion; and the United Kingdom's, US$1.43 trillion. Secondly, the percentage of digital economy is increasing. In 2016, digital economy accounted for 59.2% of the total GDP in the United States; that in the United Kingdom, 54.5%; that in Japan, 45.9%; and that in China, 30.1%. Thirdly, digital economy is growing faster than GDP. In 2016, digital economy's growth rate in the United States was 6.8% while that of GDP was 1.6%; in Japan, the two growth rates were 5.5% and 0.9 respectively; in the United Kingdom, they were 5.4% and 2% respectively; and in China, they were 16.6% and 6.7% respectively.[5]

5.2.2 The Global Economy is Increasingly Digitalized

According to *China Internet Report* released by Boston Consulting Group (BCG), in 2016, the eGDP index of major countries is as follows: 80% in Republic of Korea, 69% in China, 56% in Japan, 56% in India, 54% in the United States, 40% in Germany, 46% in Canada, 34% in France, 28% in Russia and 24% in Brazil; in that year, the top 10 countries in Internet consumption were the United States (US$1133 billion), China (US$967 billion), Germany (US$352 billion), the United Kingdom (US$335 billion), France (US$309 billion), Brazil (US$217 billion), India (US$209 billion), Japan (US$180 billion), Russia (US$156 billion) and Italy (US$74 billion).[6]

5.2.3 Internet Businesses Are the New Benchmarks of Global Businesses

By the end of 2016, among the top 10 businesses with the highest market value in the world, Apple, Google, Microsoft, Amazon and Facebook belong to the digital

---

[4]United Nations Conference on Trade and Development (UNCTAD): *Information Economy Report 2017*.

[5]ChinaInfo 100: *Digital Economy Report*.

[6]*China Internet Report* jointly released by BCD, Alibaba, Baidu and Didi Chuxing.

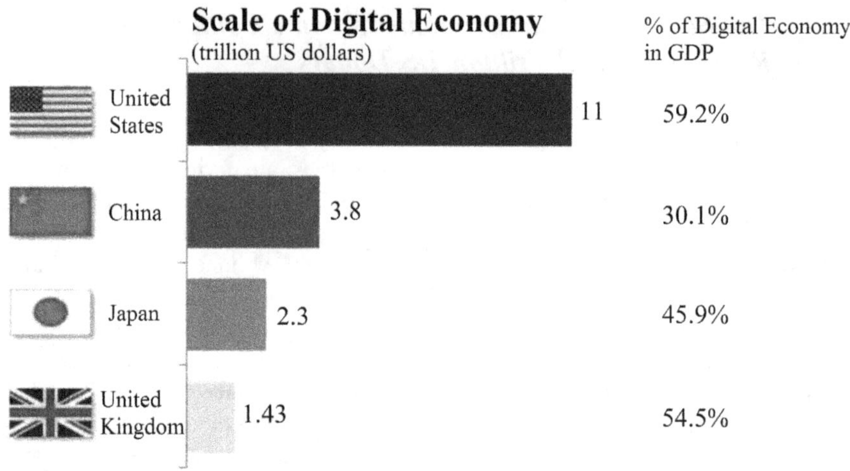

Fig. 5.3 Comparison of digital economy scale between some countries (2016)

economy category. Among the top twenty, 9 belong to the digital economy category, including AT&T, Tencent, BT (British Telecom) and Alibaba.[7] There have been different benchmark businesses in different eras. In 1996, the world's top ten businesses with the highest value were manufacturing and resource businesses. In 2006, the top ten included Citibank, BOA (Bank of America) and HSBS (Hong Kong and Shanghai Banking Corporation). In 2016, the top ten were mostly digital economy businesses like Apple, Google, Microsoft, Amazon, Facebook and Tencent, including the very top ones, namely, Google, Facebook and Tencent.[8] The rise of unicorn businesses is the feature of the era of digital economy. On the list of global top 10 unicorn businesses released by CB Insights in 2017, six are American companies and the remaining four are Didi Chuxing, Mi, LU.com and China Internet Plus Group (see Table 5.3).

5.2.4 China and the United States are the "Twin Stars" in Digital Economy

China is one of the countries with the most active investment in digital economy, ranking third in venture investment in critical digital technologies such as VR, automated driving, 3D printing, robots, drones and AI. The country is the world's largest e-commerce market, with its e-commercial transactions value accounting for

---

[7]Tencent Research Institute: *Rise of Digital Economy: the main thread of future global development.*

[8]Si Xiao of Tencent Research Institute: Internet Platform Economy: new challenges to subversive innovation and competition policies.

Table 5.3 Global top 10 unicorn companies (CB insights: global top 10 unicorn companies)

Company	Estimated value (in million US dollars)	Funding time	Country	Business
Uber	680	Aug. 23, 2013	The United States	Online car-hailing
Didi Chuxing	500	Dec. 31, 2014	China	Online car-hailing
Mi	460	Dec. 21, 2011	China	Hardware
Airbnb	293	Jul. 26, 2011	The United States	E-commerce
Spacex	212	Dec. 1, 2012	The United States	Other transportation
Palantir Technologies	200	May 5, 2011	The United States	Big data
Wework	200	Feb. 3, 2014	The United States	Facilities
LU.com	185	Dec. 26, 2014	China	Financial technology
China Internet Plus Group	180	Dec. 22, 2015	China	E-commerce
Pinterest	123	May 19, 2012	The United States	SNS

over 40% of the global total. It is also a major force in the global mobile payment, its value being 11 times that of the United States,[9] which is the country with the most mature digital economy, ranking first in digital economy scale in 2016, with its Internet consumption scale amounting to US$1133 billion, also ranking first in the world. The two countries have over 80% of the unicorn companies, 51% of which are headquartered in the United States (see Table 5.4 and Fig. 5.4).

5.2.5 *Platform Economies Have New Ecosystems of Digital Economy*

Today, the market value of the global top 10 platform economies is higher than that of top 10 traditional multinationals. The "average age" of those platform economies is only 22 while that of the 10 multinationals is 129[10] (see Fig. 5.5). Platform economies are surpassing traditional multinationals and becoming super-national economies leading a new round of globalization, such as Uber and Airbnb of the

---

[9]McKinsey & Company: *China's Digital Economy: A leading global force.*

[10]AliResearch: *Digital Economy 2.0 Report: Saying farewell to corporations and embracing platforms.*

Table 5.4 Launching time of digital technology products and applications in China and the United States (CAICT: *Report on Internet Development Trend* (2017))

The United States		China	
Time	Name	Time	Name
2002	Amazon AWS	2009	Alibaba Cloud
2008	Android OS	2011	Alibaba Yun OS
2011	Google Voice Search	2014	Baidu Voice Search
2012	Google Self-driving Car	2015	Baidu Self-driving Car
1995	MSN	1998	QQ
1996	Amazon	2000	Alibaba
1998	Google	2000	Baidu
2006	YouTube	2007	Youku
2006	Twitter	2008	Sina Weibo
2008	WhatsApp	2011	WeChat
2009	Uber	2012	Didi

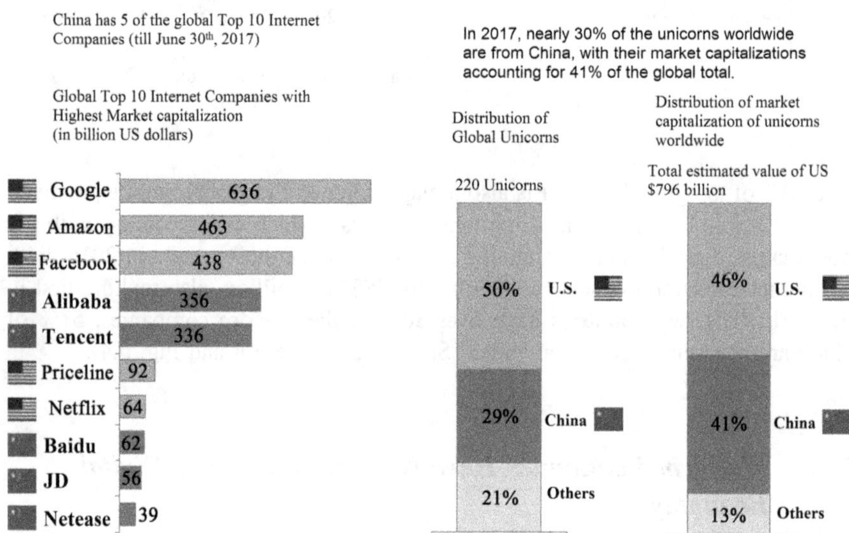

Fig. 5.4 A comparison of internet companies between China and the United States (*China Internet Report* jointly released by BCD, Alibaba, Baidu and Didi Chuxing.)

United States, and Alibaba, Didi Chuxing and OFO Bicycle Sharing of China. Airbnb has four million room sources worldwide, distributed in 65,000 cities of over 191 countries, the number being the total of the number of room sources of global five hotel groups. Airbnb provides accommodation sharing service for over 200 million people. OFO, an initiator and leader of global bicycle sharing, has over 10 million bicycles worldwide, with its daily order exceeding 25 million, providing

Platform Economies				Multinationals			
Name	Country	Market Capitalization (in 100 million US dollars)	Founding Year	Name	Country	Market Capitalization (in 100 million US dollars)	Founding Year
Apple	US	6053	1976	Berkshire Hathaway	US	4078	1956
Google	US	5482	1998	Exxon Mobil	US	3766	1882
Microsoft	US	4810	1975	Johnson and Johnson	US	3152	1886
Amazon	US	3584	1995	JP Morgan	US	3037	1859
Facebook	US	3396	2004	GE	US	2823	1892
Alibaba	China	2246	1999	Wells Fargo & Company	US	2811	1852
Tencent	China	2195	1998	AT&T	US	2525	1877
Priceline.com	US	762	1998	Procter & Gamble	US	2291	1837
Baidu	China	587	2000	Nestle	Switzerland	2220	1867
Netflix	US	527	1997	Wal-Mart	US	2200	1862

Fig. 5.5 A comparison between platform economies and multinationals (AliResearch: *Digital Economy 2.0 Report*: *saying farewell to corporations and embracing platforms.*)

over four billion travelling services for 200 million users in 180 cities of 13 countries.

Based on their advantages, leading Internet businesses of the world keep promoting the vertical integration and cross-border integration between upstream and downstream industries. First, they are extending their industrial chain control capacity, facilitating the vertical integration between upstream and downstream industries, and the establishment of collaborative business systems. Apple, an Internet giant, is establishing, with the iOS as the core, the total factor product system covering application service, chip design, terminal manufacturing and cloud platform. Secondly, by taking advantage of Internet technologies, they are accelerating cross-border integration oriented towards traditional industries and expanding the industrial space layout. Google is laying out its Internet automobile business, with its automated vehicles having covered nearly one million miles in tests, which shows to the world its promise in the automobile industry reform. Alibaba has launched the Ding Talk for businesses and developed the taogch.com platform to serve manufacturers in China, helping them to develop flexible manufacturing to accurately meet the demand of the market and to realize AI manufacturing.

A comparison between industries with capital reorganization and industries with digital reorganization is shown in Table 5.5.

Table 5.5 Comparison between industries with capital reorganization and industries with digital reorganization

Type	Industries with capital reorganization	Industries with digital reorganization
Time	Industrial economy era	Digital economy era
Sponsor	Financial institutions represented by JP Morgan	Internet businesses represented by BAT
Industries	Iron and steel, automobile and so on	Software, information and so on
Features	(1) Aiming for monopoly; (2) Merger of similar items for the increase of scale of the main industry; (3) Transfer of corporate equity (ownership); (4) Formation of large businesses	(1) Aiming to serve the whole industry and improve the efficiency of the whole industry rather than to monopolize. Providing comprehensive service for all practitioners of the main industry; (2) Connecting more similar businesses, rather than merging them; (3) Connection of more similar businesses, without change or transfer of corporate equity; (4) Pursuing joint operation instead of monopolized operation; (5) Integrating transaction markets, finance, logistics and communication into a whole; (6) Forming an industrial ecosystem

Column 1 Alibaba's Exploration in Globalization

Relying on its e-commercial platform, Alibaba invested US$249 million into Singapore Post in 2014. Till then, Cai Niao led by the corporation had had 17 overseas warehouse facilities in the countries along the "Belt & Road". In 2016, Alibaba acquired Lazada, the largest e-business of Southeast Asia, at the price of one billion US dollars, setting a new single-acquisition record of overseas investment. Ant Financial has invested in Ascend Money of Charoen Pokphand Group of Thailand and Paytm, an Indian Alipay. It has over 215 million users. Based on its large user group and relevant data, Alibaba is establishing data centers through Alibaba Cloud in Hong Kong of China, Singapore, the Middle East and Europe, and 14 regional nodes worldwide. Apsara, a large-scale computer operating system developed independently by Alibaba, connects the mega servers across the global into one supercomputer, which provides computing capacity for society in the way of online public service.

5.2.6 Digital Economy Is the New Wind Gap of Global Entrepreneurship and Innovation

The innovation capacity in the field of digital economy is one of the cores and important signs of the national innovation capacity. Digital technologies like the Internet enable main innovative bodies to collaborate with each other and organization, service and technological innovation to integrate into each other. Besides, interactions between main innovative bodies, and organization of innovative resources and application of innovative results become more networked, globalized and rapid, ushering in an innovative era characterized by integrated, systematic, mass and micro innovations. Data-driven innovation is expanding into all areas of economy, society and scientific and technical R&D, becoming the key form and direction of the national innovative development. A large number of businesses have succeeded in collaborative diversified innovation through the Internet, which has directly boosted the rapid development of e-commerce, online crowdfunding, maker initiative and mobile Internet, and accelerated the overflow of benefits to peripheral industries such as IT, marketing, logistics and design. Therefore, digital economy has become the most active basic innovative form. China is witnessing a new wave of entrepreneurship and innovation in digital economy, with the explosive growth of businesses, investment and platforms concerning entrepreneurship, the group of startups being enlarged. According to data from the China Torch Center of the Ministry of Science and Technology, there were 4298 wework businesses channeled into China Torch Program, 3255 incubators of technical businesses and over 400 business accelerators by the end of 2016, which means that the country now has 7953 platforms for entrepreneurship and innovation, ranking first in the world.

5.2.7 Digital Economy Has Brought About More Jobs

1. **Digital economy is the "stabilizing anchor" of the global job market in the post-financial-crisis era.**

The global financial crisis has slowed down the economic growth of all countries and reduced the number of jobs in traditional industries and businesses, but there is a comparative balance between supply and demand in employment, to which digital economy has contributed a lot. A survey by McKinsey & Company on 4800 small and medium-sized businesses shows that with the popularization of Internet technology in small and medium-sized businesses, the loss of one job will create 2.6 new jobs. By 2015, the productivity growth brought about by the Internet will reduce the demand for labor by 1.3–4.0%, i.e. 10–13 million jobs, but the new products and services brought about by the Internet will create 46 million new jobs. Therefore, the Internet will increase rather than decrease the employment rate.

2. **Flexible employment brought about by Internet platforms is leading to more inclusive employment.**

Digital economy makes the supply of labor in the future like water on tap, which is convenient and controllable, and individuals with flexible work time can obtain flexible jobs. According to a report by McKinsey Global Institute, in 2015, over 200 million people in the world with different talent could obtain more job opportunities and income from free-lance platforms. By 2025, all kinds of online talent platforms will be expected to contribute 2% to the total production value of the world and provide 72 million jobs. According to *World Development Report 2016*, Internet platforms can help disadvantaged groups like women, the disabled and poor people to find better jobs or to start up their own business. On the freelance platform Elance, 44% of the online workers across the world are women. India has launched Kudumbashree Project, in which information technology services are subcontracted to the co-ops of women from poor families.

3. **Digital economy is creating jobs of higher quality, so that everyone is accessible to global job opportunities without any difference.**

Though they have replaced people in some jobs of traditional industries, doing routine and repeated work, digital technologies, in general, have improved the quality of jobs and created more job opportunities with higher value of a wider range. Internet platforms like Upwork and zbj.com make long-distance work possible and help low-cost laborers from developing countries to get jobs from developed countries, which will lead to extensive global collaboration and equal treatment of workers.

> **Column 2 Impact of Technical Progress on Employment: A British Case**
>
> Take the United Kingdom as an example. Seen from the data of the past industrial revolutions, from 1992 to 2014, when technical progress was highly complementary to non-routine cognitive work, the number of corresponding jobs would increase by 365%; when technical progress made non-routine manual work replaced in a limited degree, the number of corresponding jobs would increase by 168%; when technical progress made routine cognitive work and routine manual work a significant substitution, the number of corresponding jobs would decrease by 66% and 70%, but the total number of jobs would still increase by 23%. Seen from a longer term, with the popularization of machines, the number of jobs in the United Kingdom has increased by over twice in the past 150 years.[11]

---

[11]Ian Stewart, Debapratim De and Alex Cole from Deloitte: *Technology and People: The great job-creating machine.*

5.2.8 *Digital Economy Is Ushering in a New Round of Globalization*

With the rapid development of digital economy, rapid growth of digital product and service trade and accelerated integration of regional digital markets, cross-national digital economy platforms have become the navigator and vane of economic globalization.

1. **The global digital service trade is growing rapidly.**

According to data from UNCTAD, in the statistics about service trade, the percentage of emerging digital services represented by computer and information service, communication service and technical service has increased dramatically. From 2005 to 2015, the percentage of digital service trade increased from 7.78 to 9.79%. According to statistics of McKinsey & Company, 12% of the global cargo trade was done through the Internet, and thanks to the global circulation of resources and laborers, 50% of the world's service trade has been digitalized. According to *Cross-border B2C E-commerce Market Trends 2020* jointly released by AliResearch and Accenture, the cross-border B2C e-commerce transaction volume by 2020 will reach US$994 billion and such e-commerce will benefit 943 million consumers. According to *The Transatlantic Digital Economy 2017* released by Center for Transatlantic Relations, Europe is the largest market of the U.S. digital service and digitally supported service export and the major source of the imported digitally-supported service of the United Sates.

2. **The integration of regional digital markets is being accelerated**

In May 2015, European Commission released the plan for the establishment of single digital markets and a series of measures concerning policy reform, copyright laws, consumer protection and cloud service to promote cross-border trade with EU. According to Center of European Policy Studies (CEPS), the integrated digital market within EU will bring about an economic growth of 500 billion euros (about US$689 billion) in the coming 10 years. In May 2017, the *Joint Communique of the Leaders Roundtable of the Belt and Road Forum for International Cooperation* proposed "supporting innovation action plans for e-commerce, digital economy, smart cities and science and technology parks, and encouraging greater exchanges on innovation and business startup models in the Internet age in respect of intellectual property rights."

5.3   The Information Technology Industry is Defining the Digital Future

Major countries are utilizing information technologies to redefine the economic mode and taking them as the means of improving economic development quality and effect.

5.3.1   Electronic Information Manufacturing is Facilitating the "Hard" Growth of Global Economy

1. The IC industry keeps growing rapidly.

By the end of 2016, the total output value of the global IC industry had amounted to US$339.7 billion. Intel has been on the top for 25 years with its 15.9% of the market share and Samsung has been ranking second (11.8%) for 15 years, and then come TSMC, Qualcomm and Broadcom. The IC market share of China accounts for 57% of the global total. Its IC industry market scale amounted to RMB433.55 billion *yuan* in 2016, witnessing an increase of 19.9%. The country is the largest IC market and the drive for its growth comes from automotive electronics, industrial control and communication devices and other segmented markets.

2. The sensor industry is growing steadily.

The rapid development of intelligent manufacturing, AI and IoT has promoted the application of sensors in intelligent manufacturing, smart agriculture, biology and medical science and automotive electronics. In 2016, the global sensor market scale amounted to US$174 billion. According to the Ministry of Industry and Information Technology of China, it will reach US$266 billion by 2018. In 2016, the sensor market share of the United States, Japan and Germany accounted respectively for 27.0, 20.5 and 14.0% of the global total, totally 61.5%. Giant businesses like Bosch, Broadcom, Texas Instruments, STMicroelectronics and Qorvo are the top five ones in the world.

3. The memory market scale is increasing.

In 2016, the global DRAM output value was about US$32 billion and the global NAND flash memory market value was over US$30 billion. Their total output value accounted for 95% of the global total. According to IC Insight statistics, 2017 would see a growth of 10% in the global memory market scale, which would amount to US$85.3 billion. It will grow rapidly in the coming years and will reach US$100 billion by 2020.

5.3.2 The Information Technology Service Industry Keeps Showing Its "Soft" Power

1. The global market value of mobile communication will amount to more than one trillion US dollars.

According to data from GSMA INTELLIGENCE, the number of mobile phone users amounted to 4.8 billion in 2016, and would be over 5 billion in 2017, with the compound annual growth rate of 4.2%; the mobile communication penetration rate in 2016 was 65% and it will reach 73% by 2020. In 2016, the total income from mobile communication was US$1.05 trillion.

2. The global cloud computing industry scale keeps growing rapidly.

According to data from Gartner, the global income from the cloud service market was US$ 219.6 billion, including US$48.2 billion from SaaS and US$25.4 billion from IaaS. It is estimated that the total income will reach US$411.4 billion. China's cloud computing industry scale is increasing fast. It will increase from RMB 150 billion *yuan* in 2015 to RMB430 billion *yuan* in 2019.

3. The global Internet content industry is developing dramatically.

By October 2016, the number of APPs in global mainstream APP stores had amounted to approximately 12.34 million, maintaining a fast growth. In 2016, the global online and offline entertainment and media market scale reached US$1.79 trillion, the United States, China and Japan being the top three countries. The market scale in the United States was about US$740 billion, including the online market scale of US$220 billion. The online content output value of China accounted for one fourth of that of the United States, about RMB382 billion *yuan*, equaling to that of Japan and that of Republic of Korea. It is estimated that it will surpass Japan and Republic of Korea in 2020 and rank second in the world.[12] According to the *Internet Advertising Revenue Report 2016* released by IAB, the expenditure on global online advertisements in 2016 increased by US$1.29 billion, with a year-on-year growth of 21.8%; the expenditure on mobile advertisements in the same period increased by US$1.59 billion while that on PC advertisements decreased by US$0.3 billion. According to *Interaction 2017* released by GroupM, in every increase of one dollar of investment in advertising, 77 cents would be spent on digital advertising while 17 cents on advertising on TV in 2017.

---

[12]Tencent Research Institute: *Digital Empowerment: Chinese elements promote the development of global cultural industries.*

5.4   E-Commerce is Popular Worldwide

Seen from the global economy development, e-commerce is the earliest area of Internet application with wide penetration and the most active area of digital economy. In 2016, the total transaction volume of the global e-commerce retailing market reached US$66.58 trillion.[13] According to a report by French Lengow, by the year 2016, the top 5 countries in e-commerce turnover were China, the United States, the United Kingdom, Japan and France, with their respective turnover being US$975 billion, US$648 billion, US$186 billion, US$124 billion and US$76 billion.

5.4.1   On-Line Retailing Leads the Growth of Global Retailing, Bucking the Trend

In recent years, the global retailing market has been sluggish, but online retailing is growing fast, becoming the strongest drive for global retailing growth. According to a report by eMarketer, the global online retailing turnover in 2016 was estimated to be US$1.915 trillion, with a growth rate of 23.7%, accounting for 8.7% of the global total retailing turnover. It is estimated the online retailing turnover will increase to US$4.058 trillion in 2020, accounting for 14.6% of the global total retailing turnover.

The Asia-Pacific Region remains the biggest online retailing market, with its turnover in 2016 reaching US$1 trillion, and its estimated turnover reaching US$2.725 trillion in 2020. The rise of the middle class, the increased popularization of mobile phones and the Internet, the growth of the number of e-businesses and the improvement of logistics and infrastructure will further boost the development of e-commerce in that region. China's e-commerce retailing turnover has amounted to US$899.09 billion, contributing nearly one half (47%) to the world's total. In 2016, the e-commerce retailing in North America saw an increase of 15.6% in turnover, which amounted to US$423.34 billion. Thus the region became the second largest e-commerce market. The e-commerce retailing there, according to eMarketer, will increase by over 10% by 2020, thanks to the increase of the consumers' budget in e-commerce, the popularization of the e-commerce platforms for daily necessities and the growth of the sales volume of mobile e-businesses.

The development of global e-commerce retailing is shown in Fig. 5.6.

According to *Global 1000* by Internet Retailer, the global top 10 Internet retailers in terms of transaction volume are Alibaba, Amazon, eBay, JD.com, Rakuten, Apple, Suning, Mi, Dell and Wal-mart (see Fig. 5.7), with their total transaction volume accounting for 46.2% (nearly one half) of the world's total, and with the total transaction the first three (Alibaba, Amazon and eBay) accounting for 80% of the

---

[13]Ministry of Commerce of the People's Republic of China: *Report on China's E-commerce* (2016).

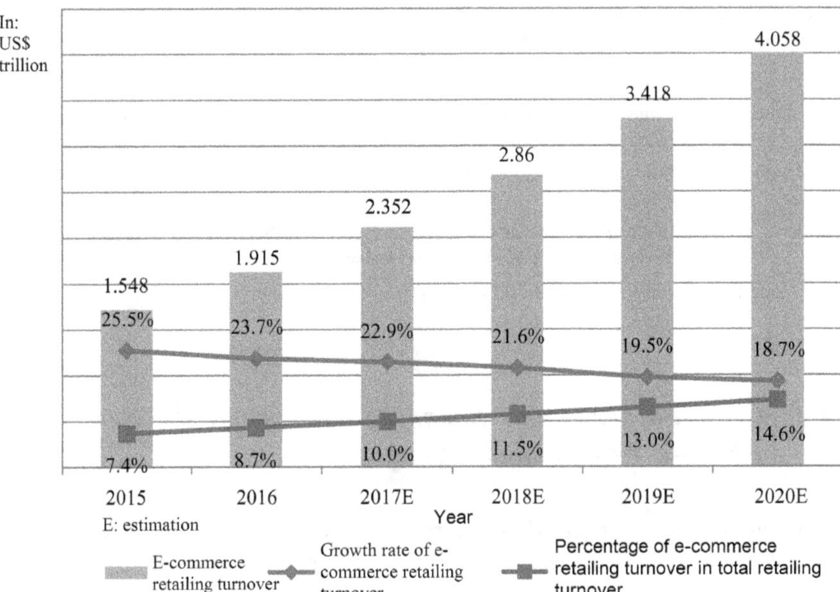

Fig. 5.6 Development of global e-commerce retailing

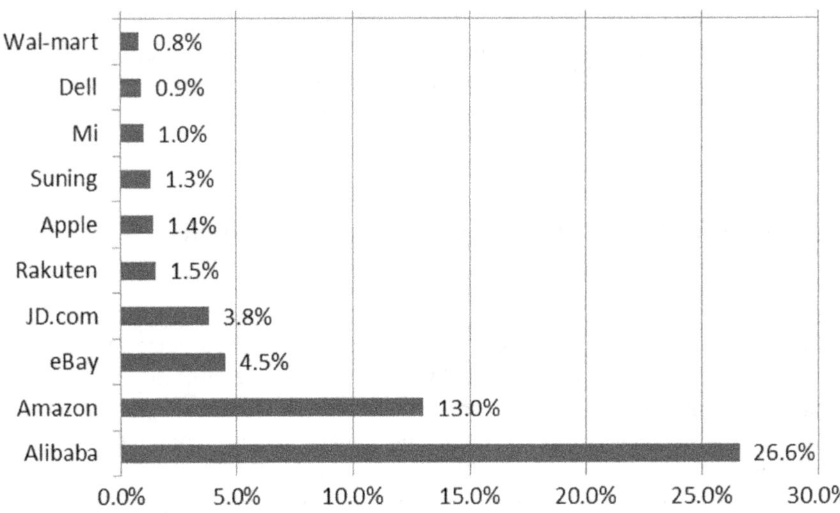

Fig. 5.7 Global top 10 businesses in percentage of internet retailing turnover (Internet Retailer: *Global 1000*)

total of the 10 businesses. The analysis of Internet Retailer shows that in 2016, the total value of the commodities sold on the top 18 platforms of the world increased by 21.5%, with the total transaction volume amounting to US$1.09 trillion.

Table 5.6 Percentage of online shoppers from major countries in cross-border online shopper (Nielsen: *Global Connected Commerce 2016*)

Region	Country and percentage
Latin America	Chile (69%), Mexico (64%), Columbia (54%), Brazil(44%), and Argentina (42%)
Asia-Pacific Region	India (74%), Australia (66%), Thailand (64%), the Philippines (61%), China (58%), Republic of Korea(50%), and Japan (32%)
Europe	Italy(79%), Germany (73%), Austria (64%), Spain(63%), France (59%), Turkey (54%), UK (52%), and Poland (34%)
North America	Canada (62%)and US (29%)
Africa	Nigeria (61%)and South Africa (50%)

5.4.2 Cross-Border Online Shopping has become the New Favorite Way of Global Consumers

According to *Global Connected Commerce 2016* by Nielsen, e-commerce is mostly confined within the country and consumers tend to buy goods from their domestic retailers, but cross-border e-commerce is developing (see Table 5.6). Consumers are paying increasing attention to retailers abroad. Among the interviewees who have done online shopping in the past six months, over one half (57%) say that they have bought something from an overseas retailer. Among the Indian interviewees, nearly three fourths (74%) say they have bought something from a retailer abroad. This is not only a trend in developing countries, since about two thirds of interviewees in Western European have also bought something from retailers abroad, for example, 79% of Italian consumers and 73% of German consumers.

5.5 Manufacturing has become the Main Battlefield of Digital Economy Development

Manufacturing is the main battlefield of digital economy development, which will promote the intelligent, personalized, networking and service-oriented development of manufacturing, the innovative reform of management and the transformation and upgrading of traditional industries. Germany, a major country committed to intelligent manufacturing, put forward in 2013 the "Industry 4.0" Strategy, ushering in a new round of industrial transformation competition worldwide. In 2016, China's manufacturing information index was 36.9 and the country's manufacturing is shifting from 2.0 to 3.0. China has made progress in "four new foundations" and cloud service and takes the lead in Internet transformation development like industrial ecosystem innovation.[14]

---

[14]ChinaInfo 100: Informatization Index of "Made in China" (2016).

5.5.1 The World's Leading Industrial Businesses are Laying Out Their Platform Economies

According to IDC, by 2018, over 50% of large industrial businesses will have their digitalization strategies and the number of digital industry platforms based on cloud architecture will increase from over 100 in 2016 to 500 in 2018.[15] Siemens began to integrate resources of software in 2007 to build its big data analysis capacity and to integrate its existing platforms since 2015 to set up the Sinalytics platform. The Predix platform of GE is committed to integrating the industrial data and systems. SAP, taking the strong application software as the core, is expanding its solutions from intelligent upper-level management to lower-level intelligent production.

Layout of industrial cloud platforms by the world's leading businesses are seen in Table 5.7.

5.5.2 The Intelligent Industrial Production Is Becoming a Trend

At present, ICT is integrating fast into the whole process of manufacturing, all links in the production chain and the whole life cycle of the product, so the new intelligent production will soon become a trend. First, the subject of production is changing from the manufacturer to the prosumer.[16] The number of prosumers in service areas and simple product manufacturing (for instance, handicrafts manufacturing) is increasing. With the emergence and popularization of 3D printing and IoT, the production which used to be complicated tends to be more simple, and it can even be done by individuals, so there will be prosumers in more areas. Secondly, production is shifting from object manufacturing to virtual manufacturing, which is defining the manufacturing mode. Virtual production lines, workshops and factories will be set up in cyberspace. With the popularization of the digital twin of plant production equipment and the digital twin of the production technique, real-time mapping can be realized between object manufacturing and virtual manufacturing whether in discrete industries or in process industries. Thirdly, the production system is shifting from closed to open. The collaboration between traditional manufacturers and upstream and downstream businesses of the industrial chain will be done online or in the network. Fourthly, the production space mode

---

[15]Accenture: *Industrial Platform Helping Digital Transformation*.

[16]The progress and penetration of ICT will lower the barriers for production, so consumers can independently produce and thus provide products and thus there will be no boundary between the producer and the consumer. Hence there has emerged the term "prosumer".

Table 5.7 Layout of industrial cloud platforms by the world's leading businesses (ChinaInfo 100: *Rise of Information Economy.*)

Platform	Equipment connection and information collection	Industrial software based on the third-party development and development environment	Industrial cloud and big data platforms
GE Predix	Gateways developed jointly with Cisco, chips jointly developed with Intel embedded GE OS and SW	Predix I.O, SDK, and API	Data analysis platform based on cloud, integrating a number of data service products
Siemens Sinalytics	Gateways and Siemens products integrated through SIMATIC	–	Big data service platform enriching data services through software tools like TERADATA IBM
IBM Bluemix	External Gateways hardware	Providing the third party with development service and function of expanding application programs for communities. Providing development integration architecture or API	Loading Waston application for big data application based on cloud platforms
Ayla	Submersible firmware on devices or gateways	Developing solutions through AMAP application for the third-party application development, and notice applied in mobile application	Providing PaaS cloud service based on Amazon Cloud and providing IoT data services through Ayla Insight
Microsoft Azure	SQL Azure Gateway and ISS	Azure SDK for NET, and Visual Studio	Cloud platform providing IaaS and PaaS services
Platform	Equipment connection and information collection	Industrial software based on the third-part development and development environment	Industrial cloud and big data platforms
PTC Thing Worx	Long distance devices connection through Thingwork edge micrlservice	Building the Thingworx marketplace application development environment through SDK and API	Setting up the cloud platform through integration with Amazon and Microsoft, and making big data prediction through Neuron
Oracle Platform	External Gateways and protocol converters on devices	JD Edwards applications for the third-party application development	Providing IaaS, Paas and Saas cloud platform services, and big data analysis based on the existing product system

(continued)

Table 5.7 (continued)

ARM Mbed	MCU based on ARM Cortex-M and mbed OS	Mbed Device Server provides mbed REST API and SDK in the way of free development + commercial authorization	Providing device cloud services based on ARM by cooperating with IBM
Intel Platform	External Gateways, Embedded OS/software and embed intel chips	Providing the third party with API management solutions	Intel® Data Center Manager, with the function of data analysis platforms
Huawei Ocean connect IOT Platform	LiteOS, an open-source IoT OS of Huawei	Providing Agent, API and SDK respectively for terminal, integration and application developers	Able to connect with the third party's cloud platform

will shift from centralized to decentralized. The superposition between additive manufacturing based on 3D printing and IoT is completely changing the production space mode, promoting the shift from centralization to decentralization of production.

Column 3 G20 Digital Economy Development and Cooperation Imitative (Excerpt)

(1) Encourage internet-based RDI and entrepreneurship through an enabling, transparent legal framework, programs to support RDI and well-functioning capital markets for innovative enterprises. Support developing and emerging countries to build capacities in digital technology and internet-based entrepreneurship.

(2) Take advantage of the internet to promote innovation in products, services, processes, organizations and business models.

(3) Encourage the integration of digital technology and manufacturing, to build a more connected, networked, and intelligent manufacturing sector. Take advantage of ICTs to improve education, health and safety, environmental protection, urban plan, healthcare and other public services. Promote the continued development of service sectors such as e-commerce, e-government, e-logistics, online tourism, and Internet finance and the sharing economy. Promote digitization of agricultural production, operation, management, and networked transformation of agricultural products distribution.

5.5.3 *Exploration of Major Countries in Intelligent Manufacturing Transformation*

The United States takes the lead in manufacturing transformation. It released in February 2016 the strategic plan The National Network for Manufacturing Innovation (NNMI), which has four major goals: (1) to increase the competitiveness of the US manufacturing; (2) to facilitate the transition of innovative technologies into scalable, cost-effective, and high-performing domestic manufacturing capabilities; (3) to accelerate the development of an advanced manufacturing workforce; and (4) to support business models that help institutes to become stable and sustainable.

Column 4 National Network for Manufacturing Innovation of the United States

Goal 1: Increase the competitiveness of U.S. manufacturing

Objective 1.1: Support the increased production of goods manufactured predominantly within the United States;

Objective 1.2: Foster the leadership of the United States in advanced manufacturing research, innovation and technology.

Goal 2: Facilitate the transition of innovative technologies into scalable, cost-effective, and high-performing domestic manufacturing capabilities

Objective 2.1: Enable access by U.S. manufacturers to proven manufacturing capabilities and capital-intensive infrastructure;

Objective 2.2: Facilitate sharing and documentation of best practices for addressing advanced manufacturing challenges;

Objective 2.3: Foster the development of standards and services that support U.S. advanced manufacturing.

Goal 3: Accelerate the development of an advanced manufacturing workforce

Objective 3.1: Nurture future workers for STEM related work;

Objective 3.2: Support, expand, and communicate relevant secondary and post-secondary pathways, including credentialing and certifications;

Objective 3.3: Support the coordination of state and local education and training curricula with advanced manufacturing skill set requirements;

Objective 3.4: Advanced-knowledge workers: researcher and engineers;

Objective 3.5: Identify the competences needed by the next generation of workers.

Goal 4: Support business models that help institutes to become stable and sustainable

The manufacturing innovation institute should not always rely on the governmental funding. It is the resource and service platform and incubator jointly set up by all governments of all levels, industry, academic community, attracting small and medium-sized business and startups to participate in it, to protect and use intellectual property right to attract industrial funds.

Germany has launched the "Industry 4.0" program. By releasing the potential of IoT, big data and AI, the country is changing the critical process of manufacturing of automobile, trains, aircrafts and machinery, making better demand prediction through big data technology, and laying out 3D printing and modelling technologies to enrich customers' choices, improve product quality, reduce cost and increase production speed.

Column 5 Implementation Roadmap of German Industry 4.0

Industry 4.0 refers to the digitalization of supply, manufacturing and sale information in the process of production to ensure fast, effective and personalized product supply. In *Hight-Tech Strategy 2020*, the German Government lists Industry 4.0 as one of the future ten projects, whose implementation can be summarized into "123,458".

"1" refers to one system, the physical system connecting resources, information, things and man-force, which is taken as the core technology and basic technology of intelligent manufacturing.

"2" refers to two themes, namely, Smart Factory and Intelligent Production, which aim to foster the intelligence of the factory, create new products, improve production efficiency and solve social problems. The focus of "Smart Factory" is to study the intelligent production system and process and how to distribute the production facilities in the network; "Intelligent Production" involves the production logistic management, interaction between man and machines and the application of 3D technologies in the industrial production.

"3" refers to three transitions, namely transition from centralized production to decentralized production, transition from common products to personalized products and transition from customer-oriented service to customer-involved service.

"4" refers to four goals, namely, developing intelligent produciton methods, optimizing automatic technologies, satisfying the new demand resulting from man force changes and forming new industrial produciton models.

"5" refers to five tasks, namely, setting up the production system integrating and networking the manufacturing process, fostering the innovation and application of ICT in the manufacturing, building standardized models, constructing new business organization models based on man-machine interaction, and enhancing R&D and promotion of security technology and know-how.

"8" refers to eight actions, namely, establishing standard technology systems, mastering the management techniques of sophisticated systems, constructing industrial broadband network infrastructure, ensuring production and information security, redesigning work positions and content, continuing professional skill training, establishing corresponding laws and supervision mechanisms, and efficiently utilizing resources.

EU has launched two-wheeled strategies to facilitate the digital transformation, namely, the leading-supplier strategy and the leading-market strategy. The former focuses on the supply side, which means that EU members should stand out in the global market competition and maintain the leading position by encouraging scientific and technical innovation and technical reforms. The latter focuses on the large-scale application of innovations with global potentials. EU tries to establish "leading markets" through legislation, labeling, standardization, identification and technology-guiding purchase. For instance, by launching the Tyre Labelling Regulation, EU stipulates that all the tyres on the market should have labels covering rolling resistance, noise and performance rate of wet and slippery road holding on them, which has facilitated the competitiveness of European tyre businesses and the increase of input of global tyre businesses in R&D to improve the quality through innovation.

India fosters its "Made in India" campaign. In September 2014, the country released a series of policies concerning "Made in India" to enhance its attractiveness in terms of investment in industries, providing one-stop services for those domestic and foreign businesses who intend to invest in India. Meanwhile, it has reformed the labor laws and taxation and simplified procedures for examination and approval to attract investment in setting up factories to increase the number of job opportunities. "Made in India" campaign involves 25 industries, covering automobile, chemical, pharmacy, textile, information technology, port construction, aviation, tourism, railway, renewable energy, mining and electronics. In mid-February 2016, the "Made in India Week" was held in Mumbai to showcase and test the achievements of the campaign.

5.6   Arrival of the Internet Finance Era

Digital revolution, intelligent data analysis and popularization of intelligent terminals have lowered the threshold for the Internet finance service industry, which promises a bright future. All types of services, such as third-party payment, crowdfunding, financial community, online exchange, personal finance, micro-credit and P2P, are changing the landscape of the financial service industry.

5.6.1   *Crowdfunding on the Internet is Witnessing an Explosive Growth Worldwide*

Crowdfunding on the Internet was started in 2001 and then began to grow explosively. Database of Forward Information Co., Ltd. shows (see Fig. 5.8) that from 2010 to 2016, the crowdfunding on the Internet saw the annual compound growth rate of over 80%. Its scale reached US$198.96 billion in 2016, and it is anticipated to reach US$300 billion by 2025, with US$96 billion in developing countries. Crowdfunding on the Internet is a totally new funding way and channel, which, to some extent, has reduced the funding cost of startups and improved funding

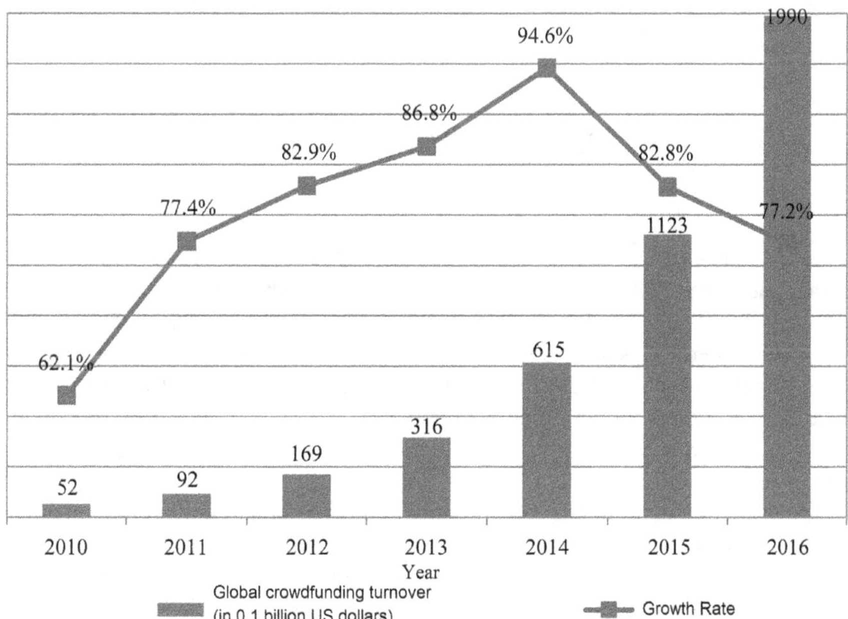

Fig. 5.8 Global crowdfunding turnover and its growth rate (Database of Forward Business Information Co., Ltd.)

efficiency. The *JOBS Act* released by the United States identified the legality of crowdfunding and authorized investors to have the equities as the returns for the investment. Besides, improved supervision and financial innovation are both promoting the rapid development of crowdfunding models.

5.6.2   Mobile Payment has become the New Trend of Global Financial Service

In 2017, Capgemini, a consulting service company, and BNP Paribas of France jointly released a report, which shows the global digital payment scale in 2020 will reach US$726 billion. In the present non-cash payment transaction, the use of debit cards account for the highest percentage (46.7%), then comes the use of credit cards, accounting for 19.5%. On the other hand, the contactless card is becoming the "new standard" of digital payment, especially in Europe. Cash is, though, the main payment way, especially for small transactions. In contrast, the cheque is being weeded out. Its use saw a decrease of 13.4% in 2015. A report released by Ipsos in September 2017 shows that mobile payment is being popularized in China, where 26% of consumers have less than RMB100 *yuan* of cash with them and 14% do not have cash at all. According to data from the Ministry of Industry and Information Technology, the scale of China's mobile payment in 2016 was over RMB81 trillion *yuan*, ranking first in the world. Baidu, Alibaba and Tencent occupy half of the third-party market share in China.

5.6.3   The Internet Insurance Market has become the Investment and Financing Hotspot

Research shows that from 2014 to 2016, nearly US$5 billion of capital across the world was invested in Internet insurance. In 2016, there were altogether 145 financing events for Internet insurance startups and about US$1.7 billion was invested in Internet insurance, with the biggest being series C-round financing of US$400 million for Oscar Health. Most financing cases of Internet insurance take place in the United States, then in Germany, the United Kingdom and other European countries. The major areas of Internet insurance financing in the world are businesses' health insurance platforms, personal insurance policy management platforms and big data application platforms. Health insurance is the most attractive area for financing. Among the businesses whose financing scale was over US$10 million in 2016, there were over 10 health insurance companies. In the global Internet insurance, there have emerged unicorn businesses like Oscar Health, Zenefits and MetroMile.

5.6.4 China Is the Leader of Global Internet Finance Development

Whether in quantity or scale, China's Internet finance ranks first in the world. According to the monitoring data from National Technology Platform of Internet Financial Risk Analysis, there are 19,000 Internet finance platforms in the country, covering 21 types of Internet finance services, including online lending, Internet asset management and online crowdfunding. There are accumulatively over 6000 online lending platforms, nearly 3500 Internet asset management ones and 800 Internet crowdfunding ones. According to data from Visual Capitalist, there are 27 unicorn businesses of financial technologies in the world, including 8 in China, who have raised the fund of US$9.4 billion and whose total estimated value was US$96.4 billion; over 14 in the United States (their raised fund being US$5.7 billon and estimated value being US$ 31 billion). The top four unicorn businesses of financial technologies are in China, with the biggest being Ant Financial head-quartered in Hangzhou, whose estimated value is US$60 billion. The second one is LU.com, a P2P loaning business in Shanghai, with its estimated value being US$18.5 billion. JD Finance, with its fund from JD.com and Tencent, is the third biggest one. The fourth is Qufenqi.com (Qudian.com), an instalment and investment platform in Beijing, with its estimated value of US$5.9 billion and raised fund of US$0.9 billion. According to the *2016 VC Fintech Investment Landscape* released by British Innovate Finance, the total financing value of China's financial technology businesses surpassed that of those in the United States for the first time in 2016 and ranked first in the world.

5.7 Sharing Economy is a New Highlight of Global Economy

Sharing economy has swept the whole world, witnessing dramatic development in Americas, Europe, Asia and Oceania and rudiment in Africa. A number of local sharing economy platforms with their own characteristics have emerged in different countries and regions.

5.7.1 The Wave of Sharing Economy has swept the World

North America takes the lead in sharing economy. According to statistics of Justpark website, over a half of sharing economy businesses are in North America, with the United States as the leader. In 2016, there were over 400 such businesses in the country, with the financing scale amounting to nearly US$20 billion. There have emerged unicorn businesses of sharing economy in different industries of the

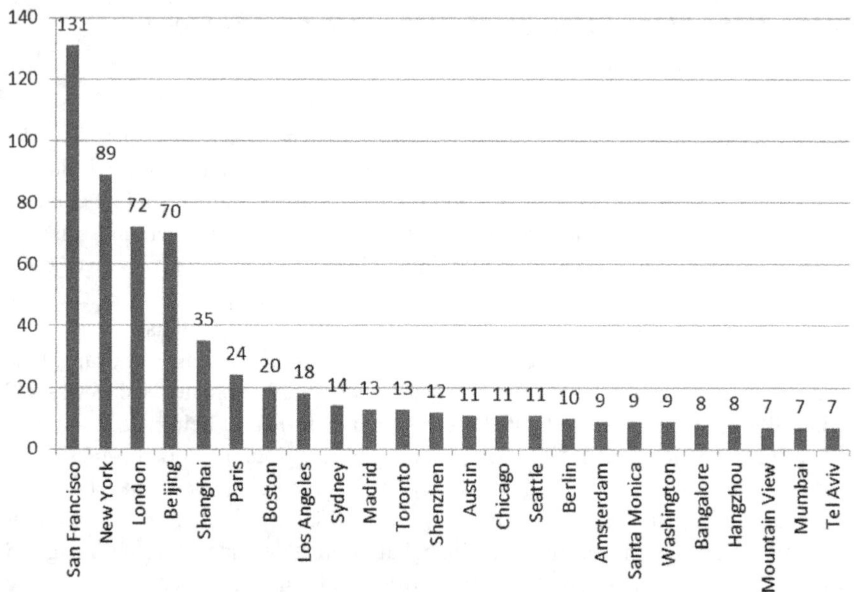

Fig. 5.9 Startups in sharing economy in representative cities of the world in 2016 (Tencent Research Institute: *Overseas sharing Economy Development Report* (2016))

United States, where San Francisco and New York host the most startups in that area, the former being called "the capital of global sharing economy". There are the most participants in sharing economy in the Untied States. In 2015, 21.7 million grown-ups used sharing economy services and the number is expected to reach 37.3 million in 2019.[17] The country also leads the trend of sharing economy. C2C personal sharing has penetrated areas like agriculture and energy, for instance, SWIM in water resource sharing, Machinery Link in agricultural equipment sharing and Yeloha in power sharing. B2B sharing is expanding into medical care, building and warehouse logistics, such as Cohealo in medical equipment sharing, Yardclub in building equipment sharing, Cargomatic in transport capacity sharing and Flexe in warehousing sharing.

The number of startups in sharing economy in representative cities of the world in 2016 is shown in Fig. 5.9.

European countries see rapid development of sharing economy. According to EU Sharing Economy Briefing, the sharing platforms of EU earned a total income of approx. 28 billion euros in 2015. Eurobarometer (2016) shows that over 50% of EU citizens have heard about sharing economy and one sixth of the citizens are participants in it. In the United Kingdom, the government proposed in 2014 the plan of building the country into a "global center of sharing economy", and London is

---

[17]*How the United States Responds to the Rise of Sharing Economy*, http://3g.ifeng.com/news/ sharenews.f?aid=107115412&mid=.

the city in Europe with the most sharing economy businesses. According to the data from UK Ministry of Commerce, about 25% of British adults have shared their idle resources like assets and skills online. In Italy, over 25% of Internet users had used sharing economy services and 74% say that they are willing to participate in sharing economy. According to Italian Parliament, the GDP owed to sharing economy will amount to 90 billion euros in 2025. A survey made by Deloitte in March 2015 shows that in Switzerland, 18% of consumers there had participated in sharing economy and 55% said that they would participate in it in the next year.

The Asia-Pacific Region is catching up and taking the lead at a high speed. A survey on the global Internet use made by Nelson Company shows that Internet users in the Asia-Pacific have a strong wish to share, 78% willing to share what they have and 81% willing to rent something from others. According to data from Ministry of Commerce, Industry and Energy of Republic of Korea, in 2015, the country's sharing economy scale was about US$425–658 million, accounting for 2.8–4.4% of the world's total. In Republic of Korea, sharing economy witnesses rapid development in areas such as automobile, housing, travelling, office, socializing and catering, know-how and personal things. It is developing fastest in transportation, accommodation and second-hand goods transaction. According to a report by Yano Research Institute, the sharing economy scale of Japan grew by 134.7% on a year-on-year basis in 2014 and 129% in 2015. In the country, sharing economy has begun to cover accommodation, car pooling and parking, medical care, housekeeping and logistics and is going to cover fashion and animation. In Australia, the capital region and West Australia have legalized care sharing. The Word of the Year 2015 selected by Australian National Dictionary Centre was "sharing economy".[18] In that year, the scale of sharing economy in Japan increased by 129%.

The African market has become a new opportunity for sharing economy. Airbnb and Uber are active in tapping the sharing economy market there. Fund sharing represented by the P2P financing platforms is popular. For example, in South Africa, Rainfin, a social credit platform, was established to meet the financing demand of small and medium-sized businesses and private startups. In the very first month after its establishment, a credit application of over 540,000 *rand* was approved. In 2014, German Lendico and British Wonga entered the South African market. Thus there are three giants of fund sharing in South Africa.

The development of sharing economy in different countries is shown in Table 5.8.

---

[18]Source: Australian National Dictionary Centre's Word of the Year 2015, http://ozwords.org/?p=7561.

Table 5.8 Development of sharing economy in different countries

Country	Development	Representative platforms
The United Kingdom	The future global sharing economy center. According to Nesta, there are in the country 16 million sharing economy participants, accounting for 25% of the total population of the country. According to statistics, the sharing economy scale of the country in 2013 amounted to 2.24 million pounds, accounting for 1.3% of its GDP. Sharing economy will become a pillar sector of British economic development. Unicorn businesses of digital economy have emerged in the financial area	Funding circle (the world's first P2P platform allowing the financing value to exceed one hundred million pounds), Zopa (the world's first P2P credit platform), Just park (a leading business in parking sharing), frn (second-hand furniture platform) and Future (a MOOC platform)
Germany	According to Bitkom statistics, those interested in car sharing account for one half of Germany's total population and those under the age of 30 account for two thirds. According to bcs statistics, the number of car sharing users in the country increased from 1.04 million in 2015 to 1.26 million in 2016 and the number of cities and towns with car sharing service stations grew from 490 in 2015 to 537 in 2016. German auto manufacturers have begun to transform their business into car sharing	Daimler car2go (having had 1.1 million members by December 2015, and having become the largest car sharing project of the world), BMW Drive Now Car Sharing Program (having established its service seven cities including Berlin and London by 2015, with over 240,000 users), and Volkswagon Quicar sharing project
France	There have emerged giant businesses of local car sharing and second-hand goods sharing in France. According to a Mediaprism survey, second-hand goods transaction is popular there. 49% of French people say that they seize all the opportunities to buy second-hand goods. For the idle things in the house, 77% of them say they would like to give these things "a second life"	Blablacar (a French long-distance car sharing giant, with 20 million active users from 19 countries, and a unicorn business of sharing economy in France), Leboncoin.fr, Top 1 website of second-hand daily goods transaction, and Ulule in crowdfunding
Israel	Israel's scientific and innovation capacity facilitates the development of sharing economy	Gett (providing high-level car sharing service in 57 cities worldwide), Fiverr (a global task releasing and outsourcing platform with the initial task price at 5 US dollars. The price increases with the credit), Ourcrowd (one of the crowdfunding businesses with the fastest development in the world) and LaZooz (an innovative carpooling APP using block chain advantages for decentralization)

(continued)

Table 5.8 (continued)

Country	Development	Representative platforms
Singapore	Singapore is universally recognized as a "garden-city country", so the ecological effects produced by sharing economy can facilitate the country's sustainable development	Grab taxi (a transportation sharing platform and unicorn business, covering taxi calling, private cars and motorcycles), Icarsclub (a P2P car renting APP), Blockpooling (a things sharing platform based on community social networks) and Rent ycoons (the first P2P renting service platform in Singapore for idle things)
Canada	High participation in sharing economy. According to a survey by Crowd Companies, 39% of Canadian population ha s participated in sharing economy, higher than that of the United Kingdom and the United States, on the top list of the world	Varage sale (a second-hand sharing platform for the neighborhood, having over one million users in North America), Breather (a P2P space sharing platform), Borrowell (an online credit platform) and Jobblis (a platform acting as a bridge between free lancers and businesses)
Republic of Korea	According to ROK Ministry of Commerce, Industry and energy, the sharing economy scale of the country by March 2015 was US$425–658 million, accounting for less than 0.5% of the total GDP, but in 2025, it will amount to US$7650–11,850 million, so it boasts great potential	Car sharing: Socar and Greencar, two domestic representatives. The former is the leader of the country's car sharing business. By August 2015, it had had over one million members with the accumulative funds of US$18 million; House sharing: Kozaza, a domestic representative, providing foreign tourists with traditional ROK housing, with the accumulative funds of US $500,000; Second-hand transaction: Open Closet, a domestic representative, a second-hand transaction platform
Japan	Japan is a typical aging society, but with the growth spurt of the tourist industry there and the government's loosening control over sharing finance, the Japanese people accept sharing economy more and more and have an increasingly high demand for it, so it enjoys a rapid start	Mercan (a C2C second-hand transaction platform), Cogicogi (a car sharing platform), Akippa (a parking place sharing platform), Crowdworks (an online task crowdsourcing platform) and Aqush (connecting individuals with idle fund and those with the demand for it)
Brazil	Sharing economy has saved the recession of the country's economy. The number of Internet users in Brazil ranks fourth in the world, laying a foundation for sharing economy development through the Internet and large-scale match between supply and demand information. In the country,	Rocket (travelling sharing, with its services distributed in South America, Asia, the Middle East and Africa), Shippify (logistic crowdsourcing, based on the public logistics system on the online platform), fleety (P2P car renting) and Buzzero (individuals' online courses provider)

(continued)

Table 5.8 (continued)

Country	Development	Representative platforms
	the economy is sluggish, so there is an urgent need for a development drive, which can help to reduce the cost and increase the income, bringing opportunities for the development of sharing economy	
Holland	Amsterdam is Europe's first sharing city, which tries to become a world-class sharing city	Snappcar (car sharing), mobypark (parking place sharing), peerby (a things renting platform, encouraging renting in the neighborhood), thuisafgehaald (meals and snacks sharing in the neighborhood) and vandebron (energy sharing, buying electricity from the neighbor)

5.7.2 China Enjoys the Most Colorful Sharing Economy

After rudiment, start and rapid development, China's sharing economy has entered a new era embracing comprehensive innovation and become the biggest highlight of the world in sharing economy. According to a research done by the State Information Center, the year 2016 saw the sharing economy turnover of about RMB3.452 trillion *yuan*, with a year-on-year growth of 103%, and the financing value in the same year was RMB171 billion *yuan*, with an increase of 130% and the number of participants reaching 600 million. Sharing economy has contributed a lot to the implementation of China's strategy of "priority of employment". In 2016, the number of economy providers and users amounted to about 60 million, including 5.85 million platform workers. Take sharing economy platforms as an example. In 2016, Didi Chuxing platform created 17.50 million flexible jobs. Among the employees, 2.384 million came from excessive capacity-reducing industries and 875,000 were army men demobilized or transferred to civilian work. It provided directly 2072,000 drivers with an average income of RMB160 *yuan* every day. There are innovative business models emerging constantly and the application area keeps expanding. Following knowledge payment, live streaming and bicycle sharing whose first development year was 2016, umbrella, charger pal, KTV, refrigerator, bed and book store sharing has emerged and sharing economy in education and medical care is witnessing the beginning. In the coming five years, the annual growth rate of sharing economy in China will be around 40% and the scale of sharing economy will account for over 10% of the total GDP by 2020. To sum up, sharing economy in the country has entered an era in which everyone can participate and everything can be shared.

5.7.3 The Government is guiding the Development of Sharing Economy

1. **Promoting the organization and construction of the sharing economy industry.**

In the United Kingdom, SEUK was set up in 2014 sponsored by sharing economy businesses, with the Ministry of Commerce as the coordinator. Today, it has developed into a self-disciplinary organization with 36 members and cooperative businesses. In Spain, 26 sharing economy businesses have set up the Sharing Economy Association of Spain. In September 2013, EU countries set up Europe Sharing Economy Coalition to facilitate the implementation of sharing economy policies at the EU level and member country level. In Republic of Korea, the municipal government of Seoul released in September 2012 the declaration of Sharing City Seoul and set up Seoul Sharing Economy Promotion Committee, who is responsible for giving advices on promoting sharing economy and reviewing and identifying sharing groups and businesses. In Singapore, representative sharing economy businesses from different areas have set up Singapore Sharing Economy Association to create a reliable and positive environment for the members' development.

2. **Facilitating the reform of governmental procurement and opening of public resources**

Different countries are reforming the governmental procurement framework and embracing sharing business. Take the United Kingdom as an example. In 2010, Croydon Council passed a resolution on cooperation with Zipcar in replacing the governmental cars with the ones from the automobile club. British officials can choose house sharing and travelling sharing on their official trip. In Brazil, Airbnb signed an official agreement with Rio 2016 Summer Olympics and became one of the accommodation providers recommended officially. Different countries are constructing the sharing public facilities, such as car pooling, car renting and public bicycle sharing in the United States and France, "open data" projects in some cities of France, the United Kingdom, Republic of Korea and the United States, and sharing street facilities like public libraries, wardrobes and cabinets in some German cities.

3. **Fostering the wider participation in sharing economy**

All governments are promoting and supporting sharing economy. In Republic of Korea, the identification system has been adopted in Seoul for sharing groups and businesses. The identified businesses have the right to use BI (Brand Identity) of city sharing and are supported by the government in cooperation with the sectors of the city. EU is facilitating the education philosophy of sharing economy, promoting the concept and principle of sharing economy in primary, secondary, tertiary, adult and occupational educations, to enhance the understanding and awareness of participation in sharing economy. Besides, different countries have launched policies to close digital divides. The United States and Australia show concern about special groups of people. For example, some cities of the United States require Uber to provide additional service for disadvantaged groups and Australia encourages sharing economy platforms to provide service for the disabled and requires Uber to provide car service for the disabled. EU stresses the elimination of information barriers and has launched Europe 2020 to support direct financing programs for local and regional high-speed broadband projects and thus foster the construction of network facilities.

4. **Launching preferential taxation policies to support the development of sharing economy**

In the United States, Chicago, Boston and Portland differentiate car sharing and traditional car renting, reducing the taxation rate for car sharing, and the street parking places in San Francisco offer a discount for the parking of sharing economy businesses. The UK Government offers tax-free treatment for short-term accommodation with the annual rent below 4250 lb and appropriates funds for the development of P2P platforms. The municipal government of Paris appropriates funds to support Bollore's operation of car sharing. The municipal government of Seoul provides offices, consulting service and activity funds for sharing economy entrepreneurs.

Major countries' measures supporting sharing economy are shown in Fig. 5.10.

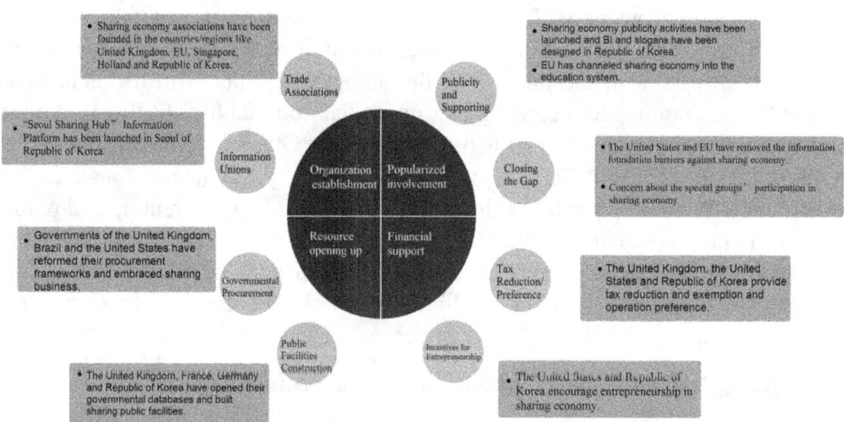

Fig. 5.10 Major countries' measures supporting sharing economy (Tencent Research Institute: *Overseas Sharing Economy Development Report* (2016))

5.8 Challenges to and Future of Digital Economy Development

Challenges to digital economy development cannot be neglected, such as the obvious digital divide, imbalanced development of digital economy, difficulty in cross-national collaborative governance, insufficient digital know-how and hazards in digital security. Despite all of these challenges, the era of digital economy has come and it will have a brighter future.

5.8.1 Challenges that Cannot be Neglected

1. **The digital divide has made global digital economy development more imbalanced.**

Digital economy is witnessing imbalanced development worldwide. In developed countries and regions like the United States and Europe, the percentage of digital economy in GDP is high, but in most developing countries, digital economy has just begun. The digital divide is the major cause of imbalanced digital economy development. According to *The State of Broadband 2017* released by UN Broadband Commission for Sustainable Development, the penetration rate among Internet users in the least developed countries would be only 17.5% at the end of 2017 and 3.9 billion people worldwide had no access to the digital world. The poverty trap is the biggest risk in the imbalanced digital economy development. As Mukhisa Kituyi, Secretary General of UNCTAD, says, "Most of the benefits from digital economy flow to quick actors. The major beneficiaries are those who have owned the know-how and opportunities. Therefore, we have to control the widening gap, the gap between nations and the gap between different groups within a country." Across the globe, the imbalance in digital economy development is increasing and thus superposes the imbalance in the industrial age. To sum up, imbalance in digital economy development is the root cause of the widening gap between nations, regions and the rich and the poor.

2. **Difficulties in cross-national collaborative governance of digital economy globalization**

In the field of global digital economy supervision and governance, the inconsistency, conflict and lack of coordination concerning policies and laws between nations have become the important issue during the digital economy globalization. In the field of trade and investment, though most developed countries have cancelled the unilateral tariff on digital technology products, there still exist trade tariff barriers for the circulation of these products in developing countries. In the field of knowledge product protection, every country has its own intellectual property right protection system, which hinders the free flow of digital technology products and services and hence the development of digital businesses. The cross-national collaborative governance over digital economy is weak. At present, there is no effective mechanism in the world for tackling the problems that have occurred in digital economy development, so there is an immediate need to innovate the multilateral, bilateral, regional or global coordination and collaboration mechanisms. For instance, in terms of property right, who should own and govern the data, how shall we govern them and use them, and how do we define the right of the owner, the user and the governor? All these questions are beyond the function and capability of existing international institutions.

3. Laborers' insufficient digital literacy hindering digital economy development

Digital economy development urgently demands to improve laborers' digital literacy, but at present, their knowledge is out-of-date and their skill mismatch is obvious. According to a report by the WEF, a survey among about 9000 young people aged from 16 to 25 from Australia, Brazil, the United Kingdom, China, France, Germany, India, the United States and South Africa shows that nearly 80% of them say they have to acquire new knowledge that they have failed to master at school since technologies are progressing so fast. Seen from the global employment distribution, the replacement of work force by applications of digital technologies will give rise to less accumulation of human resources and thus developing countries with insufficient input in education and training will more easily be faced with the risk of "technical unemployment". In the present education system and model, digital learning, ubiquitous learning and lifelong learning based on man-machine integration and online platforms cannot meet the demand of digital economy development, which reinforces the inadequacy of the laborers' digital literacy and skills.

4. Digital security as a serious challenge to digital economy

Security derived from digital economy is a global challenge and no country can be an exception. Critical infrastructure, from industrial control system and important information system to basic information networks, is directly or indirectly connected with the Internet, where there are an increasing number of threats like viruses, Trojans, hacker intrusions, and DDoS attacks. Finance and energy have become the severely afflicted areas. Cloud data, wearable devices, businesses' mobile devices, smart home devices and smart cars are key targets of attack. Some Internet businesses have some non-standard operations in data sharing, opening and transaction, which has affected national security and the users' rights and interests. In particular, the uncontrolled cross-border flow of data threatens the national data sovereignty.

5.8.2 Bright Future of Digital Economy

1. Universally benefiting and sharing digital economy

The future digital economy will promote everyone into participating in its activities and enjoying the results of universally beneficial and sharing digital economy. The governments of different countries will put more stress on the digital demand of different groups of people, so that children, women, senior citizens, the disabled and low-income groups will share the results of digital economy development without any barrier. The Internet will play a more important role in poverty relief and reduction and more poor people will be able to access to the Internet, through which they can seek to get rid of poverty. The e-government of developing countries will see a new peak of development and information networking will facilitate services of high-quality medical care, education and culture to be put-online so that everyone will enjoy equitable public services. Micro and small businesses will get more equitable opportunities of participating in market competition.

2. Interconnected digital space

Future information infrastructure will be further interconnected, so it will become a reality to establish a global village on the Internet and the information gap between nations, regions and groups of people will be narrowed. Cloud computing centers and big data centers will be improved. Smart perceptive terminals will be fixed to public infrastructure in transportation, energy and environmental protection and this infrastructure will be more intelligent. 4G technology will be applied and popularized at a higher rate and the commercialized use of 5G technology will see great breakthroughs. Wireless networks will witness wider coverage and the IPv6 popularization rate and Internet access coverage rate will be increased. More and more poverty-stricken areas and people will have equitable digital opportunities to access to the Internet. One of the focuses of the world is to provide the least developed countries with universal and affordable access to the Internet.

3. Globally integrated digital market

The future digital markets will be more globally integrated and the barriers in all countries against trade and investment access of digital economy will be removed. More and more Internet platform businesses will get involved in the establishment of global trade and economy cooperation organizations and rules. The mechanism for collaboration in digital economy will be more mature, and international organizations will play a more prominent role as the coordinator. All countries will reach more and more consensuses on information communication, principle formulation, dialogue mechanism establishment and foreign investment examination restriction. Besides, they will keep improving the coordination in key areas such as network infrastructure, digital education, e-commerce, IoT application and cross-border data flow. The conditions will be more mature for facilitating globally integrated digital market.

4. Strongly protected digital security

More and more stress will be put on the protection of digital security. Protection capability for digital economy development will be continuously enhanced. All countries will foster the threat perceptibility and continuous defense capability concerning core technology and equipment of critical information infrastructure in the fields of finance, energy, hydraulic engineering, power, communication, transportation and geographical information. Protection of data security is highly emphasized and national basic data and sensitive information security will become the priority. The mechanism for cross-border data flow security supervision will be improved. To sum up, the cross-national cooperation mechanism for cyber security will keep improving, including the response mechanism for network and information security emergency and the linkage mechanism for cyber security and informatized law enforcement.

Chapter 6
The World e-Government Development

Chinese Academy of Cyberspace Studies

Abstract E-government is the inevitable result of Internet development. The history of the world e-government is also the history of integration between governmental administration innovation and information technologies. After over two decades of development, e-government has improved the administration efficiency of governmental sectors, and facilitated public administration innovation and overall government building, to meet the increasingly diversified and complicated demands of the people. All countries take the development of e-government as an important means of improving the national governance capacity to narrow the digital divide and guarantee sustainable development.

1. Development of e-government

With the improvement of soft and hardware technologies covering infrastructure, computers, the Internet and program development, the leading edge of e-government has developed from the electronic stage to networking, data, and intelligent stages. Correspondingly, the computer-aided handling of governmental affairs has developed into online multi-agent interaction, governmental data application and AI application. The purpose of e-government has shifted from improvement of office work efficiency to facilitation of governmental transparency and service, innovation of digital governance models and means, and building of intelligent government and smart society.

2. General situation

The world e-government is generally developing steadily, but there exists difference between regions. According to an e-government survey made by the UN, from 2003 to 2016, the e-government development index (EGDI) generally kept growing steadily, from 0.402 in 2003 to 0.4922 in 2016, with an annual growth rate of about 1.6%. The overall trend of e-government development in Europe, Asia, Americas, Oceania and Africa is consistent, but there exists obvious difference. Most

Chinese Academy of Cyberspace Studies
Beijing, China

© Publishing House of Electronics Industry, Beijing and Springer-Verlag GmbH Germany, part of Springer Nature 2019

Chinese Academy of Cyberspace Studies (ed.), *World Internet Development Report 2017*, https://doi.org/10.1007/978-3-662-57524-6_6

low-index countries are from Africa, accounting for 81.25% while European countries all became high-EGDI countries in 2014.

3. Development strategies

E-government development has become an important means of different countries to meet the public demand, improve administration efficiency, and build transparent and open government. They have fully realized the importance of e-government, which is supported and led by the central government, who makes scientific and explicit plans for it and raises the implementation of the plans to a strategic level. Such a characteristic is quite obvious in the United States, the United Kingdom, Singapore and Republic of Korea, who have high-level e-government.

4. Transparent and open government

It has become a common practice of all countries to publish governmental information and data on the Internet. According to an e-government survey made by the UN, 193 UN members had set up their governmental websites by 2014, and 106 countries had established open governmental data directories by 2016, the number having increased by 130% in comparison with that in 2014. Meanwhile, a number of governmental sectors and political figures have opened their official accounts on Blog, Weibo, Twitter, Facebook, and WeChat, so that citizens can easily get the dynamic information of their government, and thus their right to know can be more fairly and equitably guaranteed.

5. Service-oriented government

Different countries are utilizing Internet technologies to improve their service system, innovate their service philosophy, integrate service resources, optimize public service and build cross-regional and around-the-clock service-oriented government. Service channels have been diversified; service content has been more practical; service forms have been personalized and service means have been more intelligent.

6. Public participation in networking

It has become an increasingly universal form of interaction between the government and the public to carry out online voting, communication and election campaigns. According to an e-government survey made by the UN, 50% of UN member countries had set up platforms for public opinion feedback by 2014. The number of countries providing first-hand public feedback on public policies through social media, BBS and Internet poll respectively accounted for 71, 51 and 39%. In 2016, 152 of 193 UN member countries (about 80%) had opened social network function on the national portals.

7. Intensive and mobile construction application

The world e-government tends to be intensive and mobile. Repeated construction and investment waste are avoided through cloud computing, and e-government is shifting from discretization to intensive construction, which has helped to reduce the cost of construction and operation and maintenance. Public service channels are shifting from PC terminals to mobile phones, tablet devices and wearable devices.

8. **Major challenges**

Challenges to the world e-government development are digital divide, information security, privacy protection and information islands. Major barriers hindering the exertion of the potential of e-government include digital divide resulting from lack of access and know-how, imperfectness of systems, cyber security resulting from occurrence of new situations and problems, privacy protection problems resulting from frequent incidents and imperfectness of laws, and information islands resulting from system isolation and fragmented administration. All of these challenges must be overcome in realizing the goal of sustainable development.

6.1   General Situation of e-Government Development

The invention and popularization of the computer and Internet have led to the information wave across the world and provide technical support for the electronic and networking development of e-government. The concept of "governance-oriented government", which is different from the traditional concept of "management-oriented government", has been proposed and widely accepted and it has become the theoretical basis and practical goal for innovating the governmental work mode. Against this backdrop, e-government has emerged.

6.1.1   History of e-Government

The concept of "e-government" originated from the United States. In 1993, the Clinton Government launched the report *Reengineering Through Information Technology*, in which the e-government program was put forward. Then the concept of "e-government" was accepted by other countries and became a terminology describing the integration between the Internet and information technology and governmental work. According to David L. McClure, an Associate Director of the U.S. General Accounting Office at that time, e-government refers to government's use of technology, particularly web-based Internet applications, to enhance the access to and delivery of government information and service to citizens, business partners, employees, other agencies and government entities. It has the potential to help build better relationships between government and the public by making interaction with citizens smoother, easier, and more efficient.[1] In the UN *Global*

---

[1]Layne K, Lee J. Developing fully functional E-government: A four stage model. *Government Information Quarterly*, 2001, 18(2), pp. 22–136.

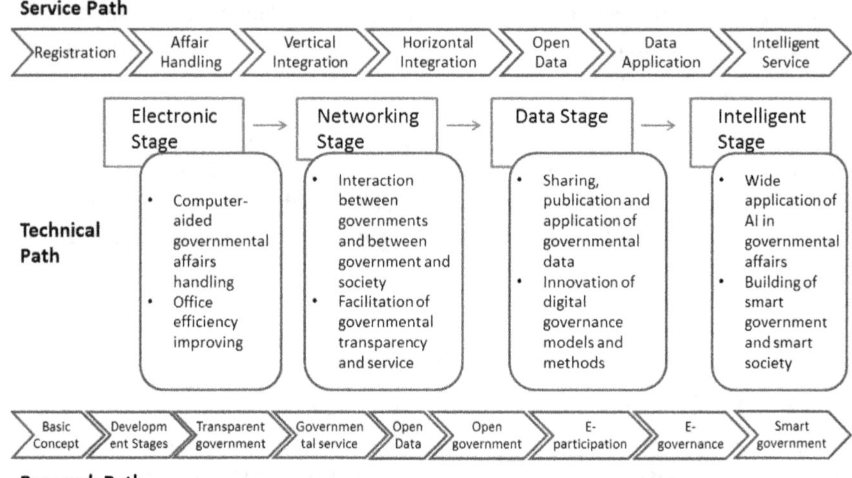

Fig. 6.1 Evolution of the world e-government

E-government Readiness Report 2004: Towards Access for Opportunity, e-government is defined as government's use of information technology to provide the public with information and basic public service, which is an early but complete and concise definition of e-government.[2]

With the improvement of software and hardware technologies covering infrastructure, computers, the Internet and program development, the leading edge of e-government has developed from the electronic stage to the networking, data and intelligent stages (see Fig. 6.1). Correspondingly, the computer-aided handling of governmental affairs has developed into online multi-agent interaction, governmental data application and AI application. The purpose of e-government has shifted from improvement of office work efficiency to facilitation of governmental transparency and service, innovation of digital governance models and means and building of intelligent government and smart society. Through the analysis of the four stages, we can trace the detailed history and general situation of e-government.

1. Electronic stage

In 1980s, with the progress of semiconductor technology and the application and popularization of the Internet, PC was invented and governments of developed countries represented by the United States began the computer-aided handling of government affairs, which marked the beginning of the world e-government development. At the electronic stage, e-government was only adopted in a few developed countries, where the traditional government affairs were handled by

---

[2]Meng Qingguo & Fan Bo. *Theory and Practice of E-government*. Beijing: Tsinghua University Press, 2006.

computers. The computing capacity higher than that of man force and the data storage density higher than that of paper files helped to improve the office efficiency.

2. **Networking stage**

The birth of World Wide Web (www) suggested that e-government entered the networking stage. As the office computers of the government were connected into the Internet, governmental employees began to do much work on the Internet, such as cross-departmental information transmission, instant messaging, development of online government portals, online publication of governmental information and online services. Thus the world e-government developed from the electronic stage to the networking stage, when governmental sectors in developed countries were able to succeed in overall Internet coverage of all office work and more and more developing countries began to know about and adopted e-government, thanks to the maturity of the Internet and the rapid development of computer technology. At this stage, there were a lot of innovations in e-government, almost covering all of its areas. The innovations could be categorized into two kinds: governmental information publication and online governmental work, including information publication, business collaboration, online service and social involvement in policy making, all based on the network. In general, the e-government development in most countries today is at this stage. For developed countries, and a few developing countries represented by China, e-government networking is already very developed.

3. **Data stage**

E-government has witnessed rapid expansion and adoption since the beginning of the 21st century and all countries have accumulated rich governmental and social data. Against this backdrop, the countries with mature networking e-government have launched the sharing, openness and application of governmental data, seeking for the data governance model of governmental affairs, which indicates that the world e-government has entered the data stage. As for data sharing, the data barriers between governmental sectors and between government and society have been removed and the "information islands" within the government has been changed, which has fostered the accuracy, timeliness and comprehensiveness and laid solid foundation for the social application of governmental information. As for data publication, more and more governmental data are published and social information is absorbed by the government, so the transparency and service of the government are being enhanced. As for data application, the great commercial and public value in governmental data has been verified by more and more practices. Through the efficient use of governmental data, governmental agencies have not only improved their e-government capability but also promoted the economic growth, and the progress of the social governance model.

4. **Intelligent stage**

At the intelligent stage, e-government is provided with governmental and social data and their application methods. The progress of information technology has improved AI technology and the development of the world's economy and society has brought about higher demand for the working capacity of the governments of all countries. The combination big data, AI and governmental has enabled the practice of e-government to develop from the thriving data stage to the intelligent stage represented by the application of AI in governmental work, with the purpose of building smart government and society. E-government at the intelligent stage covers intelligent decision making, smart management, precision service and production innovation. The government at this stage will undergo dramatic reforms in the management system, promoting the governmental process reengineering and organizational restructure and even the system revolution. E-government of the United States, the United Kingdom, France, Japan, Republic of Korea, Singapore and China is developing into the intelligent stage and manifesting their own characteristics in different fields. These countries are seeking to make leading-edge innovations in e-government in their own political system and on their own development road.

6.1.2   Development Trend and Comparative Analysis

In the world, e-government has witnessed two decades of development. From 2001 to 2016, UN Department of Economic and Social Affairs released nine survey reports on the global e-government development. The analysis of the data from the surveys can reflect systematically the trend of global e-government development. According to the surveys, from 2003 to 2016, all countries are proceeding their e-government building, so the e-government development index (EDGI) keeps increasing generally, from 0.402 in 2003 to 0.4922 in 2016, with an annual growth rate of about 1.6%,[3, 4] (see Fig. 6.2).

Since 2010, in the UN e-government survey indicators, the E-government Readiness Index (EGRI) has been shifted into E-government Development Index (EGDI), in which the e-government development is assessed from three aspects, namely, online service, telecommunication facilities and HR cost. From 2010 to 2016, the result of development in different regions (including Europe, Asia, America, Oceania and Africa) tended to be consistent (see Fig. 6.3). E-government level in Europe, Americas and Asia was higher than the world's average and that in Oceania and Africa is lower than the average. In 2012, the EGDI of Europe reached 0.7 for the first time and that of other regions was less than 0.6, so there is an obvious difference.

Since 2010, online governmental service of different regions has been dramatically improved. The annual growth rate of the online service score has amounted to

---

[3]United Nations. UN Global E-government Survey 2003: E-Government at the Crossroads. 2003.

[4]United Nations. *United Nations E-Government Survey 2016: E-Government in support of sustainable developments.* 2016.

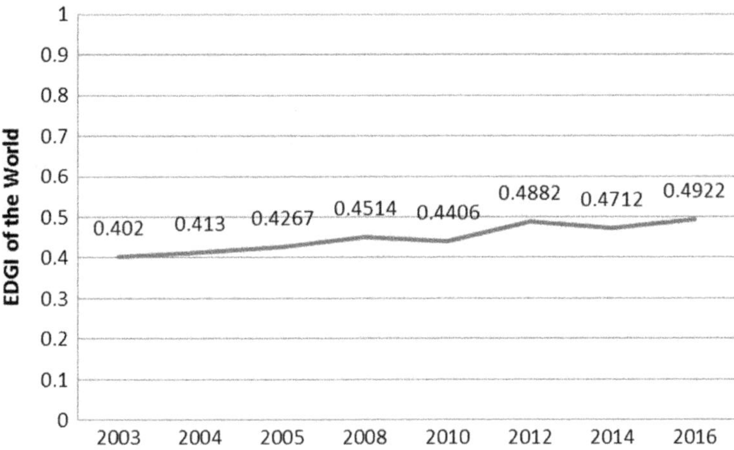

Fig. 6.2   The World e-government development (2003–2016)

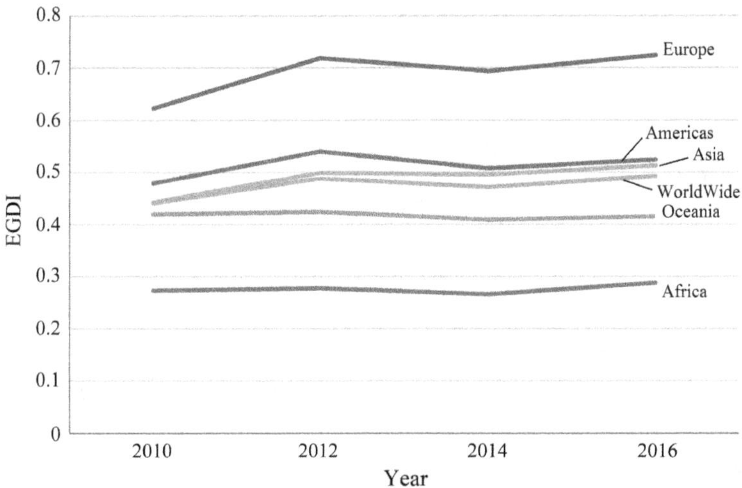

Fig. 6.3   EGDI of Different regions in the world (2010–2016)

29%. From 2014 to 2016, Africa and Americas were obviously catching up (see Fig. 6.4).

Compared with online service, the score for telecommunication facilities was lower. Africa saw the most rapid increase of the score, with the annual increase rate amounting to 41%, followed by Asia and Oceania, whose annual increase rate of the score was 30% (see Fig. 6.5). In 2016, the score for Asian telecommunication facilities surpassed the world's average level.

The index of Human Capital in different regions was higher than that of online service and telecommunication facilities. It was over 0.6 (see Fig. 6.6) on average

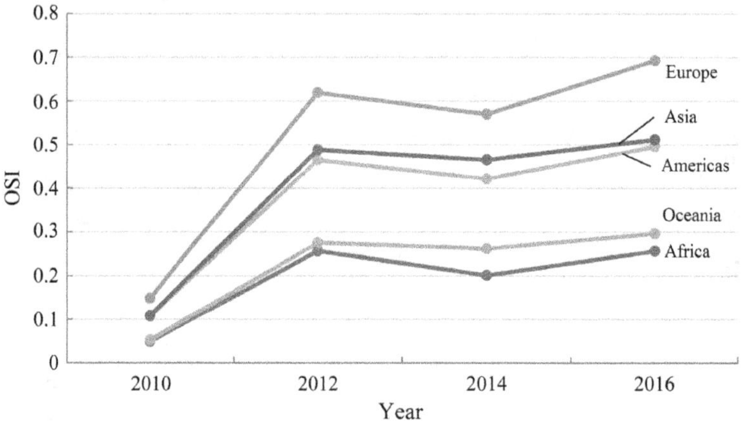

Fig. 6.4 Online service index (OSI) of different regions in the world (2010–2016)

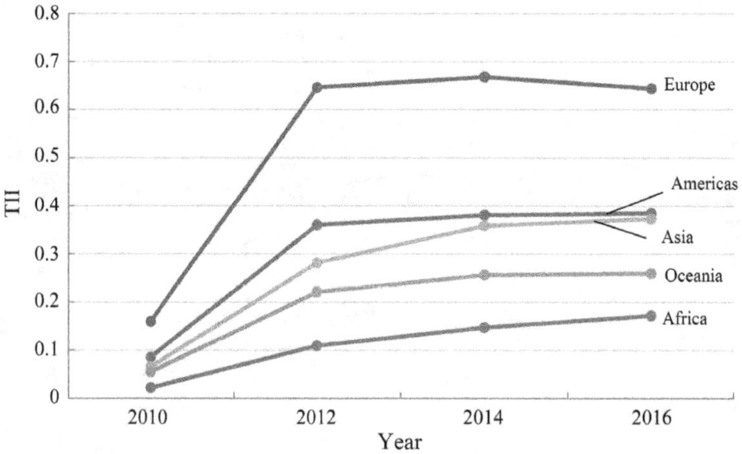

Fig. 6.5 Telecommunication infrastructure index (TII) of e-government in different regions of the world (2010–2016)

from 2012 to 2016. The increase in different regions was consistent. The annual growth rate in Europe was comparatively high, amounting to 18%; that in Africa was low, amounting to 13%. From 2014 to 2016, only the score in Africa was lower than the world's average.

Seen from the change in the past over 10 years in the ranking of top five countries in terms of EGDI in 2016, the leading countries were not always the same. In 2003, the first two countries were the United States and Denmark but in 2016, they only ranked 12th and 9th (see Fig. 6.7).

Due to the high marginal cost of e-government input in countries with a large population, their EGDI is, to some extent, lower than that of other countries with

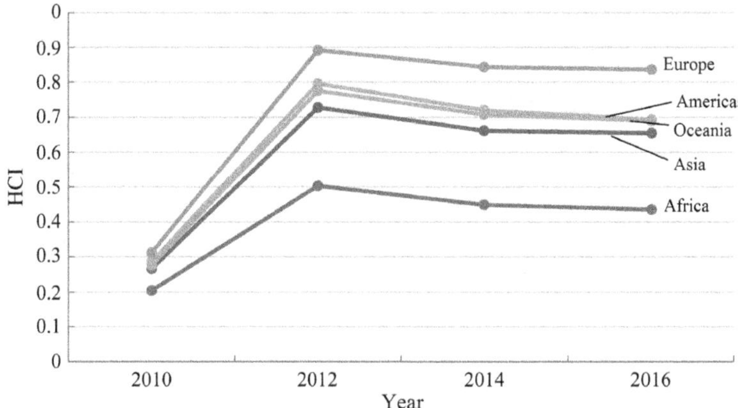

Fig. 6.6 Human capital index (HCI) of e-government in different regions of the world (2010–2016)

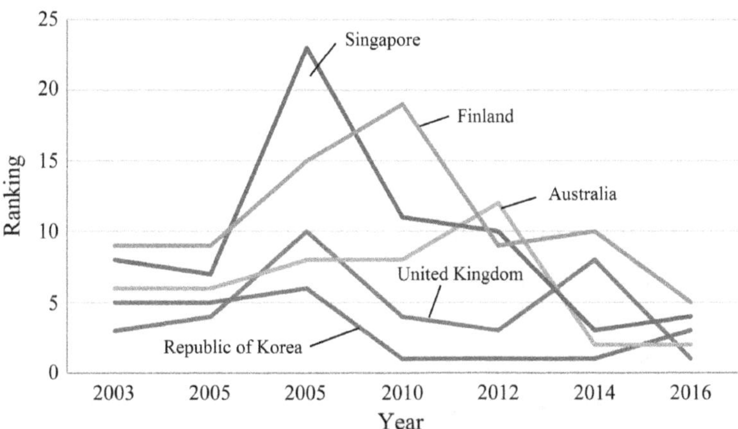

Fig. 6.7 Diachronical change in ranking of top countries in terms of EGDI (2003–2016)

the same economic development level (see Fig. 6.8). For instance, Japan and the United States rank 11th and 12th, lagging behind other major developed countries. Russia has seen rapid progress in e-government development in recent years. In 2010, it ranked 59th but in the recent three surveys, it rose to the 20–30th. China, Mexico and Brazil were in the same range, from 50th to 70th while Indonesia, India, Pakistan, Bangladesh and Nigeria rank beyond the 100th.

The analysis of the countries' distribution at different stages of e-government development (see Fig. 6.9) shows that from 2014 to 2016, there was no structural change. Take the year of 2016 as an example. Among the countries with low EGDI, African ones accounted for the most, about 81.25%; those of the middle level of EGDI were mostly countries from Africa, Asia and Americas, with almost the same

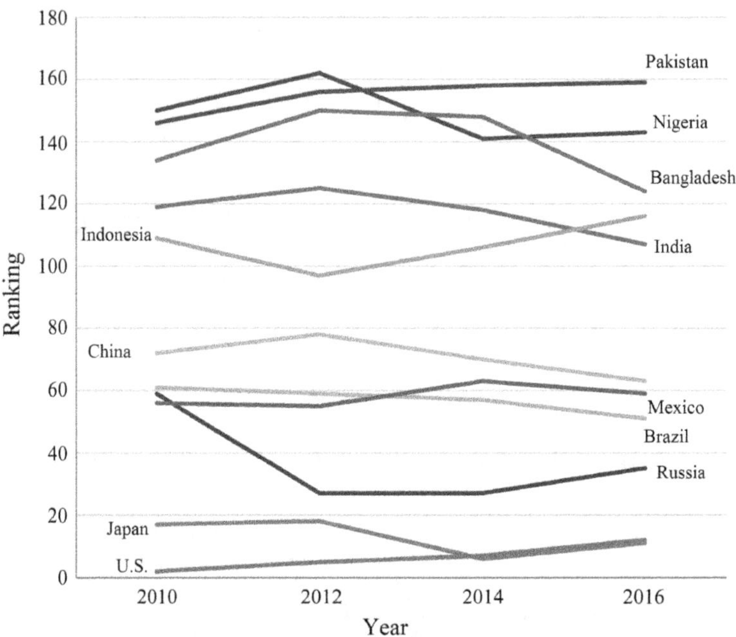

Fig. 6.8 E-government ranking of countries with a large population (2003–2016)

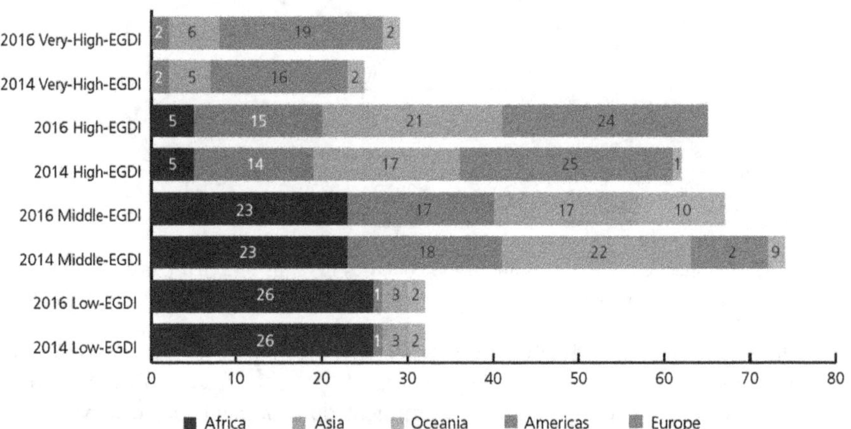

Fig. 6.9 Distribution of the continents in high EGDI (2014–2016)

percentage; those with high EGDI were mostly European and Asian countries, accounting respectively for 36.92 and 32.31%; those with extremely high EGDI were mostly European countries, accounting for the most, about 65.52%. Notably, since 2014, all European countries have been among those with high EGDI.

The development of e-government is directly related to a country's income. The analysis of three first-level EGDI indicators, namely, Online Service Index (OSI), Telecommunication Infrastructure Index (TII), Human Capital Index (HCI), shows that high-income countries and medium and high-income countries got basically the same scores in the three indicators (see Fig. 6.10). It is concluded that when the countries' income reaches a certain standard, their e-government development tends to be at the same level while there exists obvious difference in the three indicators among high-income countries, low and medium income countries and low-income countries.

In the UN e-government survey, the countries' e-participation is assessed from three respects, namely, e-information, e-consulting and e-decision-making. The result shows that from 2010 to 2016, the e-participation index of different regions kept increasing (see Fig. 6.11). Europe, Asia and Americas had an index higher than the world's average. Europe took the lead. In 2016, its e-participation index

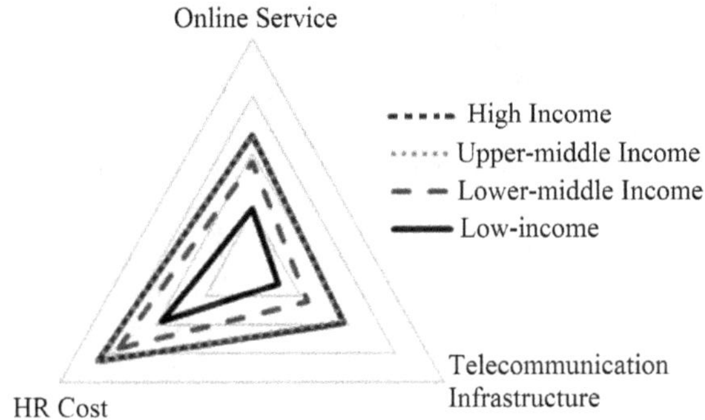

Fig. 6.10 Distribution of countries of different income in the e-government indicators

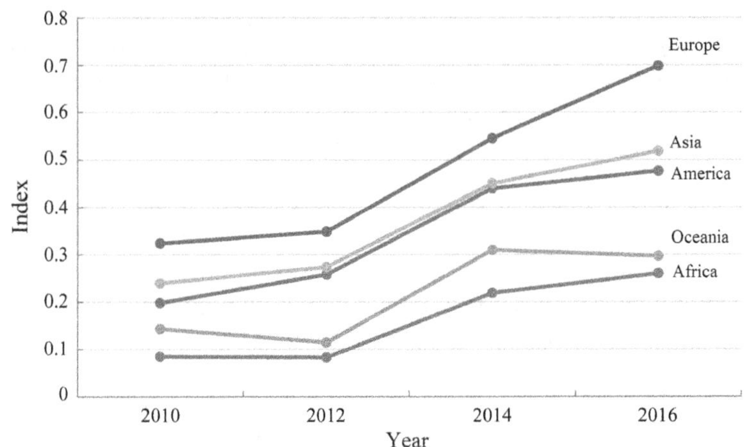

Fig. 6.11 E-participation index of different regions (2010–2016)

was 151% of the world's average. That of Africa and Oceania was always lower than the average, but the difference tended to decrease. In 2012, the e-participation index of the two continents was respectively 37 and 52% of the world's average, but it rose to 56 and 64% in 2016.

6.1.3   e-Government Development Strategies of Major Countries

The analysis of the development of countries with leading e-government shows that their governments have all realized the importance of e-government, which is universally supported and led by the central government, who makes scientific and explicit development plans for e-government and raises the implementation of the plans to a strategic height. For example, in the United Sates, the United Kingdom, Singapore and Republic of Korea, the e-government level is high.

The United States is the earliest and the most mature country in terms of e-government. In early 1990s, the country began its e-government building. In 1994, the federal government defined the objective for e-government building as "to build the customer-oriented e-government". In 2001, Office of Management and Budget announced the establishment of the E-government Special Working Unit and issued *E-Government Strategy* and signed *E-Government Act* in 2002. In 2006, the country issued *Privacy Act, Freedom of Information Act* and *Federal Information Resources Management Act*. The Obama Government signed *Memorandum on Transparency and Open Government* and issued *Open Government Directive* in 2009, and *Federal Cloud Computing Strategy* in 2011, *Digital Government:Building A 21st Century Platform To Better Serve The American People* in 2012 and Big Data Research and Development Initiative in 2014. In May 2017, President Trump signed the *Presidential Executive Order on Strengthening the Cybersecurity of Federal Networks and Critical Infrastructure* and the House of Representative passed the *Modernizing Government Technology Act* for the building of new digital service fundamentally enhancing the taxpayer' experience and improving the overall image of the government.

E-government in the United Kingdom began later than in the United States. The UK Government proposed the laboratory plan "Government Information Service" in 1994, and issued *Data Protection Act* in 1998, which identifies the citizens' legal rights to obtain information and data. It issued *Modernizing Government White Paper* in 1999 and formulated the strategy of facilitating the government reform through technology in 2005. The ICT strategy launched in 2010 was aimed to integrate and build the network platforms of public sectors, providing them with seamless speech and data transmission services. In 2012, the UK Government issued the *White Paper on Open Government* to implement the policies on open data, and in the same year, it released *Open Government Partnership: UK National Action Plan 2013–2015*, which takes the initiative opening of the government-owned data set as the action goal of the country. In 2015, the Cabinet Office's the National Information Infrastructure (NII) Implementation Document[5] further pointed out the re-use of data of strategic significance. In May 2016, the Office issued the third *Open Government Partnership: UN National Action Plan 2016–2018* and other policy documents to make governmental data transparent, and easy to find, store and visit[6]. The e-government of the United Kingdom now is on the list of leading e-governments thanks to the strong leadership, complete organization and cross-sector collaboration and information sharing.

As one of the earliest countries implementing "governmental informatization", Singapore launched the Nation IT Program in 1986, which was aimed to introduce advanced network technologies to provide the public with one-stop services. It launched the IT2000 Program in 1992 to build itself into a global IT center, and the Infocomm 21 Program in 1999 to facilitate the government's use of ICT to better serve the public. In 2006, the country launched the Intelligent Nation 2015 Program to enhance its competitive strength and innovative capability; in 2014, it issued the Intelligent Nation 2025 Program to construct the infrastructure of data collection, connection and analysis covering the country so as to predict the public demand and hence provide quality service based on data. In 2016, Agency for Science, Technology and Research of Singapore was founded, aiming to lead digital reforms of governmental sectors with ICT[7] and to cooperate with other agencies in launching new citizen-centered service. By setting clear and long-term development goals, Singapore has facilitated the cooperation between the government and businesses and is building the integrated government based on a full understanding of the public demand to provide information for the public in a collaborative way.

Republic of Korea has witnessed the rapid development of e-government. The government formulated the *Basic Plan for Information Promotion* in 1996 to facilitate the development of e-government and launched the Cyber Korea 21 Program in 1999 to improve the governmental service. In 2001, the E-government Special Committee was founded accountable for the promotion and application of e-government throughout the country. The GEA was launched in 2003 for the building of the integrated service platform serving the public, businesses and governmental agencies. In 2004, the country proposed the u-Korea Strategy to make e-government available everywhere and in 2009, announced the *Program of Comprehensive Revitalization through Cloud Computing*, stressing the prior use of cloud computing in governmental sectors. The Republic of Korea Public Administration and Security Department formulated in 2011 the *Implementation Plan for the Intelligent Government*, aiming to build the intelligent government in 2015. The plan covered four aspects, namely, openness, integration, collaboration and sustainable green growth. In 2012, the National Science and Technology Committee released the *Big Data Strategic Planning*, which received active response from the government. In 2013, the Committee proposed the national-level big data integration and development plan, including the support for big data businesses, so Republic of Korea has become the earliest Asian country in launching big data strategies.[5]

---

[5]Wang Jingxuan & Yang Daoling: Development Trend and Experience of International E-government Development, *Electronic Government*, 2015(4), pp. 24–30.

6.2 Innovations in e-Government of Major Countries

According to *United Nations E-Government Survey 2016*, e-government has become an important indicator of the sustainable development capacity of all the countries.[6] In the innovations of different countries, they tend to pursue the public-centered philosophy and construct transparent and open online government and cross-regional and around-the-clock service government model. Besides, they enhance the interaction between the government and the public as well as the public participation. It has become the consensus and goal of all governments to keep innovating the e-government building and application model.

6.2.1 Online Government that is Transparent and Open

1. Openness facilitating the performance of duties

According to the United Nations E-Government Survey 2014, the 193 UN members have all set up their governmental websites. Over 70% of these websites put stress on the optimization of information resources, organizing resources in accordance with the targeted users, application themes, hotspots and organizational structures, and thus improving the transparency of governmental affairs.[7] www.gov. cn in China integrates the open directory system of governmental information and releases all documents formulated by the Central Government of the People's Republic of China. Each document has the simultaneous link with relevant interpretations in the form of graphics, audio and video materials introducing the background and problems to be solved, and contents related to the interests of businesses and the public (see Fig. 6.12). Meanwhile, the website releases the power list of the departments of the Central Government, so that the administrative licensing power of every department can be found quickly (see Fig. 6.13). For every item of administrative licensing, the basis, service guide and examination and approval procedure are provided, so the website plays a monitoring role in transparent and standard administration in accordance with laws.

Republic of Korea, Japan, France, Finland, Demark and New Zealand, centering on the user's demand, provide information on key affairs for all groups of users. For example, they provide guidance for individuals, covering passports, driving licenses, tax payment, job hunting, social security, education aid, occupational qualifications verification. They also provide corporations with information concerning regulations, covering registration for business, recruitment, foreign trade, and application for permits. Through the governmental website, the government's

---

[6]United Nations. *United Nations E-Government Survey 2016: E-Government in Support of Sustainable Development*. 2016.

[7]United Nations. *United Nations E-Government Survey 2014: E-Government For the Future We Want*. 2014.

Fig. 6.12 Open Directory of governmental information on www.gov.cn

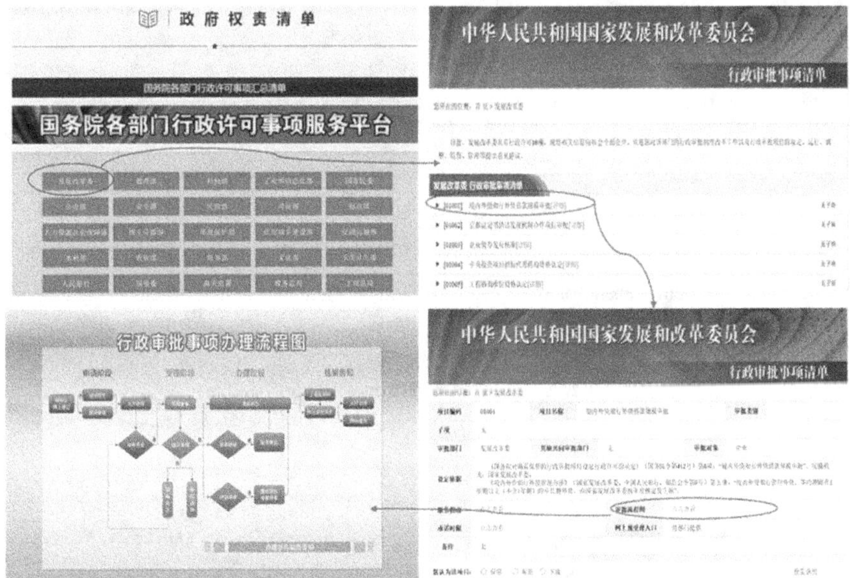

Fig. 6.13 Power List of the departments released on www.gov.cn

work and all policies are shown to the public, so that the work and policies can be assessed and monitored by society (see Fig. 6.14).

2. **Transparency ensuring equity**

Many governmental departments and political figures have opened their accounts on official blogs, Weibo, Twitter, Facebook and WeChat, so that citizens can learn about the dynamic information more easily and timely, and thus their right to know is guaranteed more equitably.

The White House of the United States has opened the official account on Flicker, YouTube, MySpace, Twitter and Facebook, copying or linking to www.white-house.gov (see Fig. 6.15). Thanks to the social media platforms, the important work

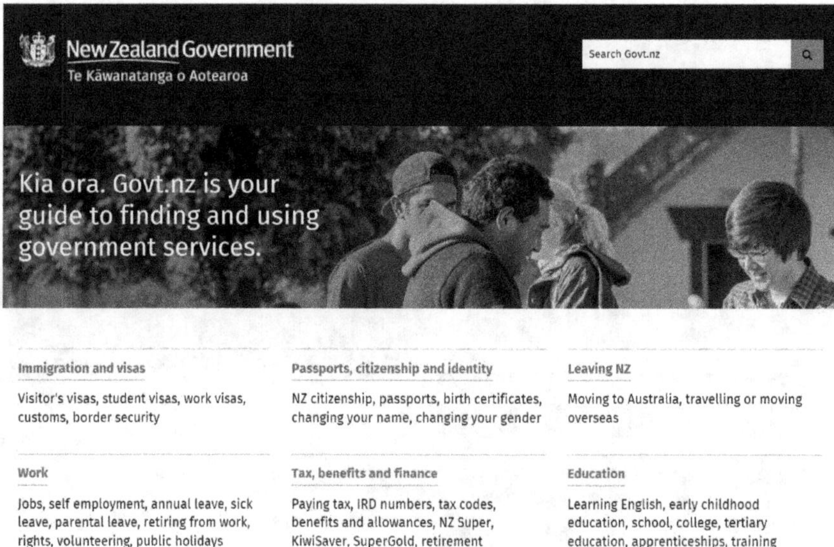

Fig. 6.14 Government website of New Zealand

Fig. 6.15 Website of the U.S. White House

of the White House can be spread across the globe every day and the latest information can be sent automatically to the users' email addresses.

Russia, Germany, France, and Chile use Twitter or Facebook to release governmental information and announcements. Some governmental agencies also take Twitter and Facebook as the tools of communication with the public.

Governmental websites of the United States, the United Kingdom, Malaysia, Singapore and Hong Kong (China) foster their connection and sharing with social platforms, so that the users can reprint and share the content that they are interested into the third-party platform and thus the governmental information can be spread widely. The users can also create their own homepage by using the sharing function of the governmental websites and thus gather the information and service on different governmental websites into their own account on the social platforms, which has helped the public to obtain and use the governmental information more conveniently. For instance, the Singaporean Government provides thousands of services through the eCitizen Platform and the users can share the pages that they are interested into scores of other platforms like Baidu, Bebo, Blip, Blogger and Diigo (see Fig. 6.16).

3. **Openness facilitating application**

Data have become the strategic property of all countries, who have carried out the exploration in opening their data. By setting up data opening platforms, the governments can meet the demand of the public and businesses to obtain, handle and use information and thus foster the transparency of government affairs and the public trust in their governments, which, in turn, boosts the innovation potential of society. According to the *United Nations E-Government Survey 2016*, 106 countries offer Open Government Data catalogues in 2016 (see Fig. 6.17), accounting for 54.9% of the 193 UN member countries, the number increasing by 130% in

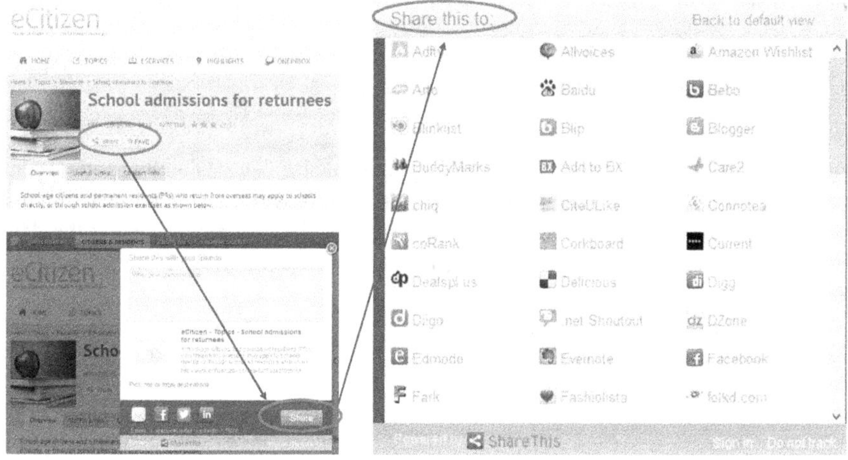

Fig. 6.16 Content sharing of the eCitizen platform of Singapore

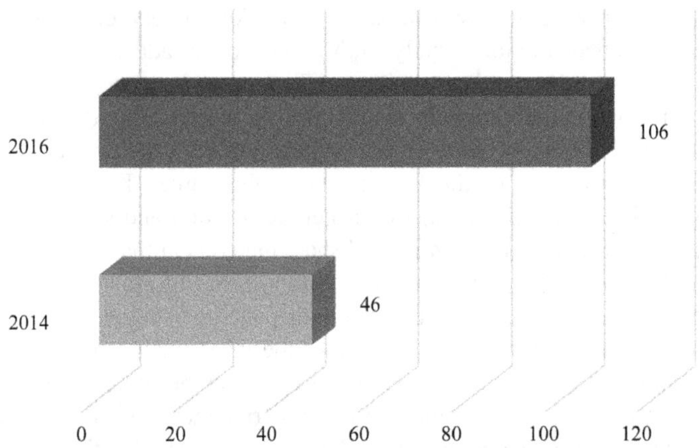

Fig. 6.17 Number of countries having set up open directories of governmental data

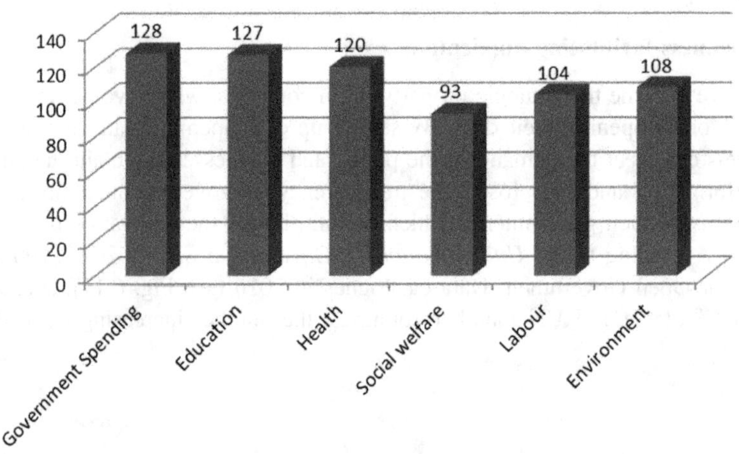

Fig. 6.18 Number of countries offering datasets, by sector

comparison with that in 2014; and 128 countries release datasets about government spending, which enabled the public to monitor the government's budget making and implementation. Besides, 127, 120, 93, 104 and 108 countries opened the data sets of education, health, social welfare, labour and environment respectively[8] (see Fig. 6.18).

The U.S. open data website (Data.gov, See Fig. 6.19) has released 196,904 governmental datasets, covering Raw Data, Geography Data and Tool Directory

---

[8]United Nations. *United Nations E-Government Survey 2016: E-Government in Support of Sustainable Development*. 2016.

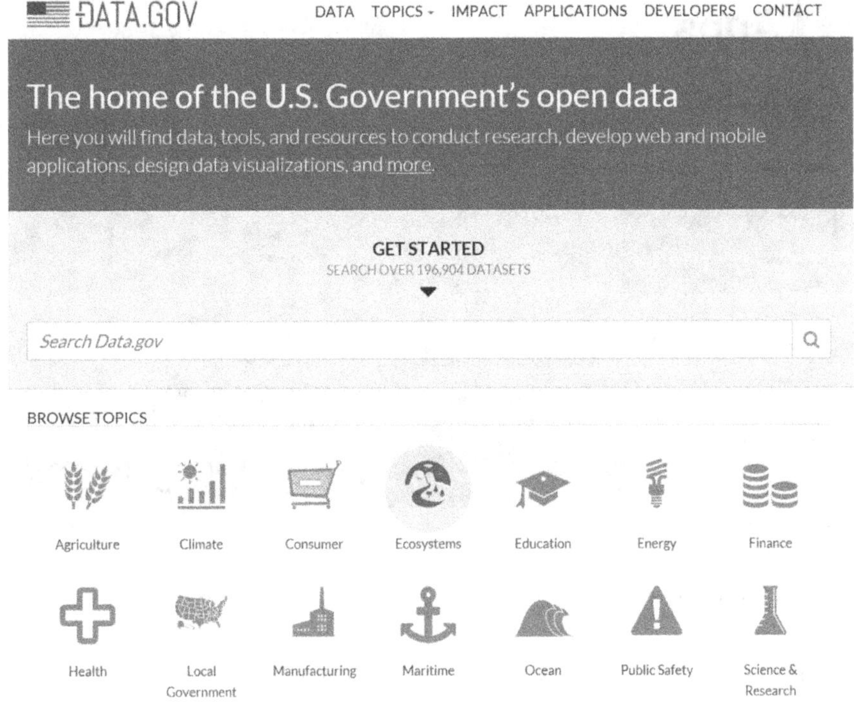

Fig. 6.19 Open data websites in the United States

with the topics of agriculture, climate, consumer, education, energy, health, manufacturing and the ocean. The website also provides tools like virtual interaction, data mining and extraction and RSS customization, which provide convenience for the users in searching for and customizing the data resources that they are interested in. These data, to some extent, can guide the businesses' investment, inspire innovation, increase job opportunities and facilitate economic growth. For instance, businesses can use them to improve their operation, expand their markets and develop new products and services; financial companies can use them to assess credit, guide investment, and quantize investment risks; and transportation and logistics businesses can use them to improve their transportation efficiency and security.

Data.gov.uk releases over 40,000 datasets, involving all areas related to the production of businesses and life of the public. For example, based on traffic accident data, it has developed and released traffic accident search service (see

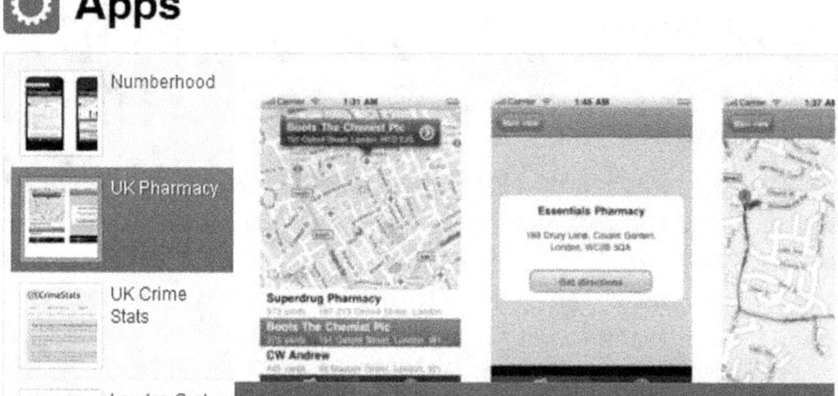

Fig. 6.20 Traffic accident search service on the UK governmental website

Fig. 6.20). The spots and frequency of traffic accidents can be shown in the e-maps of cities to provide aid for transportation security.

Canada's centralized web portal of open data is opendat.gc.ca, which releases basic urban information resources covering demographic statistics, financial budget data, immigration trend data, and employment market data as well as professional data resources covering air pollution distribution, trash recycling, water resources quality inspection report, real-time traffic and road maintenance. The information system in Fig. 6.21 integrates all effective information about the projects aided by relevant research funds available in all health departments of Canada for the convenience of professional institutions and staff in carrying out their research.

The national data platform (Data.gov.sg, See Fig. 6.22) of Singapore provides thousands of types of data of different fields, covering transportation, tourism and environment and other social services to encourage governmental agencies, social groups and individuals to use them to develop applications and offer services to society. Therefore, social organizations and the public enjoy not only the right to browse and search the data, but also the right to use the data to make valued-added development.

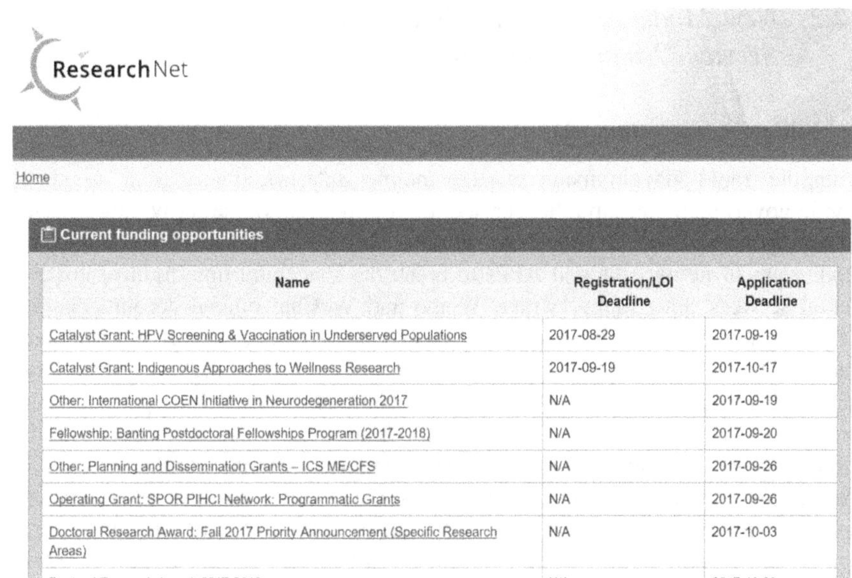

Fig. 6.21   Projects aided by relevant research funds available in all health departments of Canada

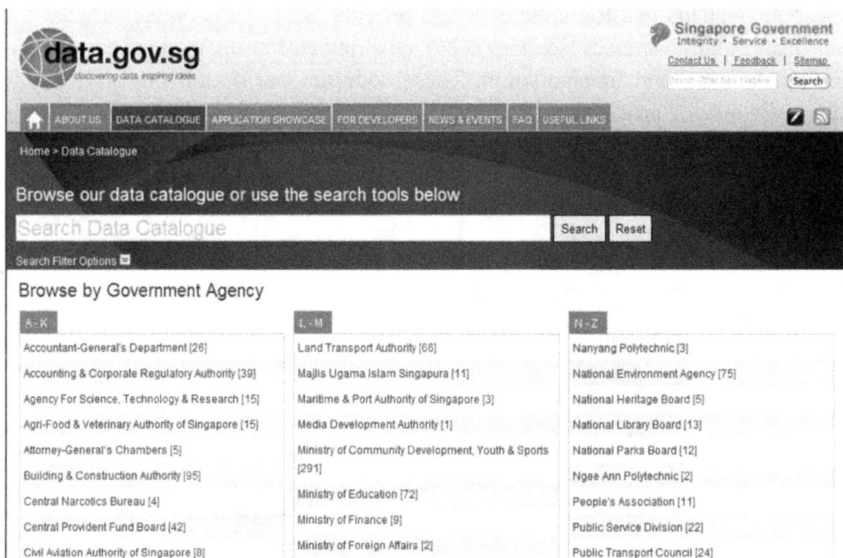

Fig. 6.22   National data platform of Singapore

6.2.2   New Model of Cross-Regional and Around-the-Clock Service-Oriented Government

1. Multiple service channels

With the rapid development of the mobile Internet, there is a trend of mobile-government and public service in the world. While providing service through governmental websites and hotlines, all governments in the world are making use of new media and APPs to break the space and time limit of governmental services. Facebook, Twitter, Weibo and WeChat official account play an increasingly important role in releasing official information, spreading governance ideas and communicating with young Internet users thanks to the fast spread, large scale and strong interaction, so they have become new channels of governmental services. For instance, American President Trump directly communicates with the public on Twitter, which then has become an important channel for the president to produce his influence and perform his duties. China has witnessed rapid development in mobile government. By the end of 2016, the number of verified governmental accounts on Sina Weibo had reached 164,522, having increased by 8.0% in comparison with that in 2015, the accounts covering government agencies, judicial departments and public servants (see Fig. 6.23).

2. Practical service content

Countries leading in e-government such as the United Kingdom, Australia, Singapore, Canada and Republic of Korea provide full-process online services for their people and businesses (see Fig. 6.24), covering education, medical care, social security, housing and transportation. These countries are optimizing and simplifying their service procedures, promoting online service of all government affairs and providing one-stop cross-sector, cross-level and cross-sphere online services

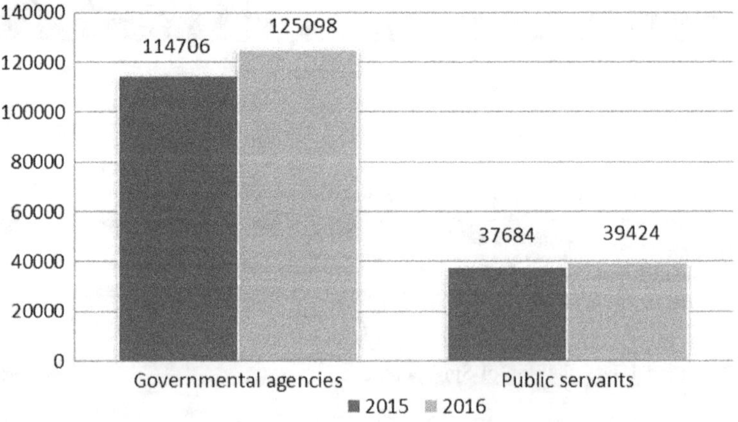

Fig. 6.23 Distribution of China's governmental weibo accounts (*Source* Sina Weibo)

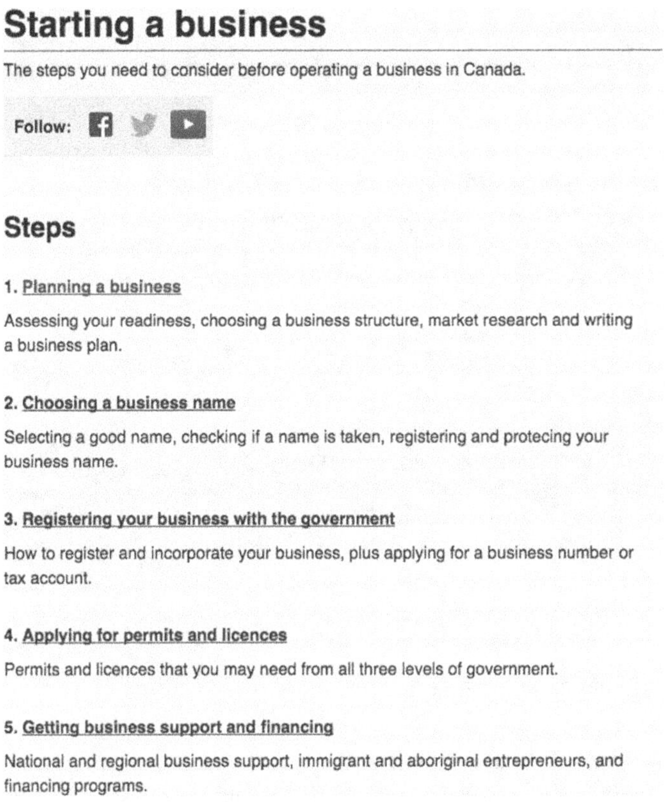

Starting a business

The steps you need to consider before operating a business in Canada.

Follow:

Steps

1. Planning a business

Assessing your readiness, choosing a business structure, market research and writing a business plan.

2. Choosing a business name

Selecting a good name, checking if a name is taken, registering and protecing your business name.

3. Registering your business with the government

How to register and incorporate your business, plus applying for a business number or tax account.

4. Applying for permits and licences

Permits and licences that you may need from all three levels of government.

5. Getting business support and financing

National and regional business support, immigrant and aboriginal entrepreneurs, and financing programs.

Fig. 6.24 Full-process online services of Canadian governmental website

concerning ID card application, driving license application, vehicle registration, environmental approval application, online payment of penalty, and social security application. In 2016, the Chinese Government issued *Guidelines on Speeding up the Internet Plus E-government*, which proposed that the integrated online governmental service platforms should be built by the end of 2017 and thus all the governmental services would be opened to the public. By the end of 2020, the country will have built one-stop Internet Plus Governmental Service System that is interconnected, collaborative and provincially coordinated to improve smart governmental service (Fig. 6.25).

3. **Personalized service forms**

With the deepening of e-government, the governments of different countries are innovating their governmental service technologies, concepts and thinking and providing increasingly personalized services, such as scenario service, customized services and personal homepage, to meet the diversified demands of the public. The Singaporean Government has set up the E-Citizen Center based on the SingPass ID

2017年各地区开展"互联网+政务服务"工作主要目标任务汇总表

| 单位\指标 | 一体化政务服务平台 | | | | 政务服务事项目录 | | | | | | | | 绩效考核机制 | | | 是否在省（区、市）政府门户网站开设专题 |
	政务服务平台是否唯一权威	覆盖层级	省本级部门入驻数	是否与省（区、市）政府门户网站前端整合	是否在省（区、市）政府门户网站集中发布	是否统一本省（区、市）目录编制标准	是否实现同一事项同一名称	是否提供全部事项办事指南	办事指南是否与政务服务事项关联	是否统一办事指南要素标准	是否开展评估考核	是否纳入年度政府绩效考核体系	是否公开考核结果	
实现地区数	31			30	31	30	25	31	31	31	28	28	24	28
实现占比	96.88%			93.75%	96.88%	93.75%	78.13%	96.88%	96.88%	96.88%	87.50%	87.50%	75.00%	87.50%
北京	是	2	40	是	是	是	是	是	是	是	是	是	否	是

Fig. 6.25 Goals of China's internet plus governmental services

Verification Control System, so that the users can have services customized, including work, life, entertainment, learning and business. The Nottingham Government in the United Kingdom has opened "#My Nottingham", containing personalized service areas of five aspects, namely, "My city council", "My leisure time", "My job", "My news" and "My sports", to make it convenient for the users to obtain the targeted information and service (see Fig. 6.26). Nanjing Municipal Government of China has developed "My Nanjing" APP, which integrates governmental and public services resources, providing the public and businesses with a large number of practical and personalized services (see Fig. 6.27), which cover social security, housing fund, regulation breach, driving license, utility, road monitoring, real-time traffic and medical appointment registration.

4. Intelligent service

With the rapid development of cloud computing, big data, IoT and smart city, e-government tends to be more intelligent. Governments of different countries taking the improvement of the citizens' life as the core goal are developing and applying new public service products to improve their service quality, starting with the areas like transportation, environment and social security. Through the sensors installed on lamp posts on the roads, the Chicago Government of the United States can collect the city road information and environment data, including air quality, noise, light intensity, temperature and wind velocity and thus provide data support for public

Fig. 6.26   #My Nottingham of the Nottingham government in the United Kingdom

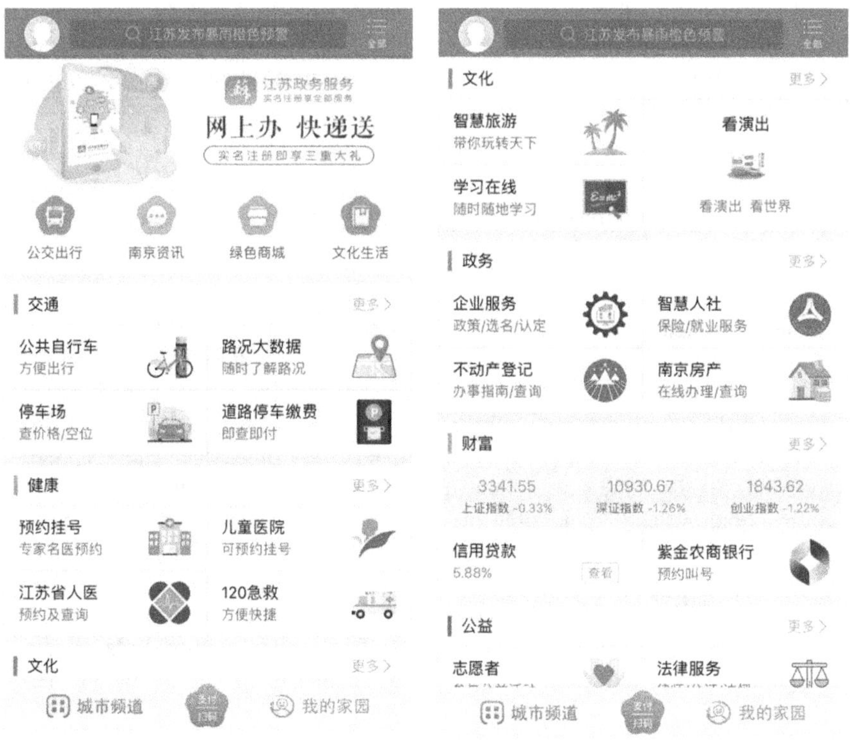

Fig. 6.27   My Nanjing APP

service. The Gloucester Government of the United Kingdom has carried out the "Smart House" pilot projects, in which smart household devices can be controlled automatically in accordance with the environmental conditions and inhouse activities can be monitored through technologies like the inductive IR cushion to meet the demand of senior citizens, whose heart rate and blood pressure data can be obtained through wearable devices and transmitted to medical institutions.

6.2.3 Government-Citizen Interaction and Public Participation in the Internet Era

1. Online voting

Online voting is a common form of e-democracy, including online election and online obtaining of the public opinions on a policy or action, which has enriched the channels for the public participation in politics and for information sharing, and thus meet the demand of the public and make the policies and actions more scientific and efficient. South Australia State of Australia has launched the website YourSAy, which collects the public opinions on the government. The public can participate in online discussion, vote, and decide how to spend the governmental funds. Since its establishment, 34,298 users have registered, 1.2 million Australian dollars have been appropriated to the communities in need and 172 governmental decisions have been made.

Column 6 Discussion on YourSAy about the Time Zone Change

On February 5th, 2015, YourSAy called on the public to give their opinions on whether South Australia should change its time zone (see Fig. 6.28). The State is located at the time zone UTC + 9.5, one of the seven "half time zones" in the world, which causes some inconvenience for the local people's life and production. Whether to change the time zone had long been a problem disturbing the people there. They were faced with three choices: to change to UTC+ 10 time zone, change to UTC + 9 time zone or to keep the original. So the State government used the website YourSAy to gather the public opinions and the possible effects of the time zone change, and released the report "what we heard" after sorting out the opinions. The report came from 908 comments of 412 users on YourSAy. Among these people, 42% chose to keep the original time zone, 41% chose to change it to UTC + 9, and 15% chose to change it to UTC + 10. The public opinions played an important role in facilitating and deciding whether to change the time zone.

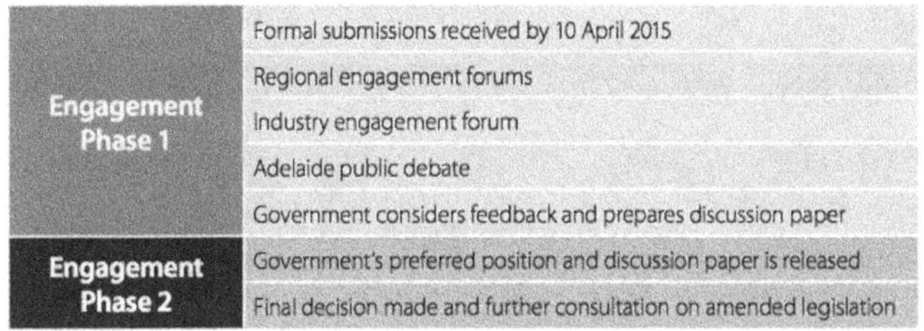

	Formal submissions received by 10 April 2015
Engagement Phase 1	Regional engagement forums
	Industry engagement forum
	Adelaide public debate
	Government considers feedback and prepares discussion paper
Engagement Phase 2	Government's preferred position and discussion paper is released
	Final decision made and further consultation on amended legislation

Submissions

The first phase of consultation will continue until 10 April 2015. A report will then be produced outlining the Government's preferred option and a further two-month period of consultation will take place.

Written submissions   Time Zone
                              Department of State Development
                              GPO Box 320 Adelaide 5001

Electronic submissions   dsd.timezone@sa.gov.au

More information   www.yoursay.sa.gov.au/timezone

Fig. 6.28 YourSAy called on the public to give their opinions on whether South Australia should change its time zone

2. Online communication

Online communication is an increasingly common interaction between the government and the people, covering online consulting, online complaint and BBS. According to the United Nations E-Government Survey 2014, 50% of the UN members had platforms for feedbacks on public opinions. The number of countries using social media, BBS and online public survey to provide first-hand public opinions for public policies accounted respectively 71, 51 and 39%.[9] In 2016, among the 193 interviewed countries, as many as 152 (about 80%) of them provided social network function (such as "thumb-up" button) on the national web portals and links with Facebook, Twitter, Sina Weibo (China) and Odnoklassniki/VK (Russia).[10] In 2014, the Chinese Government set up the column "I would like to have a word with the Premier" to collect the Internet users' opinions (see Fig. 6.29) and now there are already over one million messages, each of which the governmental personnel treat and study seriously. Some good suggestions are sent to the Premier's Office. Among the over 400,000 messages in 2016, 2071 were

---

[9]United Nations. *United Nations E-Government Survey 2014:E-Government For the Future We Want*, 2014.

[10]United Nations. *United Nations E-Government Survey 2016: E-Government in Support of Sustainable Development*, 2016.

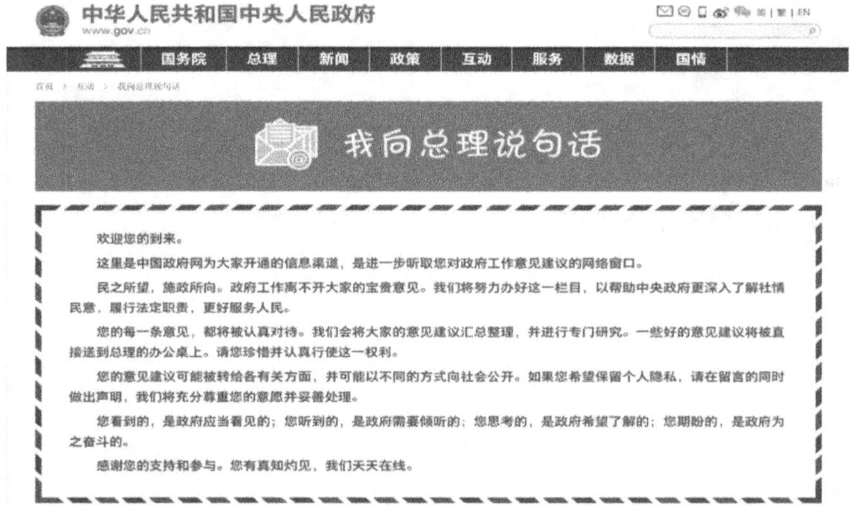

Fig. 6.29 "I would like to have a word with the Premier" on the Chinese government's website to collect public opinions

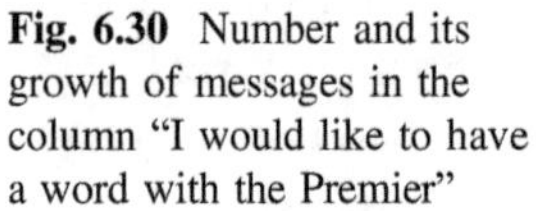

submitted to the State Council leaders and the draft group of the *Report on the Work of the Government*, and 101 of them were highly consistent with the Report. So the column has become an important bridge between Internet users and the Government (see Figs. 6.30 and 6.31).

Fig. 6.30 Number and its growth of messages in the column "I would like to have a word with the Premier"

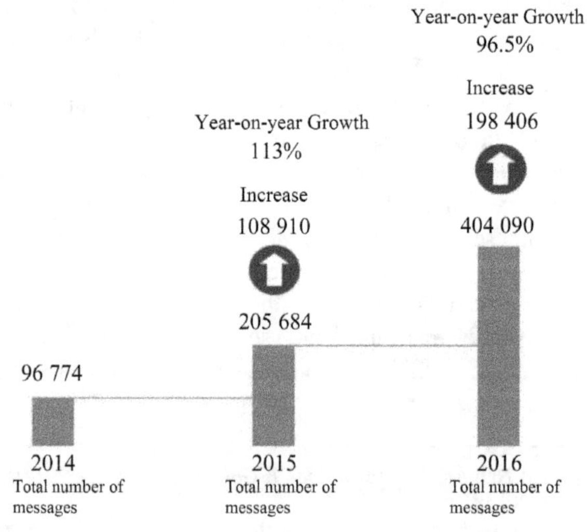

Fig. 6.31 Some messages and feedbacks in the column "I would like to have a word with the Premier"

3. **Online election campaign**

It is a trend in the Internet era to use the network platforms to facilitate election. The candidates can directly communicate with the public on the Internet and thus shorten the distance between them and facilitate the interaction between the statesmen and the public. When interviewed by CBS, Trump said that he had a large number of fans on Facebook, Twitter and Instagram, which he held as a big force. In his opinion, such an advantage had helped him to win the presidency. He also said that his rivals had spent more money but social media were more powerful and that he himself had proved

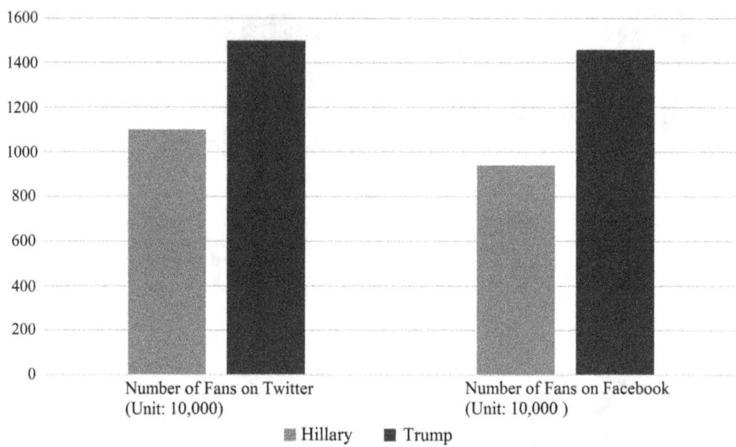

Fig. 6.32 Number of fans of trump and hillary on social media

that.[11] On Twitter, he had over 15 million fans while Hillary Clinton had 11 million; on Facebook, he had 14.59 million fans while Hillary had 9.4 million (see Fig. 6.32). Besides, Trump frequently spread his electoral manifesto and conception of government through short video and live streams[14], and every piece of the content was reprinted. Hope Hicks, press secretary of Trump's presidential campaign, said that Trump managed his accounts in person, so almost all the content on his social media had been sent by himself. She also said that his strategy was to tell the truth[15]. Through social media, Trump propagandized his conception of government, and thus enhanced the attention from media and shortened the distance between him and the public.

6.2.4 Constantly Innovated e-Government and Application Models

1. **From separate building to government cloud platforms**

It is a trend in the e-government development to build integrated e-governmental platforms. Countries leading in e-government all hold that e-government should be intensified through cloud computing to break the barriers against information

---

[11]Trump: Facebook and Twitter Helped Me to Be the Host of the White House. http://tech.sina.com.cn/i/2016-11-15/doc-ifxxsmic6289829.shtml (November 5th, 2016).

sharing and collaboration, reduce cost of building and operation and maintenance, and thus to support the efficient operation of the government.

The United States started the cloud building of government earlier than other countries, so it now has complete policies and institutional systems in that respect. In December 2009, the country launched the Federal Cloud Computing Initiative (FCCI) and founded the Executive Steering Committee (ESC) in the Federal Chief Information Officer Committee (CIOC) and Project Management Office (PMO), a daily office of General Service Administration (GSA), to facilitate cloud computing. In February 2010, the country proposed the Federal Data Center Consolidation Initiative to prevent the increase of the number of governmental data centers and to shift funds to intensified and efficient cloud computing platforms.[12] In September 2011, the United States launched the Federal Cloud Computing Strategy. So far, cloud computing has been widely used in all levels of government: the federal, the state and the county.

Column 7 Apps.gov of the United States

Apps.gov is a representative project of the Federal Government in governmental cloud building (see Fig. 6.33). It is a one-stop cloud service resource portal built by the Federal Government and an online application store providing cloud IT public service. Governmental agencies can buy their needed cloud services on Apps.gov and businesses can develop and release relevant applications for the government to buy. At present, services and applications provided on Apps.gov cover infrastructure, office software, industry applications and social media software.

EU also puts stress on the effect of cloud computing on public sectors. The European Commission has listed cloud computing in its Digital Agenda 2020, aiming to adopt government cloud computing within EU to enhance the internal market management in Europe. EU member countries have released their national cloud computing development strategies and carried out the general planning for government cloud layout and application. The United Kingdom, Australia, Canada and Japan have successively released the *UK Government Cloud Strategy*, the *Australian Government Cloud Strategy,* the *Canadian Government Cloud Computing Practice* and *Digital Japan Innovation Planning*, taking cloud computing in e-government as a systematic project.

2. **From Internet to mobile Internet**

In recent years, mobile Internet plays an increasingly important role in facilitating the transformation of governance and service modes and transparency of

---

[12]Gao Qiqi, Impact of Cloud Computing on Public Administration. *Journal of Southwest Minzu University*, 2015 (6), pp. 111–118.

Fig. 6.33 Apps.gov of the United States

governmental information and in solving practical problems of the people. It is a
new bridge between the government and the public. Throughout the world, the
development of mobile government has had a history of over one decade. Malta
announced in 2003 that it had entered the "mobile government era", with the
earliest services covering citizen complaints enquiry, updated notice concerning all

kinds of certificates and notice of examination scores. In 2013, Republic of Korea proposed that it would shift its public services to mobile terminals and build personalized and user-friendly service platforms not limited by time or space. According to the United Nations E-Government Survey 2016, many member countries are using mobile Internet to provide public services, with the number increasing from less than 50 in 2014 to nearly 100 today. The mobile Internet has eliminated the space barriers against public services and enables the public of different backgrounds and from different regions using different media to obtain governmental services timely and effectively. It is promoting governmental services to leap forward.

6.3   Challenges to e-Government

The challenges to the world e-government development are digital divide, information security, privacy protection and "information islands", which are the major barriers hindering the exertion of the potential of e-government and the difficulties that must be overcome in realizing the goal of sustainable development.

6.3.1   Digital Divide to be Closed

At present, there exists obvious difference in obtaining e-government services between regions and groups of people, which may affect these people in paying attention to public policies, obtaining public services and information and participating in public affairs through the Internet.

1. **Divide caused by lack of access means**

To obtain e-government services, the public need some carriers, mainly the information transmission networks and terminal devices, but many regions in the world are still lack of fixed broadband access and wireless communication equipment, and many low-income groups of people are universally lack of computers and smart phones. According to the UN data, 3.9 billion people have never accessed to the Internet, accounting for 52% of the world's total population, and the penetration rate of the Internet in the least developed countries was expected to be only 17.5% by the end of 2017, which is the objective barrier for the popularization of e-government.[13]

---

[13]UN Broadband Commission for Digital Development. *The State of Broadband: Broadband catalyzing sustainable development.*

2. **Divide caused by lack of know-how and skills**

The public education background, know-how and Internet literacy are the factors affecting the overall effect of e-government. To enhance the public know-how, some countries design their governmental information into conventional and simplified versions, make e-government system operation videos or compile operation coursebooks and diagrams to improve the availability of e-government. Besides, multilingual countries take the influence of language barriers on information and service availability and thus promote multilingual governmental services.

6.3.2 Serious Situation of Cyber Security

Since e-government entered the networking stage, cyber security has been a serious challenge inevitable in e-government development. In particular, the e-government system stores the core data of government affairs, provides government office work and public services, and protects the national confidential information, so its influence keeps expanding and it plays an increasingly important role, having a higher demand for cyber security assurancecapacity.

1. **Incomplete cyber security protection systems**

At present, government systems' capacity of resisting cyber attacks is not strong. Website distortion, attacks on business systems, theft of governmental data and intrusion on governmental networks frequently take place. For example, the ransomware attack that took place in 2017 paralyzed the e-government systems in many countries, severely affecting the governmental systems, financial systems and public facilities management systems of the Untied States, the United Kingdom, Russia, Sweden, Portugal and Ukraine. Incomplete management mechanisms, limited technical level and weak security awareness restrain the e-government system security protection of all countries.

2. **Striking problems with information security in emerging areas**

The application of cloud computing, IoT and big data and other new-generation information technologies in e-government has changed information obtaining, storage, transmission and use and caused new challenges to information security. For example, there are constant vulnerability, virus and network attacks on and threats to cloud. Information security protection bound becomes increasingly obscure and the responsibilities for the security needs to be redefined.

6.3.3   *Increasing Risks of Privacy Disclosure*

While providing public services and managing economy and society, the government collects and produces a large amount of information, which may involve citizens' privacy and national secrets, and whose disclosure will give rise to serious results. As different countries facilitate transparent government building and data opening, data security is attracting more attention.

1. Frequent breaches and thefts of confidential information

In recent years, there have been frequent breaches and thefts of personal information in public areas, so there are striking problems concerning data security and privacy protection. In the United States, the United Kingdom and Japan, there have been such breaches, involving sensitive information of employees, taxpayers, pensioners, telecommunication users and bank account openers. With the accelerated upgrading of technologies, citizens' information may be extracted from desensitized data through new data mining technologies, which should arouse the attention of competent authorities.

2. Privacy protection laws to be perfected

The lagging privacy protection legislation in the field of the Internet may aggravate the information breach risks. Legislation can help to clarify the rights and liabilities of governmental agencies, businesses and social organizations in collecting and using public information and provides legal guarantee for privacy protection. So far, some countries leading in e-government have formulated pertinent laws to protect their citizens' privacy, such as *Freedom of Information Act, Privacy Act* and *Internet Protection of Privacy Policy* in the United States; *Freedom of Information Act, Data Protection Act* and *Public Information Act* in the United Kingdom; *Governmental Organization Information Disclosure Act* and *Personal Information Protection Act* in Japan; and *Cybersecurity Law* in China, which have made clear requirements for the government in protecting citizen's privacy and businesses' trade secrets.

6.3.4   *Universally Existing "Information Islands"*

In the process of e-government development, information islands always exist. Isolation between systems and lack of exchange or sharing of data have given rise to non-collaboration in government work, redundancy and errors of data, high cost of operation and maintenance and difficulties for the public to obtain governmental services. Faced with the increasingly complicated economic and social administration, all governments need to remove the information islands and realize sharing of government data and collaboration in operation.

1. Technical causes of isolation between systems

Different building standards, different system structures, isolated transmission networks and no mutual recognition of data are the manifestations of e-government information islands caused by technologies. There are objective historical reasons and causes of IT development paths. First, different sectors have different starting points and technical stages in building e-government systems, and they are lack of coordinated planning, so their systems are not compatible with each other and their networks are not inter-connective. Secondly, there are different technical standards and the operation systems are built in a scattered way and laid out independently, which has also resulted in the isolation of one information system from another.

2. Fragmented operation of different organizations

Information islands are also caused by fragmented operation of different organizations. First, because of their different responsibilities, different organizations have built e-government systems with different functions, so their systems are relatively independent. Secondly, since e-government systems and data concern the core interests of relevant sectors, some sectors refuse or resist the building of inter-connective e-government systems for the sake of their own interests. Thirdly, the rights and obligations after the interconnection of e-government systems are not clearly allocated, which hinders all sectors from initiatively connecting their systems with others'. With the maturity of data integration technologies, it is the key solution to the information island problem to enhance the top-level design, clarify rights and obligations, and accelerate the intensification.

6.4 Prospect of e-Government Development

After years of development, e-government is playing an increasingly important role in fostering the governmental governance capacity, public service quality and social democracy. It is the common goal of different countries to facilitate the transparent, service-oriented, democratic and smart government through e-government. In the future, influenced by governmental reform and technical innovation, e-government models will be innovated and the government service and governance will be improved. The government information resource development and utilization will be constantly deepened and e-government building will be more intensive.

6.4.1 Openness, Sharing, Development and Utilization of Governmental Data Resources will be Constantly Deepened

Data resources have become a decisive factor measuring a country's social development, and their quantity, quality and utilization are important signs measuring a country's position in the international community. All countries take them as their basic resources and assets, and the development and utilization of governmental data as an important means of releasing the digital bonus and facilitating economic development as well as an inevitable choice in building open and transparent governments, promoting equity and justice and facilitating inclusive development. In recent years, with regards to government data resources, all countries have been doing explorations and practices concerning the re-use limit, open obtaining model and policy and regulation guarantee. In the future, the classification system of these resources will be more scientific, their ownership, use right and transfer right will be further clarified and their use scope and boundary will be clearer. The organizational collaboration mechanism for their development and utilization will be perfected to provide continuous drive for exerting their potential.

6.4.2 Innovation of Government Applications Will Facilitate Governance Capacity

With its development, e-government is no longer the simple governmental document flow or internal office work and operation application. It is not only the means and tool supporting governmental administration, but the means of promoting administrative system reforms in which the government and society both play their roles, and modernizing the national governance systems and capacities. In the future, with the technical progress and economic and social development, e-government will get involved more in market supervision. There will be more public services through e-government, which will help to innovate service modes, reduce institutional transaction cost, facilitate economic development, solve the asymmetry in government and market information, and improve the pertinency and effectiveness of supervision over market entities. It will help to know the power better and trace the power exercise and optimize the operation process so that the power can be exercised in a more transparent way. To sum up, it will become an important means of facilitating administrative system reforms in which both the government and society play their roles, and it will help to modernize the national governance systems and capacities.

6.4.3 Personalized, Customized, Standard and Intelligent Government Services Will Keep Emerging

With the development of the new-generation information technology and the deepening of e-government building, it has become possible for the government to provide personalized, customized, standard and intelligent services for different targets. E-government will help to enrich public services. It will shift application construction focus highlighting line management functions oriented to governmental agencies to integration of resources, optimization of service and improvement of satisfaction oriented to the demand of businesses and the public. Specifically, in the future, the government will attach importance to the satisfaction of the users' demand, provide in priority information and services that they need urgently, show concern about individual differences and provide personalized and customized services for different groups of users. The service quality will be highlighted and the service standard and personalization will be improved.

6.4.4 Big Data and AI Will Facilitate the Innovations of e-Government Models

The process of e-government development is the process in which governmental administration innovation has been integrated and developed with information technologies. In recent years, rapidly developing new technologies such as big data, AI, blockchain, VR and IoT have inevitably produced impact on e-government development and facilitated the constant innovation of e-government models. By using AI technology we can integrate and use the data resources in all areas of economy and society to govern the country through data. For instance, we can predict the macroeconomic turning point, monitor and analyze the industrial development in real time, stop up vulnerabilities in supervision in time and effectively, remove asymmetry in information, and reduce social transaction cost. With AI technology, government services can shift from man-to-man services to machine-to-man services, from face-to-face services to long-distance services. That is, it can help to change the traditional government service model. By using blockchain technology, we can solve problems existing in ID identification and information completeness and reliability of online government services. Some countries have made explorations in areas such as government fund spending, public digital ID identification, targeted welfare payment, and health e-file keeping. By using IoT technology, we will change the traditional data collection and supervision models, so that e-government will play a greater role in urban administration, environmental protection, traffic control and public security.

6.4.5 e-Government Building and Operation and Maintenance Models will be More Intensified

The development of Internet technologies will produce continuous influence on the e-government technology architecture. With the maturity of business models of cloud computing and thriving of software information service, new-type e-government building and management models and a number of professional information service organizations are emerging. In the future, intensified and professional building models like cloud computing will be adopted for e-government. Through outsourcing and purchase service, we can make full use of the funds, technologies and talents of the market, and attract professional social forces into e-government building to improve the use efficiency and management level of infrastructure, reduce operation and maintenance cost of e-government, and shorten the project construction period. Meanwhile, we can give full play to basic role of the market in resource allocation, create markets for businesses and facilitate the cultivation and development of information service markets in all countries.

Chapter 7
Development of the World Internet Media

Chinese Academy of Cyberspace Studies

Abstract (1) Internet media, the application with the most users, the highest use frequency and the widest scope, are the means of obtaining information, acquiring knowledge, socializing and enjoying entertainment and leisure time, playing an important role in enriching Internet users' cultural life and facilitating cultural prosperity and civilization progress. (2) Compared with mass media, Internet media are characterized by digitalization, multi-media, hypertext and strong interaction. They have experienced three stages, namely, Internet media 1.0, Internet media 2.0 and next-generation Internet media, developing from one-way communication in the direction of interaction, diversified ecosystems and smart media. (3) By product or service type, Internet media can be categorized into new media, social media, search engines, knowledge platforms and audio-visual media. North America and Europe have more media discourse power and technical advantages, with their products available worldwide; and Asia, Latin America and Africa have local products and services with great potential, with room for improvement. (4) Internet media are channels for Internet users to obtain news and information. Comprehensive portals and aggregated news websites enjoy wide coverage. For example, the number of monthly accesses to Google News is 499.2 million. Traditional news media in major countries maintain strong momentum. The number of monthly accesses to the website of *New York Times* has reached 388.1 million. Europe, North America, China, Japan, Republic of Korea and Australia are witnessing digital transformation and upgrading of their traditional media and the revolution of their media industry. Africa and Latin America are weak in Internet news media, but some countries there with good Internet infrastructure have their local influential news media. (5) Social media are one of the most popular types of Internet media, including the social network, instant messaging (IM) and online forum (BBS), with the number of their users and their growth rate and penetration rate on the leading list of the world. By the second quarter of 2017, the number of active users of social media in the world had amounted to 3.028 billion, accounting for 79.29% of the total number of the world's Internet users, but there exists imbalance in the development between regions. The penetration rate in Africa is

Chinese Academy of Cyberspace Studies
Beijing, China

© Publishing House of Electronics Industry, Beijing and Springer-Verlag GmbH Germany, part of Springer Nature 2019

Chinese Academy of Cyberspace Studies (ed.), *World Internet Development Report 2017*, https://doi.org/10.1007/978-3-662-57524-6_7

only 14%, much lower than that in other regions. Social media have increasingly comprehensive functions. News feeds, payment transaction and live video have been added into social media and the mobile terminal has become an important area. (6) Google is the leader on the search engine market. By July 2017, its market share had remained 86.83% of the global total, and it had over 4.5 billion active users, as the leading search engine in most countries of Asia, Europe, Africa and Americas, but not in some non-English speaking countries, such as Russia, China, Republic of Korea and Japan. (7) Internet media of knowledge platforms include online encyclopedia and online Q&A. Wikipedia enjoys over 15 billion accesses every month. Knowledge platforms started early and have witnessed rapid development in North America, Europe and Asia, but started late in Africa and Latin America, where there are few localized knowledge platforms. (8) Audio-visual media take the lead in all Internet media in terms of traffic. It is estimated that their traffic will account for 76% of the total traffic of Internet media in 2018. YouTube has 1.5 billion active users every month. Audio-visual media will become an important growth point of the global entertainment and media industry, and tend to generate and deliver their content by themselves. They are divided into short video, live streaming and bullet screen to meet the demand of different groups of people for entertainment. The products of the United State enjoy the fastest development while those in other regions can meet their local cultural demand. Africa and Latin America have great potential demand, with South Africa and Brazil witnessing the greatest. (9) Forms of Internet media are undergoing profound changes. Traditional media are transforming into new media, with their delivery channels expanding to social platforms. News generation flows represented by "Central Kitchen" witness constant innovation and an obvious trend of paid subscription. Internet media embody open scenario communication, participation and feedback communication, layer communication, and social interaction communication. Internet media are becoming diversified, personalized, professionalized and data-oriented. (10) The necessity and importance of Internet content governance have become the consensus of governments, businesses, trade associations, media and the public, harmful information supervision being the core. The focus of governance is on information threatening national security, information of pornography and violence, information violating privacy and online rumors. In recent years, to curb the spread of terrorism online has been another focus. Legislative and administrative supervision has been combined with civil self-discipline, self-governance of media platforms and special actions. Varieties of technical means have been adopted for comprehensive governance. With the deep combination of advanced technologies like AI, VR and intelligent algorithm with media development, news content generation and information communication will be more diversified, personalized and target-oriented. To sum up, the Internet is a big social information platform, on which billions of Internet users obtain and exchange information. Internet media are the most common Internet application and service type and an important tool of obtaining information, acquiring knowledge, socializing and enjoying entertainment and leisure time. They have changed the pattern of mass communication, and they are influencing and reshaping production and social life. They can also reshape

Internet users' knowledge acquirement ways, thinking and values. They are carriers spreading positive energy and excellent cultures, playing an important role in enriching the Internet users' cultural life and facilitating cultural prosperity and civilization progress.

7.1 Concept and Characteristics of Internet Media

Internet media are a type of new media obviously different from traditional media. In a broad sense, they cover all information carriers with media properties such as the Internet, the mobile Internet and terminal devices. In a narrow sense, they refer to media products or services providing news, socializing, information or entertainment through the Internet.

Compared with mass media and traditional media, Internet media are characterized by digitalization, multi-media, hypertext and strong interaction. Digitalization refers to the fact that in Internet media, information is stored and transmitted in the binary code, so the information capacity of Internet media is tremendous. Today, the quantity of new data every day in the world is equal to the number of printed collections of 250,000 libraries of the U.S. Congress. Multi-media indicate that Internet media can transmit simultaneously all types of media including texts, pictures, audio broadcasts, video shows, animations and so on, providing the users with diversified experiences. Hypertext means that text information from different cyberspace is organized into reticular texts through hyperlinks, allowing the users to have "skipping" browsing in any order in different space according to their own demand and interest. Strong interaction suggests that every information node in Internet media has double identities (the sender and the receiver), allowing the users to generate and spread the content and interact with each other, which cannot be done in traditional media.

7.2 Development of Internet Media

Internet media have seen evolution of their technologies, forms and models. They have experienced three stages: Internet media 1.0, Internet media 2.0 and next-generation Internet media. With the rapid development of Internet technology, Internet media forms tend to be more diversified, their content richer and their models more mature. They are developing from one-way communication (from the website to the user) to interaction, and are becoming more open, interactive and decentralized.

7.2.1 Internet Media 1.0

The age of Internet media 1.0 covered the period from 1980s to the early 21st century, represented by news websites, search engines and web portals. At this stage, traditional media expanded into the Internet and formed the central node of information production, and information began to flow from professional producers to common users, marking the completion of traditional media digitalization.

A large number of news websites emerged. Traditional news media set up their websites and started their exploration into the integration between them and new media, represented by CNN, *Chicago Tribune* and the *Wall Street Journal* in the United States, and BBC and the *Daily Telegraph* in Europe.

The pattern of the world's search engines was formed. In 1980s and 1990s, traditional media's news websites and homepages were dominant and their number kept increasing, leading to the dramatic increase of the demand for information classification and retrieval, which, in turn, led to the emergence and rapid development of search engines, such as Yahoo! (in 1995), Google (1998) and Baidu (2000), the top three search engines of the world.

Web portals saw their rapid development. They could gather different sources and common functions into one homepage through search engines, emails and forums to occupy the first entrance of the users' access to the Internet. Early web portals (see Table 7.1) were American Online (the United States), Yahoo! (the United States), Excite (the United States), Netvibes (France), Google (the United States), MSN (the United States), Naver (Republic of Korea), Lycos (Spain), Prodigy (the United States), Indiatimes (India), and Rediff (India). Influenced by the wave, some representative web portals also came into being in China. They are still active today, including Sina, NetEase and Sohu.

7.2.2 Internet Media 2.0

Around the year 2000, Internet media entered its 2.0 age, when UGC was the core, mass media shifted to we media, and SNS, BBS, online encyclopedia, blog, microblog and Q&A website became mature (see Table 7.2). Its difference from Internet media 1.0 is as follows: in terms of participation, the users witnessed the shift from their one-way "reading" to two-way "reading + writing" and Internet content generators were not only the webmasters and programmers, but common users; in terms of receiving ways, browsers were replaced by subscribed RSS readers; in terms of application type, primary applications developed into multiple Internet services.

At this stage, the popularization of mobile Internet and smart devices boosted the rapid development of mobile applications. According to statistics from APP Annie, a third-party survey corporation, in the second quarter of 2017, the number of downloads on iOS and Google Play reached nearly 25 billion across the globe. The

Table 7.1 Representative media in the age of internet media 1.0

Website	Type	Headquarter location	Domain name	Founding time	Operation status
CompuServe	Website portal	The United States	compuserve.com	1969	In operation
American Online	Website portal	The United States	aol.com	1983	In operation
Yahoo!	Website portal	The United States	yahoo.com	1995	In operation
CNN	News website	The United States	edition.cnn.com	1995	In operation
BBC	News website	The United Kingdom	bbc.co.uk	1997	In operation
Yandex	Search engine	Russia	yandex.ru	1997	In operation
NetEase	Website portal	China	163.com	1997	In operation
News24	News website	South Africa	news24.com	1998	In operation
Sina	Website portal	China	sina.com.cn	1998	In operation
Sohu	Website portal	China	sohu.com	1998	In operation
Google	Search engine	The United States	google.com	1998	In operation
Baidu	Search engine	China	baidu.com	2000	In operation

number witnessed a year-on-year growth rate of 15%, covering almost all industries, which shows that mobile applications have become an indispensable part of consumers' life.[1] Application delivery platforms attract more developers in a pro-rating style for their win-win development.

7.2.3 Next-Generation Internet Media

The world Internet media technologies are developing fast and becoming more closely related to people's production and life and producing deeper influence on social changes. Short video and live streaming have stirred up new movements of making stars; Facebook, YouTube and Tencent and other super platforms keep up with the movements; VR, AR and MR have improved the Internet users'

---

[1]App Annie Summary of the Second Quarter of 2017, https://www.appannie.com/cn/insights/market-data/google-apple-app-stores-q2-2017.

Table 7.2 Representative media in the age of internet media 2.0

Website	Type	Headquarter location	Domain name	Founding time	Operation status
Tianya	BBS	China	bbs.tianya.cn	1999	In operation
Blogger	Blog	The United States	blogger.com	1999	In operation
Wikipedia	Online encyclopedia	The United States	wikipedia.org	2001	In operation
LinkedIn	SNS	The United States	linkedin.com	2003	In operation
MySpace	SNS	The United States	myspace.com	2003	In operation
Facebook	SNS	The United States	facebook.com	2004	In operation
Reddit	BBS	The United States	reddit.com	2005	In operation
Twitter	Microblog	The United States	twitter.com	2006	In operation
Baidu Baike	Online encyclopedia	China	baike.baidu.com	2006	In operation
VK	SNS	Russia	vk.com	2006	In operation
Tumblr	Blog	The United States	tumblr.com	2007	In operation
Sina Weibo	Microblog	China	weibo.com	2009	In operation
Quora	Q&A	The United States	quora.com	2009	In operation
Zhihu	Q&A	China	zhihu.com	2011	In operation
QQ	IM	China	–	1999	In operation
WhatsApp	IM	The United States	–	2010	In operation
WeChat	IM	China	–	2011	In operation
Line	IM	Japan	–	2011	In operation

experience; and the boundary between reality and virtual space is becoming obscure and poses a challenge to the traditional definition of "media". AI, big data and algorithm revolution have completely changed the channels and experiences of obtaining news information, with a number of aggregated news applications represented by toutiao.com having become the favorite of capital. Payment for knowledge and outsourcing news are becoming new ways of info-cashing. Social media have been deeply involved in political election and social movement, and have become an important tool of all stakeholders' contacting users, producing influence and convening activities. What has happened, is happening and will happen in all these areas constitute the prospect of next-generation Internet media development.

7.3 General Situation of Internet Media Development and their Representatives

By product or service type, Internet media can be categorized into news media, social media, search engines, knowledge platforms and audio-visual media (see Table 7.3). Generally speaking, since the age of Internet media 1.0, the world Internet media have been developing fast, with their forms, contents and communication ways becoming richer and richer. North America and Europe have more media discourse power and technical advantages, with their products available worldwide; and Asia, Latin America and Africa have local products and services of great potential, with room for improvement.

7.3.1 News Media: Aggregators of Internet News

Internet news media have become high-frequency basic network applications and important channels of obtaining news information. In China, by June 2016, the number of their users had amounted to 625 million and that of mobile news media had amounted to 596 million.[2] Despite the wide coverage of comprehensive portals and aggregate news websites like Google news, Yahoo! News and Baidu News, the news websites of the traditional national news media websites maintain their momentum. The three largest news media websites in Europe–BBC, the Guardian and Daily Mail–are all traditional British media; France 24 as the international and current news TV station is the most popular in France; and bild.de, the news website of the German newspaper Bild, enjoys the largest page view among all German news websites.

Europe, North America, China, Japan, Republic of Korea and Australia are witnessing the digital transformation and upgrading of their traditional media, the thriving of their aggregate news websites and the revolution of their media industry. Africa and Latin America are weak in Internet news media, but some countries with good Internet infrastructure there have their local influential news websites or applications. For instance, the local news websites globo.com and uol.com.br are the most influential two in Brazil; the five most influential news websites in Chile are their domestic ones except Yahoo!; News24 in South Africa, and Hespress and globalnovas in Morocco are the most influential news websites, only next to Yahoo!, in the two countries.

1. *The New York Times* (The United States)

The New York Times in the United States, well known for its serious news, is the leader of the traditional media transformation. While investing in its core news

---

[2]CNNIC. *Report on China's Internet Journalism Market Development* (2016). January 2017, http://www.cnnic.net.cn/hlwfzyj/hlwxzbg/mtbg/201701/t20170111_66401.htm.

Table 7.3 Types and representatives of internet media

News media	News agency media	People.cn, nytimes.com, The Guardian, BBC, Russia Today, CNN, and News24
	Web portals	America Online, The Huffington Post, Tencent, NetEase, and Sohu
	Aggregate news	Google News, Yahoo! News, Buzzfeed, News Break, Dailyhunt, G1, and Toutiao
Social media	SNS	Facebook, Twitter, VK, Sina Weibo, Instagram, Reddit, Tumblr, Badoo, and Tianya
	IM	WeChat, Line, Whatsapp, and Hikemessenger
	We media	WeChat Official Platform
Search engines		Google, Baidu, Yandex, Bing, and Naver
Knowledge platforms	Online encyclopedia	Wikipedia, and Baidu Baike
	Q&A	Quora, and Zhihu
Audio-visual media	Online video	YouTube, Netflix, Hulu, Youku Tudou, Niconico, and Bilibili
	Music and broadcast stations	Spotify, Podcast, Pandora, and QQ music
	Short video	Instagram, and Kuaishou
	Live streaming	Twitch, and Douyu

business, it has opened up the new era of newspaper digital subscription, which has been made an indispensable part in its users' life. The number of subscribers on its website and its client terminal has amounted to over 2.2 million,[3] the number of total accesses, 1164 million; that of monthly access, 388.1 million, and that of monthly uniques, 125.6 million. The subscribers are from the United States, Canada, the United Kingdom, France and Australia, with the largest number from the United States, accounting for 72.41%.[4] The Internet news of the *New York Times* has the following characteristics: first, it has launched pay walls and native ads, the income from e-subscription being higher than that from ads, and diversified income sources having been formed; secondly, it has used big data, AR and VR to develop data analyzing tools to collect and analyze the users' data and use them for decision making concerning edition, while acquiring technology businesses to develop AR and VR-related news products for the users to experience; thirdly, it is deeply integrated with social media, having opened official accounts on social media and making full use of the data from social media to develop interactive news products.

---

[3] xinhuanet.com, *the Sixth Layoff by The New York Times within Nine Years*. June 11th, 2017, http://news.xinhuanet.com/world/2017-06/11/c_129629964.htm.

[4] similarweb, https://www.similarweb.com/website/nytimes.com.

2. *The Guardian* (The United Kingdom)

The website of *The Guardian*, founded in 1999, is only next to the news website of BBC in influence in the United Kingdom. By March 2017, the newspaper had 230,000 paying members and a weekly use rate of 14% within the country, only next to that of BBC.[5] *The Guardian* is the world's first medium having made crowdsourcing as one of the core pillars and transformed itself into a medium with digital sharing platforms. Its digital transformation has been acknowledged by the world media industry as the most open and the most complete. The philosophy for its transformation is "open news" in four levels: "Open Comment Platform", on which news is open to comments, according to which the layout and report focus can be adjusted; "Open Data Platform", on which news is open to the public, who can download and analyze the data for free and who can participate in the training and competition provided on the platform; "Open Technology Platform", on which the content is open to the third party, who can achieve win-win development with the newspaper; "Open News", which enables the readers to get involved in the whole process of topic selection and development and to provide their news works.

3. Russia Today (Russia)

Russia Today (RT) is Russia's first completely digitalized TV network. At the end of 2013, it merged with RIA Novosti and The Voice of Russia into Russia Today International News Agency, thus becoming a media colossus in Russia and taking presentation of unbiased Russian prospect as its strategic positioning. Now RT covers 700 million audiences in over 100 countries,[6] as an emerging multilingual media agency with TV broadcasting as the main business, together with other business of the Internet and social media. Its networks include three global news channels respectively in English, Arabic and Spanish and three HD channels, namely RT America, RT UK and Documentary Channel. It also has online platforms in Germany and France, and its video news center PUPTLY is quite influential on social media.

4. People.cn (China)

People.cn is a network news medium belonging to *People's Daily*. It was founded on January 1st, 1997. Positioned as a large network information platform, with news reporting as its core, it is one of China's and even the world's most prestigious network media. As the leader of the country's news websites, people.cn persists in its authority, credibility and popular style and takes spreading positive energy and mainstream theme as its mission, having become a representative of domestic news websites. It keeps releasing news to the world around-the-clock in 14 languages and 15 versions, with its readers from over 200 countries and regions. It is a window for the global Internet users to know about China. People.cn went public in 2012 and

---

[5]Nic Newman. *Reuters Institute Digital News Report 2017*, pp. 24, 55.

[6]Su Xiaochun. The Latest Development of Russia Today and Its Enlightenment, *TV Research*, 2015 (6), pp. 75–77.

since then, its operation revenue and net profit have been increasing. In 2016, the former was RMB1, 432 million *yuan* and the latter was RMB 106 million *yuan*.[7]

5. News24.com (South Africa)

News24.com is the largest news website of South Africa, founded by Naspers (a media multinational) in October 1998, providing news service in English and Zulu. As an influential news website in Africa, it founded its substations in Kenya and Nigeria in 2011. The content on the website covers local and international news of politics, business, entertainment, science and technology, and sports. By 2016, the number of its monthly visitors had surpassed three million.[8]

6. HuffPost (formerly the Huffington Post, in the United States)

Launched in May 2005, HuffPost is an aggregate medium of current affairs, with original content from the Internet. It is well-known as the first Internet newspaper, the first profit news media website having won the Pulitzer Prize, and the most influential news website of politics blogs. Half of the traffic of HuffPost is from the international market and most users are in the United States, the United Kingdom, Canada, Holland and India. According to Similarweb, 76.58% of the HuffPost users are from the United States, and its India market share is growing. In the third quarter of 2017, the total number of visits to HuffPost was 556 million, and the number of monthly independent visits was 68.09 million.[9] In terms of development strategies, HuffPost focuses on distributed news mining and socialized news communication models based on Web2.0, which makes it unique. It has launched "Off the bus" to distribute news gathering to the public and use social media to interact with the users.

7. News.google.com (The United States)

Since it was launched in 2002, news.google.com has become the most influential aggregate news website of the world, made up of investment, user products and research. Content on the website comes from over 50,000 news media worldwide. It provides news information through hotspot aggregation and personalized recommendation. "Suggested for you" on APP recommends topics and news in which the users may be interested according to the record on Google network and application and YouTube and according to the immediate preference settings. According to Similarweb data, in the third quarter of 2017, the total number of visits to news.-google.com was 1497 million, 71.27% of which is from PC and less than one third from mobile terminals; the number of monthly visits was 499.2 million and that of monthly independent visits was 44.42 million, and the average visit duration was

---

[7]Annual Report of people.cn (2016), http://www.sse.com.cn/disclosure/listedinfo/announcement/c/2017-04-20/603000_2016_n.pdf.

[8]Source: https://en.wikipedia.org/wiki/News24.

[9]https://www.similarweb.com/website/huffingtonpost.com.

over half an hour.[10] Most visits to news.google.com are from the United States, France, Japan, India and Germany.

8. Toutiao.com (China)

Founded in July 2012, toutiao.com is the news APP of Bytedance, providing personalized information services based on advanced recommendation engine technology and thus revolutionizing the traditional information distribution ways and becoming the leader of China's news APPs. By December 2016, the number of the monthly active users of toutiao.com had amounted to 121 million. In early 2017, its market penetration rate was 20.7%, ranking first in China, and its user loyalty was 67.5%, ranking second.[11] The products on toutiao.com have two features. First, they are technology-driven products. Machine learning, algorithms and data mining and other advanced technologies have been adopted for targeted content recommendation. Through analysis of the users' accounts, the users' "Interest graph" has been established, and thus the cold start was launched. Then order directories have been set up about the content that users of similar interest like, and the users' personal characteristics are sub-categorized according to their information. Secondly, information aggregation is adopted for the common prosperity of UGC and PGC. Toutiao.com has launched we media projects to enrich the content, such as "RMB10,000 *yuan* for one thousand people and one hundred groups respectively", "Toutiao Creation Space Incubator" and "RMB 1000 million *yuan* supporting short videos".

9. Dailyhunt (India)

Dailyhunt, an aggregate news platform in India, pushes personalized news to the users through news aggregation and allows the users to create their own news page by adopting key words of the topic. Since India has a large language market, which has not been completely developed, localization is an advantage of Dailyhunt, which supports 15 local languages, such as Hindi, Marathi, Bengali and Urdu, and updates over 100,000 pieces of news, with 6.5 million daily page viewers, 29 million monthly page viewers and 4.5 billion monthly page clicks.[12] Through cooperation between native ads and copyright in the form of primary information flow and primary application wall, Dailyhunt sells e-books, which is its main source of income. The rapidly developing Internet market and large population of India bring major opportunities for Dailyhunt, of which big data will be the core of development and audio and video show will be the major direction.

---

[10]https://www.similarweb.com/website/news.google.com.

[11]Big Data Report on News. Competition between toutiao.com and new.qq.com, http://www.askci.com/news/chanye/20170331/09375394737.shtml.

[12]Source: https://en.wikipedia.org/wiki/Dailyhunt.

10. **G1 (Brazil)**

G1, the biggest news website of Brazil, is affiliated to Orgaizacoes Globo, the biggest media company of the country. Its APP ranks fifth in news applications of Brazil, as the most influential news APP in the country. In the third quarter of 2017, the total number of its visits was 2513 million, and 67.38% of the traffic was from mobile terminals. It has 827.8 million monthly visits and 102.1 million monthly unique visits. Most of G1 users come from Brazil, as high as 97.23%. A small number of users are from the United States, Portugal, Japan and the Untied Kingdom.

7.3.2   Social Media: An Active Factor in Internet Media

Social media are one of the most popular type of Internet media, with the number of their users, their grow rate and penetration rate on the leading list of Internet media. Social media include SNS, IM and BBS. At present, the first two are developing fast and BBS tends to have a small number of users and to develop vertically. The function of social media is increasingly comprehensive, gradually covering news feeds, payment transaction and live video. APP has become the key area of their development. By the second quarter of 2017, the number of active users of social media in the world had amounted to 3.028 billion, accounting for 79.29% of the total number of the world's Internet users; the number of active users of mobile social media had amounted to 2.780 billion, accounting for 79.27% of the total number of mobile Internet users of the world. The geographical distribution of social media users is not balanced, 1489 million of them in Asia, 362 million in Latin America, 237 million in North America, 451 million in Europe, 170 million in Africa, and 21 million in Oceania. The penetration rate of social media in Africa is only 14%, much lower than that in other regions.[13]

The number of monthly active users of social media in 2017 is shown in Fig. 7.1.

1. **Facebook (The United States)**

Founded in February 2004, Facebook is the most influential online social medium with the largest number of users in the world. It ranks third in the 10 largest Internet businesses of the world. Actually, the number of its users is growing fast. By the second quarter of 2017, Facebook had 1320 million daily active users and 2010 million monthly active users,[14] with 160 million users in Africa, 736 million in Asia, 343 million in Europe, 370 million in Latin America and the Caribbean, 86

---

[13]Source: We are social(Global Digital Statshot Q3 2017), https://www.slideshare.net/wearesocialsg/global-digital-statshot-q3-2017.

[14]Source:Facebook Financial Report (Q2 2017), https://investor.fb.com/investor-news/press-release-details/2017/Facebook-Reports-Second-Quarter-2017-Results/default.aspx.

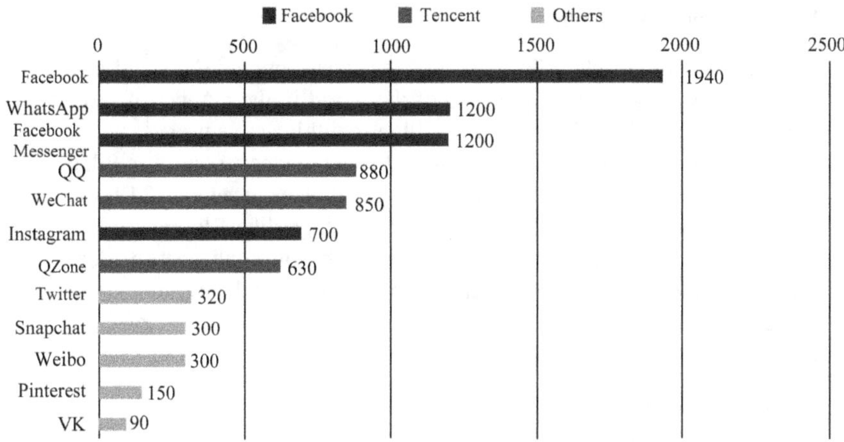

Fig. 7.1 Number of monthly active users of social media in 2017 (unit: million)

Fig. 7.2 Distribution of
facebook users in the world

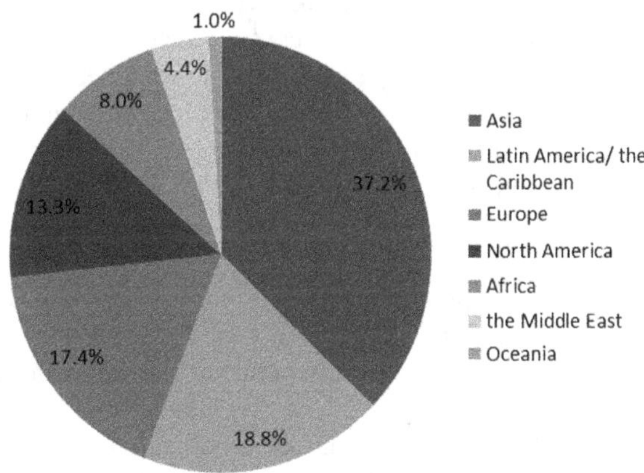

million in the Middle East, 263 million in North America and 19 million in Oceania[15] (see Fig. 7.2). The top 10 countries in the number of Facebook users are India, the United States, Brazil, Indonesia, Mexico, the Philippines, Vietnam, Thailand, Turkey and the United Kingdom.

Besides the basic functions of social websites, Facebook supports image and video uploading, sharing and browsing and has launched the live streaming channel, transaction system, blog platform, Q&A Program and open platform. Most of its profits come from ads based on AI, VR, AR and drone technologies.

---

[15]Source: http://www.internetworldstats.com/facebook.htm.

2. Twitter (The United States)

Twitter, founded in March 2006, is one of the most famous SNSs and microblog service websites, ranking among top 15 of the world in the number of visits.[16] By the second quarter 2017, it boasted 328 million monthly active users,[17] with its core function being microblog service. It launched live broadcast service in 2016. With its main direction being the digital video strategy, it plans to set up a 24-h streaming media channel, whose content will come from partners like Bloomberg News and Major League Baseball (MLB). Till now, Twitter has not only met the demand of individual and business users in sending information, socializing, brand marketing and managing customer relations, but also played an important role in global political and public opinion campaigns since 2010.

3. VKontakte (Russia)

Launched in September 2006, VKontakte is the biggest SNS, producing great influence in Russia and East Europe, popular among users speaking Russian. According to traffic statistics from Alexa, VKontakte ranks first in all Russian websites and 15th in the world. It has 452 million registered accounts and 81 million monthly active users,[18] mostly from Russia, Belarus, Kazakhstan, Kyrgyzstan, Moldova, Latvia, and some other middle Asian and east European countries. Notably, quite a few musicians promote their works on VKontakte, especially on the official VKontakte page, such as Tiësto, a music producer from Holland, and Shakira, a singer from Latin America.

4. Line (Japan)

Opened to the public in June 2011, Line is a free IM application popular in Japan and Southeast Asia, operated by Line Corporation, a Japanese subsidiary of Naver, a giant Internet business in Republic of Korea. By the first quarter of 2017, Line had 214 million monthly active users, with a total of 171 million in Japan, Taiwan (China), Thailand and Indonesia, accounting for 79.91%.[19] Line Sticker Shop is one of its most successful functions. Its other functions include Line Game, Line Pay, Line Man (take-out food application) and Line Here, all of which make it more attractive. Besides, it has set up its brick-and-mortar shops selling cartoon figures of Line Friends. Recently, its strategy has shifted onto live broadcast and video.

5. Badoo (Europe)

Founded by Andrey Andreev, a Russian entrepreneur, in November 2006, Badoo is a global SNS, influential in Europe and Americas. As a representative SNS for

---

[16]Source: Website Visits Statistics by Alexa, https://www.alexa.com/siteinfo/twitter.com.

[17]Source: Financial Report of Twitter (Q2 2017), http://files.shareholder.com/downloads/AMDA-2F526X/5691461441x0x948691/3C92BB75-01A6-441E-A910-5B8E045CAC29/TWTR_Q217_Earnings_Date_Announcement.pdf.

[18]Source: https://en.wikipedia.org/wiki/VK_(social_networking).

[19]LINE Business Stories, LINE AD CENTER, http://ad-center.line.me/mediaguide.

strangers to communicate with each other, Badoo adheres to its philosophy of helping its users to make friends in the neighboring areas. Now it has 350 million registered users in the world from 190 countries, providing services in 47 languages. It has over 700 million visits per month[20] mainly from Spain, Italy, France, Mexico and Brazil. Badoo allows its users to improve their credit points and credit heat through payment, so that it can make profits. Its future focus will be put on the upgrading and matching of the user algorithm, segmentation of user groups, exploration of more profit models and expansion of markets.

6. Hike Messenger (India)

Launched in December 2012, Hike Messenger is a popular cross-platform IM application. In January 2016, it had over 100 million users and it was sending 10 billion messages.[21] In addition to IM, it has developed graphical stickers, aggregate news, cricket playing and chat robots in accordance with the situation of the Indian market. The cricket is a popular sport in India, so Hike Messenger has launched the cricket application to meet the demand of Indians, since they can follow their favorite cricket teams and make comments on them.

7. Sina Weibo (China)

Launched in August 2009, Sina Weibo is one of the most influential microblog websites with the largest number of users. It has witnessed the growth of the number of its users and the operation performance, and become the "Hall of Public Audience". By the second quarter of 2017, it already had 159 million daily active users and 361 million monthly active users, 92% being APP users.[22] Sina Weibo always commits itself to expanding its market to third-tier and fourth-tier cities. Recently, it tries to categorize the users' interest vertically, innovate aggregate information spread models, and facilitate excellent content systems. It also facilitates video spread by seizing the opportunity brought about by small and medium-sized V-bloggers using short videos and live streams to promote themselves.

8. WeChat (China)

WeChat is an IM and social media APP developed by Tencent, an Internet business giant. First released in January 2011, it is one of the most influential IM APPs. By December 2016, the number of total monthly active accounts from domestic WeChat and WeChat International Version) had amounted to 889 million[23] and the growth rate is taking the lead in the world. Its property of socializing has brought it high user loyalty. According to statistics, 94% of WeChat users use it every day, and of these users, 61% open their WeChat more than 10 times a day and 36%, over

---

[20]Source: https://en.wikipedia.org/wiki/Badoo.

[21]Source: https://en.wikipedia.org/wiki/Hike_Messenger.

[22]Source: Financial Report of Weibo (Q2 2017), https://weibo.com/1642634100/FgkuWn9WG.

[23]Source: Financial Report of Tencent (Q4 2016), http://tech.qq.com/a/20170322/034572.htm.

30 times a day.[24] WeChat has developed from a single social APP into a comprehensive APP combining media, transaction, service and sharing. The WeChat Official Account has produced great impact on the Internet media content and become an indispensable part of we media in Chinese. The subscription account and service account meet the demand for reading long texts and receiving customized services on WeChat. By June 2017, there were over 20 million WeChat official accounts.

7.3.3 Search Engines: Index of Internet Information Resources

A search engine is an information retrieval system for finding information on the Internet and processing the information and providing it to the users with certain strategies and special computer programs. In essence, a search engine completes three-step work: retrieving information, processing information and giving the result. In 1990s, the rapid development of www led to the sharp increase of the number of homepages, so there was an immediate need to develop an information searching tool for retrieving online information resources, and thus emerged the search engine. Yahoo! launched in 1995 was the first modern search engine in real sense. Google and Baidu, the second-generation search engines, have their advantage in search speed and result by using the PageRank based on hyper-link analysis.

In general, Google is the search engine with the largest market share. It is the leading one in most countries and regions of Asia, Europe, Africa and Americas, except some non-English speaking countries and regions. According to statistics, in 2015, Yandex, a Russian Internet business, was the leader of the local market, with its market share accounting for 58%; Baidu was the leader of the Chinese market, with its market share accounting for 55%; the second in China is Qihoo 360, with its market share, 28%; the third in China was Sogou, with its market share, 12.8%; in Republic of Korea, Naver and Daum were the two leaders, with their market share increasing from 90% in 2013 and 97% in 2015; Yahoo! had 40% of the market share, but it still uses the Google algorithm to provide search results.[25]

---

[24]Source: Report on WeChat Users Data (2016), https://wallstreetcn.com/articles/252592.

[25]Source: return on now, https://returnonnow.com/internet-marketing-resources/2015-search-engine-market-share-by-country.

1. **Google (the United States)**

Google is the leader of the global search engine, launched in September 1997. Headquartered in California of the United States, it provides information service worldwide in 123 languages. By July 2017, its market share had accounted for 86.83%,[26] with over 4.5 billion active users in the world,[27] and over three billion search requests processed every day.[28] Most of its income comes from advertisements. In 2016, its global income totaled about US$89.46 billion,[29] which was the highest among the science and technology businesses, thanks to PageRank, which can priorly show quality webpages. Besides webpages, users can search for images, news websites and videos in different regions on Google. As a rendezvous point of global information, Google provides varieties of services like public information, maps, unit conversion, calculation and translation. In June 2011, it launched "voice search". Its AdSense enables small and medium-sized websites to make profits through traffic.

2. **Yandex (Russia)**

Founded in 1993, Yandex is the largest search engine and the most popular website in Russia, and the fourth largest search engine in the world, with its market share in Russia exceeding that of Google. In the fourth quarter of 2016, its market share was 55.4% of Russia's total of search engines.[30] It is influential also in Kazakhstan, Belarus and Turkey, as the second largest engine in Europe. Besides searching, Yandex has developed services and products like e-mail addresses, maps, news, videos and online payment.

3. **Baidu (China)**

Founded in January 2000, Baidu is the world's largest search engine in Chinese, with the world's largest Chinese web library. It entered Japan in 2008, and its Portuguese version entered Brazil in July 2014. 93.9% of Baidu users are from China, 2% from Japan, 1.2% from the United States, and 0.8% from Republic of Korea.[31] Based on its core technology "hyperlink analysis" Baidu assesses the quality of its linked websites by analyzing their number to ensure that the websites are arranged on Baidu according to their popularity, from higher to lower.

---

[26]Source: https://www.statista.com/statistics/216573/worldwide-market-share-of-search-engines-turns-19-company-celebrating-doodle-form/704831001.

[27]Source: https://www.usatoday.com/story/tech/talkingtech/2017/09/27/google.

[28]Source: http://www.internetlivestats.com/google-search-statistics.

[29]Source: https://www.statista.com/statistics/266206/googles-annual-global-revenue.

[30]Source: Financial Report of Yandex (Q4, 2016), http://ir.yandex.com/releasedetail.cfm?releaseid=1012130.

[31]Source: https://en.wikipedia.org/wiki/Baidu.

7.3.4 Knowledge Platforms: Collective Wisdom of the Internet

Internet media of knowledge platforms include online encyclopedia and online Q&A. They are interactive platforms with Internet users' knowledge, experience and opinions on them. The online encyclopedia is a kind of encyclopedia, the content and structure of which can be revised by the users in collaboration on the browsers. Online Q&A allows the users to raise and answer questions and edit and share the questions raised or answered by others. Knowledge platforms started early in North America, Europe and Asia and they have developed fast there while they started late in Africa and Latin America, where there are few large local platforms of knowledge.

1. Wikipedia (The United States)

Launched in January 2001, Wikipedia is the most influential encyclopedia in several hundred languages. Since 2007, it has had over 15 billion visits every month and "wiki" has even been used to refer to all the online encyclopedias. The English Wikipedia is the largest of all the encyclopedias, with over five million articles by September 2016. At the very beginning, there were only 457 users editing over 10 pieces of content, but in 2017, the number was over 2.36 million.[32] Operated by wiki engine, Wikipedia has no owner or leader of the content creation, but anyone can create the entries, add or revise their structure and content. Wikipedia is used more and more as the collaboration software in businesses, covering project communication, intranet and files. It relies on charity donations for its operation, and in recent years, the Wikipedia Foundation, a non-profit organization, has held large-scale money-donation activities to pay for the maintenance of the websites of Wikipedia.

2. Baidu baike (China)

The testing version of Baidu baike was launched on April 20th, 2006. By September 2017, it had had a record of over 15 million entries and over 6.3 million users had participated in the editing.[33] Today, it is the largest online encyclopedia of China. Baidu baike, Baidu tieba, Baidu zhidao and Baidu wenku constitute a knowledge platform array of Baidu. With comprehensive functions and modes similar to those of Wikipedia, Baidu releases the content after it is verified by the machine and the assessor. From May 2012, Baidu began to launch the brand digital museum project, which enables Internet users to appreciate the exhibits through audio guide, scenario simulation and 3D display.

---

[32]Source: https://en.wikipedia.org/wiki/Wikipedia.

[33]Source: https://baike.baidu.com/item/%E7%99%BE%E5%BA%A6%E7%99%BE%E7%A7%91.

3. **Quora (The United States)**

Based on California of the United States, Quora is a Q&A and knowledge sharing website, including the website launched in June 2010 and the application launched in September 2011, providing services in English, French, German, Spanish and Italian. Quora allows its users to raise and answer questions after their registration and revise the raised questions in collaboration while giving suggestions, comments and thumbs-up on the answers and sharing questions and answers. The number of Quora users has been growing since its establishment, with 190 million unique users in 2017. By August 2017, the largest user group of Quora was in the United States, accounting for 35.1% of the total number of its users in the world; and the second largest was in India, accounting for 20.1%. Quora obtained some D-round funds in April 2017, with the total estimated value exceeding US$1.8 billion.[34] Quora's rapid development is owed to the participation and contribution of experts and celebrities from different areas.

4. **Zhihu (China)**

Zhihu, launched in January 2011, is a Chinese Q&A website where questions are created and answered by the community of its users. It pursues the philosophy of "sharing your knowledge, experiences and ideas with the world". In May 2013, relying on the large amount of valuable content it generates every day, Zhihu released Zhihu Daily, a totally new news APP. By September 2017, it had over 100 million subscribers and 26 million active daily users.[35] At the same time, it decided to open registration to organizations, allowing legal and qualified organizations to register with it, including but not limited to research institutions, charity organizations, governmental agencies, media and businesses. Now the influential organizations on Zhihu are Science Museum of China, Microsoft Research Asia, Disney, China Banknote Printing and Minting Corp., and other businesses, organization and universities. Zhihu completed its D-round funding of US$100 million in January 2017. Its estimated value is US$1000 million and it is now a unicorn business.

7.3.5 Audio-Visual Media: A New Force of Internet Entertainment

Audio-visual media or streaming media include online videos, online music, online broadcast, live streaming, short videos and other products or services. They are one of the Internet services with the largest number of visitors, the longest duration, the highest frequency and the largest traffic. It is predicted that in 2018, its traffic will

---

[34]Source: https://en.wikipedia.org/wiki/Quora.

[35]Zhihu Announces That the Number of Its Subscribers Has Amounted to over One Billion and It Will Open Organizational Accounts for Registration, http://tech.qq.com/a/20170920/020694.htm.

account for 76% of the total Internet traffic.[36] In the coming five years, the global entertainment and media industry will continue to expand, its growth rate in 36 countries//regions will be higher than that of GDP and audio-visual media of Internet will become an important growth point. Audio-visual media in the early period mainly had the functions of media sharing and online play, but they have expanded into self-generation and self-delivery of content. There have emerged segmented varieties like short videos, online live streaming and bullet screen websites to meet the entertainment demand of different user groups.

The Untied States have strong audio-visual media products, which have a large number of users in other countries and regions. The cultural diversity of Asia makes audio-visual media colorful there, and all the products have their cultural characteristics. Africa and Latin America have great potential demand for audio-visual media content, with South Africa and Brazil taking the lead.

1. YouTube (The United States)

Created in February 2005, YouTube became the fastest-growing website in 2006. It was acquired by Google in October that year for US$1.65 billion. It is the largest streaming media service provider in the world, with its services covering video sharing and play, live streaming and music. By 2017, YouTube had launched the local versions in over 88 countries and regions, including South Africa, Nigeria, Ghana, Kenya and Uganda, with 1.5 billion monthly active users, whose video watching time added up to 1.0 billion hours and the number of watch times totaled also several billion, with 70% from mobile devices. As for the number of YouTube viewers in the world, the United States ranks first, with 167 million monthly active users; Brazil, second, with 70 million monthly active users; and Russia, third, with 47.4 million. A decade has passed since its creation, but YouTube remains the leader in the field of streaming media though it has experienced the UGC explosion, rise of social media and thriving of the mobile Internet.

2. Netflix (The United States)

Netflix is an American online entertainment company founded in California August 1997. It specializes in and provides streaming media, online video-on-demand, and film and television production and online distribution. As of October 2017, Netflix had 109 million users from 190 countries and regions, including 52.77 million in the United States.[37] Netflix's video-on-demand streaming services allows subscribers to stream television series and films via the Netflix website on PC, or the Netflix software on a variety of supported platforms, including smartphones and tablets, digital media players, video game consoles and TVs. Netflix entered the content-production industry in 2013, debuting its series *House of Cards*. It has

---

[36]PWC: Global Entertainment and Media Outlook (2016-2020), https://www.pwccn.com/en/industries/telecommunications-media-and-technology/global-entertainment-and-media-outlook-2016-2020.html.

[37]Source: https://www.statista.com/statistics/250934/quarterly-number-of-netflix-streaming-subscribers-worldwide.

greatly-expanded the production of both film and television series since then, offering "Netflix Original" content. It released an estimated 126 original series of films in 2016, more than any other network or cable channel. It plans to invest US $8 billion in producing 80 original films and 30 animations in 2018.

3. **Youku Tudou (China)**

Youku Tudou is an online video site with the most users in China. Tudou.com was launched in April 2005 while Youku.com was launched in December 2006. The two were merged into one in March 2011, and the merger was then acquired by Alibaba in October 2015. At present, Youku and Tudou cover 580 million multi-screen servers and 1180 million daily plays,[38] supporting PCs, TVs and mobile terminals, with different content forms including copyright, co-production, self-production, self-channel, live-streaming and VR. Now, Youku Tudou is shifting the traditional model dominated by hard advertisement to the "user income + content marketing" model through its quality content. Self-produced series and variety shows are an innovative measure for content production, with over 50 series (altogether 400 episodes) having been produced. In March 2017, Youku Tudou began to transform into a short video platform, expanding into the PUGC field.

4. **Kuaishou (China)**

Kuaishou is China's mobile short video and live streaming APP, with over 600 million users in the world, mostly in China. It has the largest number of users in the country, producing millions of original videos every day. By February 2016, the total number of its users was over 300 million, including over 10 million of monthly active users.[39] The number of weekly accesses by person was 165.7,[40] so it enjoys high user loyalty, higher than other APPs of its kind. Of the number of active users of short video in China in the second quarter of 2017, that of Kuaishou users accounted for 7.63%, ranking first.[41] The APP has tried to introduce the advertising system of information flow. In the future, besides the spreading-out video advertising, it will adopt models like value-added services and games transport to diversity its ways of making profits.

5. **Twitch (The Untied States)**

Twitch is a live streaming video platform by Twitch Interactive, a subsidiary of Amazon. It was introduced in June 2011 as a spin-off of Justin.tv, but was acquired by Amazon for US$970 million in August 2014. It focuses on videos, games, and live streams, including broadcasts of eSports competitions, creative content, real life streams and more recently, music broadcasts. Content on the site can either be

---

[38]Source: https://baike.baidu.com/item/%E4%BC%98%E9%85%B7%E7%BD%91.

[39]Source: https://www.kuaishou.com/about.html.

[40]Source: Cheetah Lab: China Short Video App Report, http://data.cmcm.com/report/detail/148.

[41]Source: iiMedia Research: China Short Video Market Report, http://www.iimedia.cn/51028.html.

viewed live or via video on demand. As of December 2016, the most popular video was game competition, such as Hearthstone and Dota 2 with a combined total of over 174 million hours watched.[42] Beside games, Twitch provides music dramas, music festivals, broadcasts and real-life streams.

6. Niconico (Japan)

Launched in December 2006, Niconico is the largest online bullet screen video website in Japan. Its biggest characteristic is its real-time bullet screen comments by the viewers, which fly from left to right on the screen when they watch the video to create the experience of all viewers' watching the video together. The atmosphere and culture created by Niconico are highly related to the "otaku culture" and nijigen (lit. "two-dimensional space") culture and the videos are anime, computer games and pop music. Its income comes from paying members, advertisements and the self-owned advertising platform.

7.4 Revolutions in Internet Media Forms

Reform of Internet media forms is characterized by the transformation from traditional media to new media and innovations in content delivery, generation flow and subscription model. Besides, Internet media tend to have open scenarios and focus on participation, layered structure and social interaction. Internet media content and sense of immediacy keep being enriched and improved. New aggregates and feeds, crowdsourcing and crowdfunding news, professional content generation and data-driven journalism have become the trend.

7.4.1 Transformation of Traditional Media

The Internet is producing a great impact on the content generation and spread means, promoting the content delivery of traditional media to expand onto social platforms and news production flow to innovate itself to keep up with the revolution in the users' obtaining information. Thanks to their advantages in content, traditional media keep adjusting their strategies and thus get revitalized in the Internet era.

1. Expansion of content delivery of traditional media onto social platforms

The sharp increase of the number of social platform users has brought opportunities for the resurrection of traditional media, which once fell into difficulties. It is the trend for traditional media to expand onto social platforms. In China, traditional

---

[42]Source: https://en.wikipedia.org/wiki/Twitch.tv.

media have developed in the direction of Weibo official accounts, WeChat official accounts and online news clients, the platforms of which have all been expanded. According to *Report on China's Mobile Media Communication Index (2015)* by people.cn, 1253 media, including newspapers, magazines and TV stations covered in the report, have opened 1304 Sina Weibo accounts, and 1167 WeChat official accounts. 100% of the top 100 media in the index of newspapers, magazines and broadcasts have opened their WeChat official accounts. Major traditional media in the world have expanded onto mainstream social media platforms. Influential ones like the *New York Times*, the *Wall Street Journal*, CNN and Fox News have registered with over 10 social platforms, like Apple News, Facebook, Instagram and Twitter, which are welcome by those traditional media.

2. **Innovation in news production flow of traditional media: Central Kitchen Model**

The practice of "Central Kitchen" is an important innovation in the news production flow of traditional media. In it, three layers of organizational structure, i.e. editor-in-chief's control center—management system of content release and channel distribution—collection and edition collaboration platform, is adopted. Through the platform, varieties of internal resources are integrated to form the full-process production line covering topic selection, content generation and delivery tracing, which makes the collaboration flexible and efficient. The "Central Kitchen" is nothing new. As early as in 2000, some media group built in Florida a news plaza involving paper media, broadcast stations, TVs and websites. Then, some other media from countries like Australia tried to establish the "Central Kitchen", but with no satisfactory results. In China, the representative of such a model is *People's Daily* (see Fig. 7.3). "Central Kitchen" is a new-type media production model characterized by "one collection, more generations, diversified deliveries and around-the-clock news rolling". After such a model of *People's Daily* was launched in February 2016, it soon aroused wide attention and heated discussion from both home and abroad. At present, the "Central Kitchen" of *People's Daily* can provide news products in 18 languages, offering news to 500 mainstream media and news websites in the world.

3. **Constant development of the paid subscription model of traditional media**

For a long time, advertising is the most important profit-making means of traditional media. Recently, the paid subscription model has become a striking trend. The strategy adopted for paid subscription of traditional media is that they provide only a little content on the new media platforms but much on their official websites, on which the readers can read the news only by paying for it. Media with metering pay represented by *The New York Times* release only a little content on mainstream platforms and open a small number of free articles on their official websites to attract users to pay for more content. Media with paid subscription represented by *Financial Times*, *Chicago Tribune* and *Los Angeles Times* seldom release free articles on social platforms, but more content on Apple News for which the users must pay. Take the last general election week as an example. *Chicago Tribune*

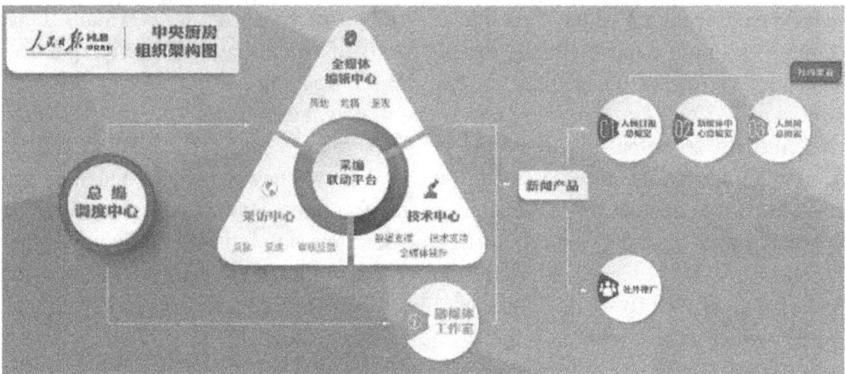

Fig. 7.3 Organizational structure of the central kitchen of *People's Daily*

released 1280 reports on Apple News, while the *Wall Street Journal* and *The Guardian* allowed the users to read the news only after they subscribed as their members and then to enjoy more benefits.

7.4.2 Revolution in the Internet Media Communication Means[43]

Openness, interactivity and participation of the Internet overturn the traditional information communication means and break the one-way information communication model. Against this backdrop, Internet media communication is an open-scenario multi-communication connecting everything, a participative communication in which "everyone can be a journalist", a layered communication bringing out a number of online communities, and a social interaction constantly facilitating advantageous opinions.

1. Open scenario communication

Internet media have transformed from a tool to a medium and then to a platform of socializing, which provides an opportunity for open scenario communication. First, they provide over three billion Internet users worldwide with more diversified channels of communication. In particular, the emergence of social media and knowledge platforms inspire Internet users to express their personal opinions openly on the Internet. Secondly, Internet media have become platforms, which gather users and discover and meet their demand in different aspects. They are

---

[43]The main ideas come from Xie Xinzhou, Dean of School of New Media of Peking University. For details, see Xie Xinzhou and Li Bing. The Main Channel Role and Its Realization Route of New Media in Building Consensus, *Journalism & Communication*, 2016, 23(5).

developing from open communication into in-depth communication, penetrating the social life. The birth and quick popularization of the mobile Internet has facilitated the emergence of scenario communication, which connects everything "anywhere anytime".

2. **Participative communication**

Internet media communication is an evolving participative communication. With new media, Internet users are no longer passive information receiver and Internet media owners are no longer the only information producer and communicator. Internet users can choose to contact the information similar with their opinions and positions and feed back their understanding and interpretations onto the Internet through reprinting and comments and thus participate in the information production and communication of Internet media. At present, a large number of Internet media not only attach importance to adjusting their content and strategies of communicating by analyzing the real-time data of their users, but also channel the users' feedbacks into their interface design, function setting and content presentation to enhance the interactivity. Therefore, the information communication of Internet media has changed into participative and feedback communication that can be selected, edited, commented and refuted.

3. **Layered communication**

With the emergence and development of social media, the "relationship" has become an important factor influencing Internet media communication. "Access from one point and sharing across the Internet" as an information communication model has developed into a multi-level diffusion model. The decentralization of Internet media communication suggests another re-centralization. The Internet has expanded the interpersonal relations from blood ties and geographical relations to relations concerning interest and career and the large Internet user community is therefore divided into sub-communities, or "layers", which are a little closed, isolated in cyberspace. On the other hand, the individuals usually have more than one "sub-community", so all the sub-communication is in layered diffusion structure with one layer extending to another.

4. **Socially interactive communication**

Since the Internet entered the era of Web2.0, its social attributes have been enhanced and the Internet users' personal characteristics and social relations have been externalized and online interaction has had more social attributes manifested in the following aspects. First, online interaction is becoming more similar with daily social interaction and the sense of immediacy is becoming stronger. The interaction between Internet media and Internet users will become more down-to-earth through writing style change and selective use of Internet slangs, and AR and VR technologies will be adopted to create more real interactive scenarios. Secondly, Internet users always avoid being isolated in the interaction. Therefore, when seeing the ideas that they support are welcome, they will take active part in the discussion, expressing and spreading their ideas boldly; otherwise, they will keep silent.

7.4.3 Revolution in Internet Media Content and Its Generation Means

The content generation means of Internet media is developing in the direction of diversification, personalization, professionalization and data-orientation. The richness and sense of immediacy will be enhanced. News aggregation and feeds will enhance personalized communication, and crowdsourcing and crowdfunding will change the news generation flow. UGC and PGC will be integrated with each other and data-driven journalism will facilitate news generation.

1. **Media content richness and sense of immediacy are improved.**

The medium forms supported and presented by Internet media have developed from monotonous texts and images to any combination of texts, images, voices and pictures. Traditional media, for instance, newspapers, TV, pictures and magazines, have been digitalized and integrated. The communication of Internet media tends to be more "real" and realistic and makes the boundary between virtual and real world obscure, so that the users have the sense of immediacy. The video telephone enables people to see each other's expression and action from afar, so that they seem to be communicating face to face. In live streams, the audience can raise questions to the anchor, give him/her a virtual gift and then receive his/her reply and thanks. VR videos and panoramic maps enable the users to control the camera lens and make them move in the scenario, which is consistent with the real experience. The 360° panoramic news has changed the reading style and transformed watching into active exploration, in which the users seem to be on the "spot".

2. **News aggregation and feeds enhance personalized communication.**

News aggregation services or products do not generate news content, but capture a large amount of news through the Internet and recommend it to the users. News aggregation services, to some extent, solve the problem that the users can hardly find the content that they are interested in and related to in the large amount of information. Key technologies of news aggregation include data capture (obtaining a large amount of news content from a large number of news websites), news clustering (putting the news of the same or similar topic into one cluster through some algorithm), content quality scoring (scoring the quality of the news of the same kind from different perspectives) and content output (feeding the news content with high quality scores to the users).

3. **News crowdsourcing and crowdfunding change the news generation flow.**

"Crowdsourcing news" can be categorized into two basic types, namely, crowdsourcing and crowdfunding news. The former refers to the content that news agencies invite the users to contribute to the news report through network platforms. In this process, the users can play the multiple roles, including the news source, text journalist, photographer and commentator. Crowdfunding news is a model in which the media worker proposes the reporting plan which should be approved by the public and be implemented through a number of small funds from the public. The organizer of crowdfunding news can be a journalist, a freelance writer, a news agency or any other organization, who aims to achieve the common journalist goal and to see quality news. At present, the major challenge to this model is the lack of effective business models. The existing business models include extraction of 4–10% as administration fee, point system, charge of subscription fees and "tipping".

4. UGC and PGC are integrated with each other.

User-generated content (UGC) is the form of content created and shared by ordinary users on the information releasing and transmitting platforms. Demand for content-oriented information, lack of quality content and diversity of profit-making ways facilitated the emergence of profession-generated content (PGC). At present, official accounts, big Vs, anchors and online celebrities in journalism, entertainment information, knowledge sharing, fun stories, live games, life shows and popular music have emerged one after another. In China, the representative knowledge talk shows are Luogic show and Papi entertainment short video; in the United States, the YouTube anchor PewDiePie's personal channel has over 57 million subscribers, with over 15.7 billion page views; all the video platforms have begun to create content by themselves and plan to support the creators in technology, promotion and profit division.

5. Data-driven journalism facilitates news generation.

Data-Driven-Journalism or data journalism is a process of digging deep into data by scraping, cleansing and structuring it, filtering by mining for specific information, visualizing and making a story. According to a survey made in September 2017, 42% of journalists are practicing data journalism (twice or more a week). 51% of news agencies in the United States now have a special data journalist.[44] The future development of data-driven journalism will be based on big data and more convenient and faster visual tools.

7.5 Internet Content Governance in Different Countries

Secure, stable and prosperous cyberspace is of significance to all countries and the whole world. However, worldwide privacy invasion, intellectual property right infringement and cybercrime happen frequently, and network monitoring, cyber attack and cyber terrorism have become the public hazards, so that necessity and importance of internet content governance has become the consensus of all governments, businesses, trade associations, media and the public. Though the target of Internet content governance varies from one country or region to another, the core

---

[44]Source: European Journalism Center, State of data journalism globally,http:// datadrivenjournalism.net/news_and_analysis/state_of_data_journalism_globally.

of the governance is the supervision over harmful information, which includes the following: (1) information threatening national security; (2) information of pornography and violence; (3) information invading citizens' privacy, reputation and portrait right; (4) online rumors; and (5) false news. So far, all countries have adopted a variety of strategies on Internet content governance, including the combination of legislative and administrative supervision, self-discipline of the public, media platforms' self-governance, and special actions, which, together with varieties of technical ways, have contained the generation and spread of harmful and false information in Internet media.

7.5.1 The United States: Mature System and Advanced Technology

As the origin of the Internet and the country with the highest number of websites, the United Sates has been paying attention to the Internet content governance since the end of the 20th century and has formed a mature system made up of supervision organizations, supervision laws, civil organizations, governance technology and means, and special actions.

In terms of supervision organizations, the United States has the system made up of traditional administrative organizations and independent committees, the former including the White House, Department of Defense, CIA, Department of Justice, FBI, Department of Homeland Security and National Security Agency. Federal Communications Commission (FCC) is an independent agency in charge of the Internet, regulating and guiding Internet information communication. To enhance the supervision over all kinds of information, the United States has established new organizations like 67th NWG, IC3, and Special Forces Combating Children-targeted Cybercrime. It has also launched Social Network Surveillance Center and Social Network/Media Capacity Project specially supervising the social media like Facebook, Twitter and MySpace.

As for legislation on supervision, the country has launched different laws on different types of information, for instance, *Children's Internet Protection Act* on online information of pornography and violence. Civil governance organizations are mainly engaged in supervision on information and rumors harmful to juveniles and in protection of privacy.

As for governance technology and means, the NSA and FBI have launched projects of large-scale monitoring including Echelon, PRISM, Upstream and Boundless Informant and others by NSA, and DCSNet,MainCore and Magic Lantern and others by FBI. Besides, a series of special actions have been taken targeting online pornography, Internet fraud and illegal transaction (see Table 7.4).

Table 7.4 Special actions of internet information governance in the United States

Competent authorities	Special action	Time	Supervised information category
FBI	Innocent images national initiative	1995	Online pornography and violence
	Avalanche	2001	Online children pornography information
	Operation site down	2003	Online economic fraud
	Cleanup and operation tovar	2013	Botnet
FBI, Department of Justice and NW3C	Cyber loss	2001	Internet fraud
FTC	Recovery	1997–1998	False information of Internet medical care
FCC	E-Con	2003	Online order fraud
National Intellectual Property Rights Coordination Center	Operation in our sites	2010	Online infringement on intellectual property right
Joint Special Action	Signature	2014	Crimes in anonymous networks

7.5.2 China: Importance Attached and Comprehensive Coordination

Since the 18th National Congress of the Communist Party of china, the CPC and Chinese Government have attached great importance to cyber content governance and have made explorations and efforts. On February 27th, 2014, the Central Leading Group for Cyberspace Affairs held the first meeting to facilitate the central leadership and coordination, proposing that unified planning and layout/should be made for cyber security and informatization, and thus opening up a new way of the coordination in Internet information governance.

Cybersecurity Law of the People's Republic of China, launched in 2016, was China's first basic law comprehensively regulating cyber security administration. It defines the target of Internet content governance, the responsibilities of competent administrators and obligations of relevant participants. Special legislations cover journalism, audio-visual programs, APPs, search engines and live streaming platforms. Specifically, these laws include *Provisions for the Administration of Internet News Information Services*, *Provisions on the Administrative Law Enforcement Procedures for Internet Information Content Management*, *Provisions on the Administration of Internet Audio-Visual Program Service*, *Provisions on the Administration of Internet Information Search Services*, *Provisions on the Administration of Mobile Internet Applications Information Services*, *Provisions on the Administration of Internet Live-Streaming Services*, and *Provisions on the Administration of Internet Forum and Community Services*. Industrial legislation covers *Interim Measures for the Administration of Internet-Based Advertising* released by the State Administration of Industry and Commerce, *Provisions on the Administration of Online Publishing Services* and *Provisions on the Administration of Private Network and Targeted Communication Audiovisual Program Services* by State Administration of Press, Publication, Radio, Film and Television and Ministry of Industry and Information Technology, and *Measures for the Administration of Cyber Performance Business Operations* by Ministry of Culture. All those laws are legal basis for cyberspace governance and promote it to develop in the direction of being ruled and regulated by law.

Cyber Administration of China and local cyber administration sectors have carried out initiatives such as Sword Net, Cleaning, Seedling-protection, Clean Sources and Clearness to counter online infringement, pornography, false information and online fraud. In 2016, 678 illegal websites were interviewed in accordance with laws, 3467 illegal websites were deprived of their licenses or records or even closed by cyberspace administration authorities together with industry and information technology authorities. 5604 cases were submitted to judiciary authorities, and 5.06 million illegal account communities were closed in accordance with relevant service agreements.[45]

7.5.3 Russia: Leading Cyber Security Technologies

As one of the biggest Internet countries in the world, Russia, faced with the increasingly complicated cyber environment and shocks and challenges from flooding harmful information, has set up a multi-level Internet information governance system made up of laws, agencies and technologies.

The Russian Government has set up special cyber supervision agencies in the traditional administrative structure, including the Federal Security Bureau (FSB), Federal Communication and Media Administration (FCMA) and Ministry of Internal Affairs (MIA), which have different responsibilities but coordinate with each other. FSB monitors the information concerning the national security on the Internet; FCMA administers the information concerning children's safety, pornography and violence; and MIA Monitoring Center is in charge of monitoring emerging media like Twitter and Facebook. To protect the data of Russian users, the Federal Government has set up the Data Protection Agency to prevent the outflow of the domestic data.

Russia has formulated the *Mass Media Law* and other pertinent laws and regulations as the legal basis for Internet governance. The *Mass Media Law* covers all major Internet governance areas, for instance, information threatening the national security, online pornography and violence, Internet fraud and privacy invasion. Special Internet laws include *Law of New Rules for Prominent Bloggers, Wi-Fi*

---

[45]Source: http://www.chinanews.com/gn/2017/01-20/8131002.shtml.

Real Name Certification and *Protecting Juveniles from Disturbance of Information Harmful to Their Health and Development.*

The country has the most advanced online data monitoring system, i.e. operation monitoring system, a comprehensive system monitoring telephone calls, Internet telephone and Internet communication. The Federal Government requires all telephone and Internet service suppliers to have the system installed. FSB can get data without informing those suppliers, so all the telephone and Internet communication data can be recorded.

Russia has carried out a number of special actions targeting the spread of harmful information, including enhancement of monitoring over all media information and closing of websites spreading harmful information, for instance, about children pornography and suicide methods.

7.5.4 Europe: Terrorism Combat as the Priority of Internet Governance

Since middle and late 1990s, EU has made a lot of legislations on racial discrimination, terrorism combat, juvenile protection and users' privacy protection (see Table 7.5). After Nazi memorabilia was sold on Yahoo! auction site in 2000, fight against racial discrimination became a key area of Internet governance. Since the beginning of the 21st century, terrorism has been a serious threat to European security. The terrorist attack in London in 2011, and that at the office of *Charlie Hebdo* in 2015 have helped to increase the speed of legislation and administration in relevant areas.

7.5.5 Singapore: Participation of the Public and Mild Governance

As one of the first countries using the Internet and one of those with the highest Internet popularization rate, Singapore has been successful in cyber content governance. To ensure national security and social stability and the citizens' freedom and right to the greatest extent, the country pursues the principle of "light interference with content supervision" and announces that the government "will not exercise excessive control over the Internet". Singapore Broadcasting Authority, Films and Publication Department and Singapore Film Commission were merged into the Media Development Administration (MDA), which exercises mild control over the Internet, which is of Singaporean characteristic.

MDA and the Police Force are the governmental agencies responsible for Internet governance. In 2015, the country founded Cyber Security Administration (CSA) affiliated to the Prime Minister's Office. CSA is in charge of the coordination

Table 7.5 Internet Governance of Europe

Country/Region	Type	Content
EU	Polices and laws	*Internet security action plan 1999–2008, cyber crime convention, additional protocol to the convention on cybercrime concerning the criminalization of acts of a raciest and xenophobic nature committed through computer systems, security action plan 2015–2020,* and *e-Europe action plan and directive 2002/58 on privacy and electronic communications*
	Supervision agencies	ENISA and EC3
	Special actions	Action plan for secure access to the internet, odysseus, circamp project combating internet children pornography, and establishment of the online censorship system against violent and terrible contents
The United Kingdom	Polices and laws	*National cyber security strategy, counterterrorism law, safety net agreement regarding rating, reporting and responsibility, privacy and electronic communications* (EC Directive) *regulations 2003,* and *suggestions on children's cyber security*
	Supervision agencies	Office of cyber security, cyber security operations centre, children cyber protection special working Group and CEOP
	Special actions	Operation ore, notarization action, rescue action against children pornography, and special action of deleting internet terrorist information
France	Policies and laws	*Counterterrorism law, protection of minors,* and *information society act*
	Supervision agencies	Cyber and information security administration
France	Special actions	Shielding five websites pampering terrorism
Germany	Polices and laws	*Law of juvenile protection in public places, state agreement on juvenile media protection, law of blocking webpage login, federal data protection law* and *federal information security law*
	Supervision agencies	Federal information technology security administration, national cyber defense center, juvenile media protection committee, and inspection office for media harmful to juveniles
	Special actions	Special investigation action against cross-national children pornography networks

among all governmental agencies concerning cyber security. As for legislation, the country has issued *Internet Behavior Law* and *Law of Internet Practice*, which make it explicit that content violating moral and political standards is the focus of supervision. As for governance means, Singapore has formed the licensing and registration system and censorship of "differential treatment". Besides, the government invites the public to participate in Internet content supervision law and policy making and has established the New Media Consulting Council, a non-official organization made up of representatives from the telecommunication industry and network service agencies and common people. The council presents the opinions from civil society to the government and proposes suggestions on Internet supervision.

7.5.6 *India: Terrorism Countering Above Everything in Internet Governance*

India enjoys developed information technologies. While bringing about new opportunities for Indian development, network and information technologies have produced some negative effects. Since the terrorist attack in Mumbai happened in November 2008, the Indian Government has facilitated the supervision on Internet content. As for supervision agencies, the country has established an administration system covering policies, administration, justice and technology, made up of Ministry of Communications and Information Technology, Network Laws Consulting Committee, chiefs of certification authorities, information decision officials, courts of appeal concerning network laws and computer emergency response groups. As for supervision laws, an Internet monitoring law system has been established, with *Information Technology Law* as the dominant and *Indian Penal Code* and regulations of different sectors as the supporting part. *Information Technology Law* contains detailed provisions and norms about online information supervision. As for supervision technology and means, India has established a complete monitoring system and launched special actions to shield malicious websites, delete harmful information and counter harmful online content or speeches. From June 2013 to December 2014, the country organized a number of special actions, in which 39 pornography websites, 472 file-sharing websites and 140 social media websites and links were shielded.

7.5.7 *Japan: Importance Attached to Privacy and Juveniles Protection*

Japan, a major country of Internet, has its own characteristics in Internet governance. Filtering of harmful information and personal rights and interests protection are the country's focuses of Internet governance.

All administrative agencies in Japan have their supervision sections, which have clear division of work. The Center for Cyber Security Strategy led by the chief cabinet secretary is responsible for coordination of the cyber security policies of different governmental agencies. The cabinet, by cooperating with different prefectures, promotes filtering software for juveniles' access to the Internet and cuts off their contact with harmful information to protect their Internet access environment. The Metropolitan Police Department has set up the Special Group Combating Cybercrime, who investigates into the cases of money defrauding in online bank transactions and provides crime information like online fraud for local police. Private Life Research Society does research on the right of network privacy.

Targeting pornography and violence, Japan has launched *Entertainment Regulation and Standardization Law*, *Law against Illegal Actions including Deceiving Children through Matchmaking Websites* and *Law of Rectifying the Juveniles' Secure Access to the Internet*. It has also issued the *Personal Information Protection Law* to ensure reasonable circulation of personal information and fundamental rights and interests of individuals.

As for civil organizations, Japanese Internet Society is an important Internet content governance agency in Japan. Its supervision and tackling of harmful Internet information covers network hotline, online expert consultation, online technology crime identification, filtering software layout, privacy patent protection and information rating.

As for special actions, Japan attaches importance to countering online pornography. On April 21st, 2011, Japanese Cyber Security Society announced that they would mandatorily block the channels for obtaining child pornographic pictures from ISP level.

7.6 Future Development of Internet Media

Since 1980s, the Internet has brought about an unprecedented revolution in media. Great changes have taken place in the pattern of media, opinion ecosystem, audience and communication technologies. In the future, the deep combination of advanced technologies such as AI, VR, and intelligent algorithm with media development will make news content generation and information communication more diversified, personalized and target-oriented.

7.6.1 AI Facilitates News Generation and Content Identification

AI is a new technology for developing theories, methods, technologies and application systems used for simulating, extending and expanding human intelligence. It is aimed to make robots capable of doing complicated jobs that can usually be done with human intelligence. It is a trend today for AI's application to be expanded into Internet media. As for news generation, Associated press (AP), Los Angeles Times, Yandex, finance.99.com, and Xinhua News Agency began to use AI for writing in

2014, but the content has been confined to sports reporting and financial information, so in the future the machines should learn to deepen their understanding of meaning and texts, so that they are capable of writing news about other aspects. As for content identification, AI can make simple identification of texts, pictures and videos, which, to some extent, helps with the content examination and management by human beings, but the machines' capability of identifying the scenarios has to be cultivated so that they can better detect the network environment.

7.6.2 VR Facilitates the Visualization of Information Communication

VR is one of the hot topics in Internet journalism. In 2015, some world-class media like *New York Times*, BBC, AP, ABC News and *The Guardian* launched their VR-based news products. *People's Daily*, CCTV and news.sina.com.cn in China soon caught up, pushing the visualization of information communication into a new stage. Around 2016, major social platforms and video websites launched 360° video and live streaming services. Experience shows that popular content covers landscape (for instance, the series of videos of National Geographic channels, and submarine diving), sports games (for instance, football games), breaking news (for example, news about storms, tornadoes, landing of SpaceX, etc.), creative commercial advertisements and music videos.

7.6.3 Intelligent Algorithm Improves Communication Efficiency of Internet Media

Intelligent recommendation algorithm is the product of response to mass information on the Internet. The rapid development of the Internet and mobile Internet has resulted in expansion and fragmentation of information and accelerated the users' demand for personalized and vertical news. At present, major news media, social media and search engines are increasing their input in algorithm technology. In the world, Google and Facebook are the pioneers of intelligent recommendation algorithm. They channeled "personalized recommendation" into their technical development strategies as early as in 2005. Toutiao, yidianzixun.com, kb.qq.com and hot moments articles of WeChat later launched personalized news services relying on intelligent recommendation algorithm. Though there are problems like low accuracy and information cocoons with it, intelligent recommendation algorithm can help to realize precise recommendation and personalized reading by learning about users' preference, so it improves communication efficiency and users' experience in the information explosion era.

Chapter 8
International Cyberspace Governance

Chinese Academy of Cyberspace Studies

Abstract Cyberspace is the product of the IT Revolution. Since 1990s, IT Revolution represented by the Internet has gone beyond national and regional borders and cultural differences and created the man-made space connecting all countries and regions, which is the global cyberspace. On the one hand, state and non-state actors have more and more interactions in the cyberspace to enjoy benefits and convenience. On the other hand, traditional politics, economy and social structures are faced with the totally new shock and challenge resulting from the cyberspace and practice proves that no individual actor can only rely on himself in tackling the shock and challenge. How to exercise effective governance over the global cyberspace and build peaceful, secure, open and cooperative cyberspace has become the key strategic issue facing all countries and all actors. (1) The antagonism between the "multiple stakeholders" model and "multilateralism" is the main thread running through the global cyberspace governance, but the two have begun to integrate with each other. China holds that the cyber sovereignty should be respected and consultation should be carried out about common issues, and the country advocates multilateral cyberspace governance involving different participants. (2) The international community has different opinions on global cyberspace governance principles, digital economy development and cyberspace militarization, but they all hold that complete governance rules should be made for cyberspace governance. (3) The global cyberspace governance mechanism keeps being improved. ITU, UN and ICANN play different roles in cyberspace governance and all countries participate in it through bilateral and multilateral mechanisms. (4) China is an active promoter of the reform of the global cyberspace governance system. The proposals on international cyberspace made by President Xi Jinping have won extensive acknowledgement in the international community. To build a community of shared future in cyberspace is becoming the common direction of responsible actors' effort.

Chinese Academy of Cyberspace Studies
Beijing, China

© Publishing House of Electronics Industry, Beijing and Springer-Verlag GmbH Germany, part of Springer Nature 2019

Chinese Academy of Cyberspace Studies (ed.), *World Internet Development Report 2017*,
https://doi.org/10.1007/978-3-662-57524-6_8

8.1 Background of Cyberspace Governance

In recent years, cyberspace governance has increasingly become an important area of the global governance system and has attracted the attention and participation of all countries and regions. Horizontally, cyberspace governance is becoming increasingly necessary and more difficult in comparison with other areas of global governance. Vertically, the order and model of global cyberspace governance keep changing with the characteristics of different stages.

8.1.1 Importance and Necessity of Cyberspace Governance

Cyberspace has become the "fifth space" after the sea, the land, the outer space and the sky. The United States released in 2005 *The National Defense Strategy of the United States of America* and China released in 2016 the *National Cybersecurity Strategy*. In all national strategies of major cyber countries, cyberspace is taken as the "fifth space", equal to the sea, the land, the outer space and the sky.[1] There is no single body who can exercise mandatory administration over the global cyberspace, but varieties of actors, including governments, businesses, non-governmental organizations and individuals, are trying to expand their rights and actions in cyberspace, so they have formulated related codes of conduct conforming to their own interest to get more resources and interests. In fact, in the global cyberspace, there exist big differences in technologies, resources and power of discourse between countries and between Internet businesses. For instance, according to IDC statistics, in the first quarter of 2017, in the ranking of global server system manufacturers in terms of market share, the top four were from Europe and Americas, and their total market share accounted for 57.6% of the worlds' total.[2] Operators of current leading root servers of top geographical domain names in the world are from ICT developed countries in Europe and Americas.[3]

Anarchy in global cyberspace governance gives rise to increasing hazards and threats facing all actors. First, with the increasingly relevance of the Internet to politics, economy, society, people's livelihood and finance, there emerge increasingly diversified, multiplied and complicated hazards in and threats to cyberspace. Cybercrime and malicious cyberattack and other activities not only threaten the security of critical infrastructure of all nations, but also affect people's daily life. In

---

[1] U.S. Department of Defense. *The National Defense Strategy of the United States of America.* March 2005, http://archive.defense.gov/news/Mar2005/d20050318nds1.pdf.

[2] The top four manufacturers were HPE/ New H3C Group, Dell, Cisco and IBM, with their market share accounting respectively for 24.2%, 20.1%, 7.0% and 6.3%. See Worldwide Server Market Revenue Declines 4.6% in the First Quarter as the Market Prepares for a Major Refresh, According to IDC, June 6, 2017, https://www.idc.com/getdoc.jsp?containerId=prUS42707717.

[3] Internet Assigned Numbers Authority. Root Servers, http://www.iana.org/domains/root/servers.

particular, some non-state actors are increasingly likely to launch terrorist attacks through cyber technologies. They keep making propagandas for terrorism, and recruit soldiers for terrorist organizations through cyberspace, so there is increasing harm resulting from cyber attacks and terrorist attacks. Some countries even use the cyberspace to launch military activities and facilitate arm race, which gives rise to greater factors for instability and insecurity. Secondly, since cyberspace is universal, without national boundary, one cyber hazard and threat may affect the whole world, and no country or actor can be free from the threat and none can tackle the situation independently. Therefore, it has become a must in guaranteeing the cyber security that all countries and actors cooperate with each other and set up an internal governance system conforming to the security and economic and social interests of all countries.

8.1.2 History of Cyberspace Governance

Since 1990s, the position on and practice in the Internet and the global cyberspace governance have undergone some changes. There have been two lines. The first is the governance dominated by technology communities, and the major organization is ICANN. The other is the governance dominated by governments, who is more and more deeply involved in it. Based on the two lines, the global cyberspace governance can be divided into four stages.

The first stage lasted from the founding of ICANN in 1990s till the opening of the World Summit on the Information Society (WSIS) in Tunis in 2005. Superficially, multiple stakeholders participated in the cyberspace governance dominated by the technology communities at this stage, but actually the U.S. Government controlled the overall situation. At that time, PCs and laptops were the major terminals and browsers were the major software for access to the Internet, so root servers, root zone files and root zone file systems supporting global TLD resolution were the most symbolic critical infrastructure and in what model to govern the root servers was the most symbolic indicator measuring the global cyberspace governance.[4] In 1997, with the Clinton government announcing the Information Highway Program, the United States had opinions different from technology communities contributing to early development of the Internet. They had heated and complicated debates on what principles to follow for governing root servers, root zone files and root zone file systems, the most important and symbolic infrastructure of the Internet. The debate resulted in the establishment of the mechanism for multiple stakeholders' governance in accordance with the position of technology communities and the founding of ICANN, but in fact, the U.S.

---

[4]Marshall Leaffer. Domain Names, Globalization, and Internet Governance, *Indiana Journal of Global Legal Studies*, 1998, pp.139–165.

Government established its unilateral governance over root servers, root zone files and root zone file systems by setting up IANA.[5]

The second stage lasted from the opening of the WSIS in Tunis in 2005 until the PRISM revealed by Snowden in 2013. At this stage, the Working Group on Internet Governance (WGIG) was founded and the U.S. Government's unilateral governance over ICANN was suspected and challenged, but the multiple stakeholders' governance remained the main model. The model, task, dynamic mechanism and actors became the focuses of all parties involved. In 2005, at the WSIS in Tunis, the international community had a wide-scope discussion on whether to exercise governance on the Internet and how to govern it. As a consequence, the WGIG was founded and research was started on the basic issues like the definition of cyberspace governance. Reports were released on the results of the research then. In 2006, the Internet Governance Forum (IGF) was set up as an important platform in the framework of the UN for open discussion on Internet governance. In 2011, countries including China and Russia submitted to the 66th UN General Assembly the *International Code of Conduct for Information Security*, advocating that the UN should play a leading role in cyberspace governance. In the same year, the global cyberspace governance conference was held with the United States and the United Kingdom acting as the leaders, and thus London Process was started.

The third stage lasted from Snowden's disclosure of the PRISM in 2013 till the U.S. Government's "delivery" of the administration over IANA in September 2016. At this stage, the reform of ICANN witnessed progress and the U.S. Government's administration over IANA shifted from administrative to judiciary. The model of multiple stakeholders' governance was challenged by the multilateral governance mechanism. In June 2013, PRISM was disclosed to the public, which shook the moral and theoretical basis for the United States to openly monopolize the jurisdiction over the critical infrastructure of global cyberspace. "Good governance of benign hegemony" could not continue to be the excuse for the U.S. Government to unilaterally govern the basic Internet infrastructure, and the stagnant ICANN witnessed progress in its reform. The scandal enhanced the attention of all countries to cyberspace governance, which became an important issue on the agenda of global governance. The international community had more different ideas on the cyberspace governance philosophy and model and the competition became fiercer, which led the global cyberspace governance into chaos and difficulty to some extent.[6] Brazil and China successively launched the cyberspace stakeholders' conference and the World Internet Conference. Through such platforms, countries can discuss core issues like cyber and national security, and cyber sovereignty. Under the pressure from China and Russia, NTIA of the United States announced in March 2014 that they were considering delivering the administration over IANA.

---

[5]Château de Bossey. Report of the Working Group on Internet Governance. WGIG, 2015, http://www.wgig.org/docs/WGIGREPORT.pdf.

[6]Lu Chuanying, Power Game, Idea Evolution and China Strategy on Cyberspace. *Global Review*, 2016(1), pp. 117–134.

The delivery was completed in September 2016, which might not be complete because in essence, the U.S. executive administration over the basic Internet resources like root servers was shifted into the judicial administration in accordance with the U.S. laws. Anyway, the delivery was of some positive significance seen from the practice of global cyberspace governance. Meanwhile, China, a major Internet country, launched the World Internet Conference in 2014, inviting governmental leaders, ministers in charge, entrepreneurs and technical communities to be present to discuss the way of cyberspace development and governance. Especially, President Xi Jinping was present at the Second World Internet Conference and put forward "four principles" and "five proposals" on promoting the cyber space governance system reform and building a community of shared future in cyberspace. The principles and proposals won the universal recognition of the international community and produced increasing influence on cyberspace development and governance. The multilateral governance mechanism dominated by governments was more and more recognized and accepted by the international community.

The fourth stage started from the U.S. Government's delivery of the supervision over IANA in October 2016. Since then, the global cyberspace governance has been at the stage of paralleled multilateral and multi-stakeholders' governance. All governments are more actively involved in cyberspace governance. With its rapid popularization and development, the Internet keeps penetrating all areas, including economy, politics, culture, society and military affairs, and multilateral governance is far from satisfying the demand of the reality. Meanwhile, all countries take the cyberspace development and governance as a new area of international cooperation and a new height for improving their competitiveness. Multilateral mechanisms like the UN, NATO, G20, APEC, BRICS and CICA, and bilateral dialogue mechanisms, such as China–U.S., China–Russia, China–UK and U.S.–Russia dialogues, take cyberspace governance as their priority of cooperation. Countering cybercrime and cyber terrorism and guaranteeing the order in the cyberspace have become the focus of intergovernmental discussion and cooperation. All countries' concern about and participation in cyberspace governance have been facilitated, which indicates that global cyberspace governance has entered the "deep-water zone".

8.2 Major Ideas of Cyberspace Governance

Global cyberspace is the field where different governance ideas compete with each other fiercely. In general, the conflict between Global Commons and sovereignty principles, and the difference between Multi-stakeholder Model and Multilateralism have been dominating the cyberspace governance philosophy. The communication and integration between conflicting ideas have been put on agenda. Actually, different actors have their characteristic ideas, for instance, China and the United States.

8.2.1 A Brief Introduction to Major Ideas on Cyberspace Governance

As for cyberspace governance, there exist two basic ideas. Traditionally ICT developed countries represented by Europe and the United States have established and are promoting the Multi-stakeholder Model, holding that cyberspace belongs to Global Commons, so its governance should be done by actors beyond governmental agencies. Quite a few developing countries and emerging economies advocate Multilateralism, holding that cyberspace has its attribute of sovereignty and that the national government should be the dominator in global cyberspace governance. The above-mentioned two ideas answer the questions of three levels, namely the essential attribute of cyberspace, the position of the national government in global cyberspace governance and the model to be adopted for the governance. On the one hand, the conflict between the two ideas is quite obvious and their competition will remain the main thread running through global cyberspace governance. On the other hand, in recent years, different countries advocating the two ideas and models have been communicating with each other and trying to seek cooperation in cyberspace.

Specifically, to ignore national sovereignty and underplay the role of the government in cyberspace and promote the Multi-stakeholder Model is beneficial to traditional major countries of ICT. Some Western countries represented by the United Sates advocate the idea of "Internet freedom". According to the *International Strategy for Cyberspace* issued by the Obama Government, the national government's control over cyberspace out of the sovereignty principle hinders the free flow of the Internet information, the global cyberspace should be completely commercialized space, and the impact of national policies on cyberspace should be reduced to the minimum.[7] This, in fact, ensures that the countries with technical advantages can output digital products and services to developing countries, which will benefit the former and thus consolidate their advantages. European countries and the United Sates also pay attention to their own cyber security, so they oppose the principle of cyberspace sovereignty superficially and guarantee their own digital sovereignty in essence.

Multi-stakeholder Model is the way adopted by ICT powers to impair the role to other national governments and guarantee their own position as the rule makers. It is the extension of the idea about multi-stakeholder governance of corporations,[8] of which the owners are not only the investors, but also employers, creditors, suppliers and consumers. To apply Multi-stakeholder Model of corporate governance to

---

[7]The White House. International Strategy for Cyberspace: Prosperity, Security, and Openness in a Networked World. May 2011, p. 5, https://obamawhitehouse.archives.gov/sites/default/files/rss_viewer/international_strategy_for_cyberspace.pdf.

[8]John E. Savage and Bruce W. McConnell. Exploring Multi-Stakeholder Internet Governance. EastWest Institute, January 2015, https://www.eastwest.ngo/sites/default/files/Exploring%20Multi-Stakeholder%20Internet%20Governance_0.pdf.

cyberspace governance is, in essence, a metaphor or analogue. According to Lawrence E. Strickling, a communication and information official from the U.S. Department of Commerce, Multi-stakeholder Model involves the full involvement of all stakeholders, consensus-based decision-making and operating in an open, transparent and accountable manner.[9] However, and unavoidable fact is that different actors have different capabilities and not all sovereign states can participate in cyberspace building in a symmetrical and equal way, which results in the imbalanced influence of different actors.

In the Internet governance after 1990s, especially in the operation of ICANN, Multi-stakeholder Model has been set up and operated, in which decisions are made on cyberspace governance in a bottom-up process. The "bottom" refers to the loose areas or professional problem committees made up of businesses, individuals and non-governmental organizations, and the "up" refers to the guiding committee made up of a few professionals, which has the highest decision-making right. In this case, representatives from sovereign states have no right to policy-making but express non-mandatory suggestions on public policies and international laws.[10] The design of the ICANN architecture stipulates that U.S. NTIA has the executive authority over root servers, root zone files and zone file systems. The U.S. Government, through procurement bidding, authorizes ICANN to manage the domain name resolution system, including domain name registration, IP address allocation and technical parameter revision, all of which are under the supervision of the U.S. Government. According to a report by WGIG, this has enabled the United States to enjoy great authority in the policy-making of ICANN, and put root servers t under the jurisdiction of the U.S. Government.[11]

In comparison with Multi-stakeholder Model, Multilateralism has emerged in recent years, which reflects the development trend of the Internet, which at the beginning, was only something new involving a few actors, such as the technical staff. With the popularization of the Internet, cyberspace now covers most people and it is closely related to economy, society and security, so there has emerged more and more overlaps between cyberspace governance and executive authority. In this change, the government cannot stay out but should perform its function to ensure smooth and secure operation of society and economy. The unilateral authority of countries with advantages in ICT over cyberspace governance has stimulated developing countries to protect their own interests and promote reform of order through multilateralism. In emerging economies, the information and communication industry is becoming increasingly important in economic development. It is becoming increasingly necessary for them to get involved in

---

[9]Lawrence E. Strickling. Moving Together Beyond Dubai. National Telecommunications & Information Administration, April 2, 2013, https://www.ntia.doc.gov/blog/2013/moving-together-beyond-dubai.

[10]Lennard G. Kruger. Internet Governance and the Domain Name System: Issues for Congress. Congressional Research Service, November 18, 2016, https://fas.org/sgp/crs/misc/R42351.pdf.

[11]Château de Bossey. Report of the Working Group on Internet Governance. WGIG, 2015, http://www.wgig.org/docs/WGIGREPORT.pdf.

cyberspace governance and they are getting more and more active in it, but limited by the Multi-stakeholder Model, their right to get involved cannot be realized. The PRISM scandal has enhanced the sense of insecurity of developing countries in the area of communication, which has stirred up the wave of enhancing cyberspace governance within sovereignty.

China plays a key role in facilitating the international community's respect for cyberspace sovereignty. As mentioned above, the applicability of the sovereignty principle to cyberspace is the natural requirement of the Internet development. However, the official proposal by a sovereign state on respecting the cyberspace sovereignty principle was first made by China. According to *The Internet in China* (White Paper) released in 2010, "The Chinese Government believes that the Internet is an important infrastructure facility for the nation. Within Chinese territory the Internet is under the jurisdiction of Chinese sovereignty. The Internet sovereignty of the country should be respected and protected."[12] In 2014, when delivering a speech at the Congress of Brazil, President Xi Jinping pointed out, "In today's world, the Internet development has brought about new challenges to the national sovereignty, security and development interests, so we must tackle them seriously. Though the Internet is highly globalized, every nation's sovereign interests in the field of information must not be violated and the Internet development must not be used to invade other countries' sovereignty. There are no double standards in that filed and every nation has the right to guarantee its information security. We will not allow it to happen that one country is safe but others are not or that some countries are safe but others are not. We should not sacrifice other countries' security to ensure our so-called absolute security."[13] He made a further illustration of this idea at the Second World Internet Conference, emphasizing that respect for sovereignty over the cyberspace should be defined as one of the four principles. He said, "The principle of sovereign equality enshrined in the *Charter of the United Nations* is one of the basic norms in contemporary international relations. It covers all aspects of state-to-state relations, which also includes cyberspace."[14] *Cybersecurity Law of the People's Republic of China*, which was put into enforcement officially on June 1st, 2017, stipulates that one of the aims of the law is to "safeguard the sovereignty over the cyberspace, national security and public interests."[15] In the framework of the UN, thanks to the constant effort of China and Russia, the report made by UNGGE identifies the position of

---

[12]Information Office of the State Council of the People's Republic of China,*The Internet in China* (White Paper), people.cn, June 8th, 2010, http://politics.people.com.cn/GB/1026/11813615.html.

[13]Xinhuanet.com, Speech delivered by President Xi Jinping at the National Congress of Brazil, July 7th, 2014, http://news.xinhuanet.com/world/2014-07/17/c_1111665403.htm.

[14]Xinhuanet.com, Remarks by President Xi Jinping at the Opening Ceremony of the Second World Internet Conference, December 16th, 2015, http://news.xinhuanet.com/politics/2015-12/16/c_1117481089.htm.

[15]Ministry of Industry and Information Technology of the People's Republic of China, Cybersecurity Law of the People's Republic of China, November 8th, 2016, http://www.miit.gov.cn/n1146295/n1146557/n1146614/c5345009/content.html .

national sovereignty in the global cyberspace governance. The report, released in June 2013, covers the following content, "State sovereignty and the international norms and principles that flow from it apply to states' conduct of ICT-related activities and to their jurisdiction over ICT infrastructure within their territory." This is an explicit declaration of the national sovereignty principle for cyberspace.[16]

On the basis of the sovereignty principle, Multilateralism does not oppose non-state actors' involvement in global cyberspace governance, but advocates that all countries' governments play a bottom-up policy-making and leading role. Then a basic framework should be set up for the rules on the involvement in global cyberspace governance, dominated by the equal cooperation between governments. China stresses the significance of Multilateralism in safeguarding the global cyberspace order, but does not put Multilateralism in absolute opposition to the Multi-stakeholder Model. Instead, the country emphasizes that we should seek for a more inclusive solution to improve global cyberspace governance. Joseph S Nye Jr., an American scholar, points out that cyberspace contains physical infrastructure and virtual assets, the administration of which should comply with different logic and principles, the former being under the legal jurisdiction under the sovereignty principle of the national government, and it being difficult to exercise sovereign administration over the latter.[17] In a recent UNGGE report, the declaration about the sovereignty principle for cyberspace is put in parallelism with the concern about human rights and freedom, which shows that both Multilateralism and Multi-stakesholder Model have been universally recognized by the international community.

8.2.2 *Philosophy and Position of Major Countries on Cyberspace Governance*

Major countries have not simply chosen between Multi-stakeholder Model and Multilateralism, but have absorbed and integrated some factors of the two models into their position in accordance with their own demand and their idea on global cyberspace order, so they have their own philosophy and position, which tend to be competitive in dialogues and cooperation (see Table 8.1).

Though a strong advocator of Multi-stakeholder Model, the United States does not pay no attention to the government's role in cyberspace governance or to sovereignty over the cyberspace. In recent national strategies, the country continues to advocate Multi-stakeholder Model on the one hand, but on the other proposes

---

[16]United Nations General Assembly. Group of Governmental Experts on Developments in the Field of Information and Telecommunications in the Context of International Society. June 24, 2013. http://www.unidir.org/files/medias/pdfs/developments-in-the-field-of-information-and-telecommunications-in-the-context-of-international-security-2012-2013-a-68-98-eng-0-578.pdf.

[17]Joseph S Nye Jr. The Regime Complex for Managing Global Cyber Activities. *Global Commission on Internet Governance*, Paper Series No. 1, May 2014.

Table 8.1 Overview of major countries' philosophy on global cyberspace governance

Country	Representative documents	Main ideas
The United States	*International Strategy for Cyberspace* (2011)	Promise on fundamental freedom, privacy and free information flow; prior policies on market opening, domestic cyber protection, law implementation, military means, Internet governance, international development and Internet freedom
China	Remarks by President Xi Jinping at the Opening Ceremony of the Second World Internet Conference (2015); and *International Strategy of Cooperation on Cyberspace* (2017)	Four principles: respect for cyber sovereignty; maintenance of peace and security; promotion of openness and cooperation; and cultivation of good order Five proposals: First, speed up the building of global Internet infrastructure and promote inter-connectivity. Second, build an online platform for cultural exchange and mutual learning. Third, promote innovative development of cyber economy for common prosperity. Fourth, maintain cyber security and promote orderly development. Fifth, build an Internet governance system to promote equity and justice
EU	*Cyber crime Convention* (2001); *Cybersecurity Strategy of the European Union* (2013); and *The Digital Single Market Strategy* (2015)	Collaboration in countering cybercrime, enhancing cyberspace defense, and reducing and even eliminating the barriers against cross-border data flow
Russia	*Doctrine of Information Security of Russia* (2000, amended in 2016)	Information security is the key aspect of national security. It covers sovereignty, political and social stability, territory integrity, human rights and freedom and critical information infrastructure
India	Statement made by India at the 66th Session of the UN General Assembly (2011); and *National Cyber Security Policy* (2013)	Proposal on establishing United Nations Committee for internet-related policies; Establishment of secure and resilient cyberspace of the people, business and government

protecting their own critical infrastructure through the government's actions to meet the challenge to cyber security. According to the *International Strategy for Cyberspace* released by the United States in 2011, the government plays a key role in cyberspace governance.[18] In the *DoD Cyber Security* released by the Department

---

[18]The White House. International Strategy for Cyberspace: Prosperity, Security, and Openness in a Networked World. May 2011, https://obamawhitehouse.archives.gov/sites/default/files/rss_viewer/international_strategy_for_cyberspace.pdf.

of Defense, serious cyberattack is taken as a cyber war, and every means can be used to counter the attack, including PGM.[19] After Trump took presidency, he signed an executive order on improving the U.S. governmental agencies' cyber security to prevent cyberattack from other governments or hackers. Then he announced that the former Cyber Command was elevated into the Unified Combatant Command, and pointed out that the elevation would consolidate the U. S. cyberspace actions and improve the country's defense capabilities.[20] Trump's actions mentioned above show that they show concern about data sovereignty and the government's role and their tendency in military preparations concerning cyberspace.

China pursues the philosophy that the respect for cyber sovereignty should be combined with the facilitation of mutual benefit and win-win cooperation in cyberspace and thus a community of shared future in cyberspace can be built. At the Second World Internet Conference held in 2015, President Xi Jinping put forward "four principles" and "five proposals" on Internet development and governance in his keynote speech. He said, "We should respect the right of individual countries to independently choose their own path of cyber development, model of cyber regulation and Internet public policies, and participate in international cyberspace governance on an equal footing. No country should pursue cyber hegemony, interfere in other countries' internal affairs or engage in, connive at or support cyber activities that undermine other countries' national security."[21] This is the foundation for mutual benefit and cooperation in cyberspace. Of course, to offer undifferentiated support to the cyber sovereignty of different countries does not only point to the demand of China itself for protecting its security and interests, or mean that countries should close their doors in terms of cyberspace. Instead, China holds that international cyberspace governance should feature a multilateral approach with multi-party participation. It should be based on consultation among all parties, leveraging the role of various players, including governments, international organizations, Internet companies, technology communities, non-governmental institutions and individual citizens. There should be no unilateralism.[22] "Four principles" and "five proposals" conform to the rule of the Internet development, fully manifesting the responsibility and mission that China would like to undertake as a big country.

---

[19]U.S. Department of Defense. The DoD Cyber Strategy. April 2015, https://www.defense.gov/Portals/1/features/2015/0415_cyber-strategy/Final_2015_DoD_CYBER_STRATEGY_for_web.pdf.

[20]The White House. Statement by President Donald J. Trump on the Elevation of Cyber Command. August 18, 2017, https://www.whitehouse.gov/the-press-office/2017/08/18/statement-donald-j-trump-elevation-cyber-command.

[21]Xinhuanet.com, Remarks by President Xi Jinping at the Opening Ceremony of the Second World Internet Conference, December 16th, 2015, http://news.xinhuanet.com/politics/2015-12/16/c_1117481089.htm.

[22]Xinhuanet.com, Remarks by President Xi Jinping at the Opening Ceremony of the Second World Internet Conference, December 16th, 2015, http://news.xinhuanet.com/politics/2015-12/16/c_1117481089.htm.

In March 2017, China issued the *International Strategy of Cooperation on Cyberspace*, which covers the opportunities and challenges, basic principles, strategic goals and action plans for China to participate in international cyberspace governance. In line with the "four principles" and "five proposals", the *International Strategy of Cooperation on Cyberspace* defines the strategic goals as: to guarantee China's cyber sovereignty, security and development interests, to ensure the secure and orderly flow of Internet information, improve the international interconnectivity, protect the peace, security and stability of cyberspace, foster the rule of law on cyberspace, promote global digital economy development, and deepen cyber culture exchange so that the Internet development achievements can benefit the world and all peoples.[23] These goals combine China's economy and society as well as its security and interests with the interests of the international community, an aim to promote the success of international cyberspace governance through practical international cooperation.

EU tends to adopt an extensive and comprehensive social governance model for cyberspace governance, putting emphasis on the protection of citizens' rights and interests, and taking the cyberspace as a democratic place governed by law instead of a place of arm race. EU also prefer Multi-stakeholder Model, but meanwhile emphasizes that governmental departments should have their position and extent of competence. By releasing the *Digital Single Market Strategy* in 2015, EU aims to focus on the public-private cooperation in cyberspace and consolidate cyber security by facilitating cooperative innovations and increasing investment,[24] Besides, through policy frameworks and practices, EU provides rich experience for multilateral collaboration in cyberspace governance. Cooperative focuses are on countering cybercrime and setting up the cyber defense system and unified digital market, showing to the international community the possibility and potential of coordinating policies and centralizing actions in cyberspace governance through treaties and strategies.

Russia's position on international cyberspace governance has been affected by its relationship with Western countries. Due to the tension in traditional political and security areas between Russia and Western countries, it tends to take cyberspace also as something political, holding that the Internet will become a tool for Western countries to subvert other regimes. In the multilateral frameworks like the UN, Russia has proposed formulating new-type codes of conduct. For instance, in 2011, it submitted jointly with China to the UN General Assembly the *International Code of Conduct for Information Security*, holding that an

---

[23]Ministry of Foreign Affairs of the People's Republic of China, *International Strategy of Cooperation on Cyberspace*, http://www.fmprc.gov.cn/web/ziliao_674904/tytj_674911/zcwj_674915/t1442389.shtml.

[24]European Commission. *Digital Single Market: Bringing Down Barriers to Unlock Online Opportunities*. https://ec.europa.eu/commission/priorities/digital-single-market_en.

international cyber security treaty it had proposed should be signed in the framework of the UN to prevent the realization of hostile political purposes through the Internet.[25]

India and Brazil are also active in advocating the reform of the order of international cyberspace governance and the establishment of a more democratic and inclusive governance system. India made a statement at the UN General Assembly in 2011, proposing that a new mechanism entitled United Nations Committee for Internet-Related Policies should be established, which should be directly responsible for the UN General Assembly.[26] This, in essence, is a model of multilateralism, which has been identified and supported in the framework of IBSA (India, Brazil, South Africa) Dialogue Forum. After PRISM was revealed, Mrs. Dilma Rousseff, then President of Brazil, postponed her state visit to the United States and openly condemned the U.S. Government for its violating international laws by monitoring other countries, appealing to the UN for the formulation of the Internet supervision norms.[27] But in recent years, the position of India and Brazil has changed. The change in the former started with the beginning of Narendra Modi's regime. In 2015, the country officially and openly declared that it supported Multi-stakeholder Model of Internet governance,[28] which resulted from Modi Government's seeking for domestic private sector's support and for cooperation with the United States in security. Brazil has quietly shifted from a country supporting Multilateralism to a country supporting Multi-stakeholder Model.[29]

8.3 Platforms of Cyberspace Governance

There are varieties of cyberspace governance platforms, global and regional and bilateral.

---

[25]United Nations General Assembly. *Letter Dated 12 September 2011 from the Permanent Representatives of China, the Russian Federation, Tajikistan and Uzbekistan to the United Nations Addressed to the Secretary-General*. September 14, 2011, https://ccdcoe.org/sites/default/files/documents/UN-110912-CodeOfConduct_0.pdf.

[26]The Centre for Internet Society. *India's Statement Proposing UN Committee for Internet-Related Policy*. October 26, 2011, https://cis-india.org/internet-governance/blog/india-statement-un-cirp.

[27]General Assembly of the United Nations. *Brazil, HE Mrs. Dilma Rousseff, President*. September 24, 2013, https://gadebate.un.org/en/68/brazil.

[28]ICANN. *Indian Government Declares Support for Multistakeholder Model of Internet Governance at ICANN53*. June 22, 2015, https://www.icann.org/resources/press-material/release-2015-06-22-en.

[29]Cameron F. Kerry. *Bridging the Internet-Cyber Gap: Digital Policy Lessons for the Next Administration*. Center for Technology Innovation at Brookings, October 2016, https://www.brookings.edu/wp-content/uploads/2016/10/internet-cyber-gap-final.pdf; Gulshan Rai. *Lesson from BRICS: Developing an Indian Strategy on Global Internet Governance*. Observer Research Foundation, September 6, 2014, http://www.orfonline.org/article/lessons-from-brics-developing-an-indian-strategy-on-global-internet-governance.

8.3.1   Global Platforms of Cyberspace Governance

Global platforms of cyberspace governance include global organizations committed to cyberspace governance, like the UN, and special organizations in Internet technology management, such as ICANN. Both kinds influence and promote the global cyberspace governance from different levels and in different degree.

1. **Global comprehensive organizations of cyberspace governance: the UN and others**

A number of global multilateral organizations like the UN promote international cyberspace governance in different ways. In the framework of the UN, platforms concerning cyberspace governance are UNGGE, WSIS, IGF, and ECOSOC. There are other comprehensive international organizations, such as WEF, World Bank, IMF, WTO, and INTERPOL, which are all involved in cyberspace governance.

Important comprehensive global organizations concerning cyberspace governance are shown in Table 8.2.

The activities of the international organizations have played an important role in promoting global cyberspace governance from different perspectives and in different ways.

Table 8.2 Overview of comprehensive global organizations concerning cyberspace governance

Organization	Overview	Achievements
UNGGE	Established for times: 2004–2005, 2009–2010, 2012–2013, 2014–2015, and 2016–2017	Expert reports to the UN General Assembly
WSIS	The first-stage conference was held in Geneva in 2003, and the second-stage conference, in Tunis in 2005	*Geneva Declaration of Principles, Geneva Action Plan, Tunis Commitment* and *Tunis Agenda for the Information Society*
IGF	Founded in 2006, and operated by multi-stakeholder Advisory Group and Secretariat founded with the authorization of the UN Secretary General	Varieties of activities, including annual meetings, focusing on international cyberspace governance
ITU	Formerly International Telegraph Union founded in 1865; main function; formulating management systems and standards for international radio and telecommunication, with 193 member states	*ITU Convention, Constitution of the International Telecommunication Union, International Communication Rules* (new version) and *Measuring the Information Society Report*, ICT Development Index, and Global Cybersecurity Index (GCI)
WEF	Founded in 1971; Davos Forum every year; a number of activities and proposals	Discussions on the IT Revolution at the Davos Forums; *Global Information Technology Report*, and Networked Readiness Index

(1) They carry out research on the status of global cyberspace and set up the framework of measuring and analysis to enhance the world's concern and comprehension about issues related to cyberspace.

ITU has released *Measuring the Information Society Report* and assessed the information and communication development and application in most countries and regions in accordance with the ICT Development Index (IDI). In 2014, ITU launched the Global Cybersecurity Index (GCI), which is the only authoritative international rating index measuring a country's cyber security to assess the member states' commitment to cyber security.

In recent years, WEF has been producing increasing impact on cyberspace governance. It released *The Global Information Technology Report* in 2001 and then the Networked Readiness Index (NRI) to measure the opportunities for development brought about by all economies' development and use of ICT. The ranking by the NRI shows that there exists a big "digital divide" between countries and regions in terms of the application of emerging Internet technologies and transformation of digital economy, taking environment, readiness, use and influence into consideration. Therefore, there is an immediate need for the international society to cope with it.[30] *The Global Risks Report 2017* released by WEF has listed cyber security, for instance, cyberattack, data disclosure, and arm race in cyberspace, as one of the major risks facing today's world.[31]

The *World Development Report* released by the World Bank in recent years also stresses the global cyberspace governance. In the Report of 2016, the World Bank holds that due to the widening gap in ICT and digital economy between countries, it is necessary to strengthen international cooperation in Internet governance, cross-border circulation of digital products and services, and relevant public products supply.[32] Besides, IMF has released the *Fintech and Financial Services: Initial Considerations*, which defines an analysis framework for fintech to respond to the users' demand, change the world competition landscape and influence supervision measures, and points out the necessity to carry out international cooperation in fintech and relevant areas.

(2) Multilateral conferences themed cyberspace governance have been held to facilitate dialogues between countries.

In the UN framework, the most important dialogue platforms are WSIS and IGF. WSIS was divided into two stages. The first-stage conference was held in Geneva in 2003. Two programmatic documents were adopted, namely, *Geneva Declaration of*

---

[30]Silja Baller, Soumitra Dutta and Bruno Lanvin, eds. *The Global Information Technology Report 2016: Innovating in the Digital Economy.* 2016, http://www3.weforum.org/docs/GITR2016/GITR_2016_full%20report_final.pdf.

[31]World Economic Forum. *The Global Risks Report 2017*, 12th Edition. 2017, http://www3.weforum.org/.

[32]World Bank. *World Development Report 2016*: Digital Dividends. 2016, http://documents.worldbank.org/curated/en/896971468194972881/pdf/102725-PUB-Replacement-PUBLIC.pdf.

Principles and *Geneva Action Plan*. At this stage, the Working Group on Internet Governance (WGIG) was established. In 2005, the second-stage conference of WSIS was held in Tunis and also two programmatic documents were adopted, namely, *Tunis Commitment* and *Tunis Agenda for the Information Society*. The former emphasizes the social challenge of closing the digital gap and the latter sets up the agenda for international cyberspace governance. The latter also proposed to the UN Secretary General the establishment of a democratic and transparent forum with extensive involvement and with no restraint. Thus the forum IGF was founded in 2006. In the past over ten years, IGF has been focusing on the five cyberspace governance topics, namely, openness, security, diversity, entry permission and critical resources. Its annual activities include annual meetings, workshops, dynamic union meetings, best practice forums, boundary meetings, host country meetings, flash meetings, open forums, inter-regional dialogue meetings, non-conference sessions, IGF villages, providing a wide range of platforms for countries and Internet organizations to get involved in global cyberspace governance.

(3) Formulation of rules and norms of cyberspace governance is advocated.

UNGGE, made up of a limited number of members, is a working group specialized in research on information security and in release of relevant reports. By far, five sessions have been set up, with the former three made up of experts from 15 countries, and the fourth and fifth made up of experts from 20 to 25 countries respectively.[33] UNGGE usually submits reports to the UN General Assembly to promote the international Internet governance, but quite a few people are pessimistic about the prospect of UNGGE, holding that the group can hardly play an effective role in formulating norms of global cyberspace. ECOSOC, UNODA and Counter-Terrorism Committee under the UN also facilitate the international cooperation and rules formulation concerning cyber security, cybercrime, cyber wars and counter-terrorism.

(4) International actions are taken in areas like cyber security.

UNESCO launches open online courses, mobile learning weeks, and international Internet education partner conferences to facilitate the international community's stress on cultivating the next-generation Internet force and promote the application of ICT in education. INTERPOL carries out and coordinates multilateral actions in countering cybercrime, covering relevant technology exploration, digital laboratory operation for investigating and detecting cybercrime, cooperation with local law enforcement agencies, and collaborative trailing and cracking of cybercrime.

2. **A special organization for global cyberspace governance: ICANN and its history**

ICANN, founded in 1998, is playing an important role in the international Internet governance, since it has the specific authority over critical Internet resources such

---

[33]Geneva Internet Platform. Members of UN GGE. https://dig.watch/processes/ungge.

Table 8.3 ICANN events

Time	Event
July 1st, 1997	The U.S. Government promised to promote the privatization of DNS management
September 30th, 1998	ICANN was officially founded in California and the first board meeting was held one month later
November 25th, 1998	ICANN signed the MOU with the U.S. Department of Commerce and identified the latter's supervisory power over ICANN
February 9th, 2000	ICANN signed with NITA an agreement, according to which the former should take up the function of the latter
June 30th, 2005	After WGIG required ICANN to reform itself, the latter published the "principle declaration" as a response
September 29th, 2006	The U.S. Department of Commerce announced that it would sign a "joint project agreement" lasting three years with ICANN
September 30th, 2009	The "joint project agreement" signed between ICANN and the U.S. Department of Commerce expired, and the two parties reached a new "obligation confirmation agreement"
March 14th, 2014	NTIA made a statement, announcing that it would quit conditionally the supervisory authority over ICANN
March 10th, 2016	ICANN signed with the U.S. Department of Commerce an agreement on the quitting of the supervisory authority
October 1st, 2016	ICANN officially broke away from the supervision of NTIA

as root servers and IP address allocation. ICANN has its special mechanism. On the one hand, though its board members are from different countries, the final veto power is in the Department of Commerce of the United States and is subject to the laws of that country. On the other hand, it cannot be taken as a U.S. organization and the international community has never given up reforming ICANN, trying to make it internationalized, which has seen some achievements in recent years.

ICANN events are shown in Table 8.3.

At the cyberspace governance platforms like WSIS in the UN framework, the reform of ICANN is always a sensitive but important issue. According to the report released at the Western Asia Preparatory Conference for WSIS in 2003, TLD and IP address assignment should fall into the scope of the national sovereignty.[34] The control power over important network resources is the most central power in international Internet governance, representing the hard power in cyberspace, and determining to a great extent, the soft power such as Internet governance rules formulation power and discourse power. On the roles of and relationship between

---

[34]WSIS Executive Secretariat. *Report of the Western Asia Regional Conference for WSIS.* February 5, 2003, http://www.itu.int/net/wsis/preparatory/regional/beirut.html.

ICANN and ITU in global cyberspace governance the parties to the series of WSIS conferences cannot reach an agreement.[35]

Quite a few actors have proposed solutions to the ICANN reform, including the solution proposed by Brazil and other countries on the Sao Paulo Conference, the solution proposed by East West Institute on Berlin Summit and the solution proposed by India on ICANN Busan Summit.[36] All these solutions comply with Multilateralism in some degree, aiming to improve the power and position of sovereign nations' governments in ICANN management. For instance, Brazil and some other countries have proposed that greater power should be given to GAC while India has proposed that the management power over ICANN should be shifted to ITU.

In March 2014, under the pressure from the PRISM scandal, NTIA under the U. S. Department of Commerce released an official statement, announcing that it would give up the supervisory power over ICANN. In March 2016, the stakeholder communities in the world submitted the suggestion on the shift of the power, which was then completed in September 2016. But the United States declared that it would not accept NTIA being replaced by a solution dominated by any government or under any governmental agency, so Multi-stakeholder Model remains the basic principle for ICANN reform and operation.[37] Therefore, the change in ICANN does not mean that the United States has given up its control over critical Internet infrastructure like the global domain name management system; instead, in a relatively long time, the internationalization of ICANN is to be furthered.

8.3.2 Regional Cyberspace Governance Platforms

In global governance, regional organizations or multilateral cooperation mechanisms like BRICS can easily launch effective cooperation initiatives and actions because of their common geo-economic and political interests or their common goal for global governance. These organizations have been involved in cyberspace governance in recent years and play an important role through their initiatives, declarations or joint actions.

An overview of regional cyberspace governance platforms is shown in Table 8.4.

---

[35]Wolfgang Kleinwachter. Beyond ICANN vs ITU? How WSIS Tries to Enter the New Territory of Internet Governance. *The International Journal for Communications Studies*, Vol. 64, No. 3–4, 2004, pp. 233–251.

[36]Shen Yi. Disputes over the Principles of Global Cyberspace Governance and China's Strategic Choice. *Foreign Affairs Review*, 2015 (2), pp. 65–79.

[37]National Telecommunications & Information Administration, United States Department of Commerce. *NTIA Announces Intent to Transition Key Internet Domain Name Functions*. March 14, 2014, https://www.ntia.doc.gov/ press-release/2014/ntia-announces-intent-transition-key-internet-domain-name-functions.

Table 8.4 Overview of regional cyberspace governance platforms

Organization	Overview	Achievements in cyberspace governance
APEC	Founded in 1989, it has 21 member countries; holding leadership summits and ministerial conferences, aiming to facilitate the Asia-Pacific economic integration and proposing the establishment of the Asia-Pacific free trade zone	*e-APEC Strategy, APEC Strategy to Ensure Trusted Safe and Sustainable Online Environment*; Asia Pacific Computer Emergency Response Team (APCERT)
BRICS	Member states: Brazil, Russia, India, China and South Africa. The first leadership meeting was held in 2008; and the new development bank and the emergency reserve mechanism have been established	BRICS Joint Cybersecurity Group; *BRICS Leaders Xiamen Declaration*, promising to coordinate their position in ICT R&D, digital economy, cyber security and international cyberspace order and to take joint actions
NATO	Founded in 1949, NATO has 29 full member states, having the military committee, which has taken joint military actions for times	NATO Cooperative Cyber Defence Centre of Excellence; NATO Computer Incident Response Capability (NCIRC); and NATO's Cyber Defence Pledge
SCO	Founded in 2001, SCO has 8 full member states and 4 observing countries; taking countering terrorism, separatism and extremism as the cooperative goal	Proposed to the UN the formation of the *International Code of Conduct for Information Security*; and Joint Counter-terrorism Cyber Exercise
ASEAN	Founded in 1967, ASEAN has 10 member states and 2 observing countries. The unique ASEAN style has been formed. ASEAN announced the establishment of ASEAN Community in 2015	ASEAN Cybersecurity Ministerial Meeting

Multilateral cooperation mechanisms like APEC and BRICS have expressed their unique ideas on cyberspace governance to the international community through declarations or similar texts. In 2000, leaders of APEC member states made their declaration through their summit to facilitate the universal access to the Internet in urban and rural areas of all countries. In 2001, at the meeting of APEC member states leaders held in Shanghai, the *e-APEC Strategy* was released.[38] In 2005, APEC released *APEC Strategy to Ensure Trusted Safe and Sustainable Online Environment* to seek to expand the cooperation within the mechanism to

---

[38]e-APEC Task Force. *e-APEC Strategy*. October 2011, http://publications.apec.org/publication-detail.php?pub_id=584.

improve information and cyber security, coordinate the policy framework for transaction and communication and to jointly counter cybercrime.[39]

BRICS Summits have been enhancing the topic about cyberspace governance in recent years. Through summit declarations, the countries try to coordinate the position on critical issues, such as digital economy, cyber security and global cyberspace governance order, to expand their cooperation. In the *Sanya Declaration* made in 2011, BRICS pointed out, "We express our commitment to cooperate for strengthening international information security. We will pay special attention to combating cybercrime."[40] They released in 2017 the *Xiamen Declaration*, in which they expressed that they would encourage the establishment of BRICS Institute of Future Networks, enhance joint BRICS research, development and innovation, increase investment in ICT research and development, and improve ICT infrastructure and connectivity in the five countries. In terms of digital economy, they pledge to take actions to create conditions for the prosperity and thriving of digital economy, and enhance the policy sharing in e-commerce and IPR. As for cyber security, they enhance international cooperation against terrorist and criminal misuse of ICTs. In terms of internal cyberspace order, they support the central role of the UN in international affairs, emphasize the sovereignty principle, and advocate the establishment of internationally applicable rules.[41]

APEC, NATO, SCO and ASEAN have shown increasing concern about cyber security. APEC established in 2003 the APCERT, which is a union made up of the computer emergency response groups of the countries in the region. APCERT aims to enhance the joint capacity of response to cyber security threats within the region. By the end of 2016, it has 28 acting members from 20 economies of the Asia-Pacific Region, and three supporting members. Among all these members, five are from China.[42] Similar mechanisms are adopted at cross-regional platforms. In 2016, NCIRC and CERT-EU reached an agreement on technical cooperation, and thus established relevant technical information sharing and best practice exchange mechanisms to facilitate the coordination and cooperation in cyber incident prevention, detection and response in accordance with their autonomy and procedures of policy making. ASEAN is committed to promoting its member states to exchange and coordinate in policies concerning cyber security through ASEAN regional forums and cybersecurity ministerial conferences. Besides, multilateral exercises in terms of cyber security are also important actions by some regional organizations. They include Locked Shields, a multi-national exercise in the framework of NATO,

---

[39]APEC. *APEC Strategy to Ensure Trusted Safe and Sustainable Online Environment*. November 2005.

[40]People.cn. *Sanya Declaration*. BRICS Leaders Meeting, April 4th, 2011, http://politics.people.com.cn/GB/1024/14391387.html.

[41]Xinhuanet.com. *BRICS Leaders' Xiamen Declaration*, September 4th, 2017, http://news.xinhuanet.com/world/2017-09/04/c_1121603652.htm.

[42]APCERT Secretariat. *APCERT Annual Report 2016*. https://www.apcert.org/documents/pdf/APCERT_Annual_Report_2016.pdf.

the Cyber Europe of EU, Cyber Atlantic 2011 jointly launched by EU and NATO, and the joint SCO counter-terrorism cyber exercise in Xiamen in 2015.

8.3.3 Bilateral Cyberspace Dialogue and Cooperation Mechanisms

Major countries including China and the United States have launched the dialogue and cooperation mechanisms (see Table 8.5), which have created the opportunity for the countries to exchange experience, coordinate policies and carry out cooperative actions.

China has launched dialogue and mechanisms of cooperation with the United States, Russia and EU member states. China-U.S. cooperation has witnessed the most rapid progress and practical achievements. Sino–U.S. Cyber Working Group was established in 2013 and the first meeting of the group was held in the same year, ushering in the dialogues between the two countries. In 2015, when visiting the United States, President Xi Jinping reached consensus with the Obama Government on jointly countering cybercrime and promoting the formulation of codes of conduct for cyberspace security. Then the U.S.–China High-level Joint Dialogue on Cybercrime and Related Issues was launched. Three such dialogues have been held since the first one in 2015. As the achievement of the dialogues, *Guidelines for Cybercrime and Related Issues* was made upon the agreement between the two countries and the U.S.–China Hotline was opened on combating cybercrime and related issues. Then desktop maneuvers of cyber security were carried out. In April 2017, when President Xi Jinping met Trump during his state

Table 8.5 Overview of bilateral dialogue and cooperation mechanisms for cyberspace security

Bilateral countries	Major achievements
China–U.S.	U.S.-China High-level Joint Dialogue on Cybercrime and Related Issues,and U.S.-China Law Enforcement and Cybersecurity Dialogue
China–Russia	*Agreement on cooperation in Ensuring International Information Security between the Government of People's Republic of China and the Federal Government of Russia*, and *The President of the People's Republic of China and the President of the Russian Federation's Joint Declaration on Collaborative Development of Information Network*
China–EU	Forum of Sino-Europe Information Society Cooperation, EU-China ICT Dialogue, and China-EU Digital Economy Dialogue Forum
U.S.–Japan	Japan-U.S. Cyber Dialogue
U.S.–India	U.S.–India Cyber dialogue, and U.S.–India Cyber Partnership Framework
U.S.–Russia	Bilateral Presidential Commission Working Group for Responding to ICT Threats, and White House – Kremlin Palace Cybersecurity Hotline Mechanism
Russia–India	Russia–India Cybersecurity Cooperation Agreement

visit to the United States, the Law Enforcement and Cybersecurity Dialogue was identified as one of the four high-level dialogue and cooperation mechanisms. The first Law Enforcement and Cybersecurity Dialogue was held in October 2017, when both countries agreed to facilitate and improve their cooperation in combating cybercrime.

China and Russia have been cooperating with each other in global cyberspace governance for a long time. For instance, they jointly advocate cyberspace cooperation in the framework of SCO and position coordination in cyber security on the platform of the UN. During President Xi Jinping's visit to Russia in 2015, the two countries signed the *Agreement on Cooperation in Ensuring International Information Security between the Government of People's Republic of China and the Federal Government of Russia*. The two countries will strengthen the cooperation in tackling cyberspace threats, countering cyber terrorism and responding to computer emergencies on bilateral and multilateral platforms, and jointly build a safe and open international Internet governance system. Then in 2016, during President Putin's visit to China, the two countries published *The President of the People's Republic of China and the President of the Russian Federation's Joint Declaration on Collaborative Development of Information Network*, which has facilitated the establishment of a just and equal information cyberspace development and security system.

China and EU have diversified dialogue mechanisms for cyberspace. They held the Forum of Sino-Europe Information Society Cooperation in 2002, and the first meeting of the Sino-EU Information Society Dialogue Working Group in 2003, having dialogues on policies, legislation and technology research and development of the information society. Another series of dialogue mechanism is the EU-China ICT Dialogue launched in 2009, eight sessions of which had been held by 2017, covering ICT policy-making and supervision, and the development of digital economy. In the first China-Australia high-level security dialogue held in April 2017, cooperation in cyber security was the focus. Both countries declared that they would set up the mechanism for information sharing in combating and preventing cybercrime and send to each other a cybersecurity team to enhance exchanges in the respect. In October 2014, China, Japan and Republic of Korea signed the *MOU on Enhancing Cybersecurity Cooperation*, set up the consultation mechanisms for cybersecurity affairs, and discussed the possible cooperation in jointly combating cybercrime and cyber terrorism and responding to Internet emergencies.

The United States is active in building mechanism for bilateral dialogues and cooperation in cyberspace. The dialogue mechanism concerning cyberspace between the country and Japan was launched in 2013, and five such dialogues had been held by 2017. Other cooperations between the two countries are U.S.-Japan Cyber Defense Policy Working Group and Japan's Cabinet Cybersecurity Center's joining the AIS Project of the U.S. Department of Homeland and Security. The dialogue and cooperation mechanism between the United States and India was launched in 2011, marked by the signing of the *MOU on Cybersecurity Cooperation*, according to which both countries are seeking to set up the mechanism for multi-channel cybersecurity information sharing. In 2016, the two countries signed a U.S.-India Cyber Partnership Framework, which defines the principles and expected forms for their cooperation in cyberspace. By 2016, they had held five United States-India cyber dialogues. The U.S. bilateral cooperation in cyberspace goes beyond its allies. In 2013, the country and Russia set up a special ICT security working group, and announced that the computer emergency responding teams of the two countries would share information on threats and use the Nuclear Risk Reduction Center set up during the Cold War period in building trust in cyber security. Since the general election of the United States in 2016, cyber security has been a key issue in the competition between the United States and Russia.

It can be seen that China and the United States are the dominating forces in building bilateral cyberspace dialogue and cooperation mechanisms. Other countries have also launched some dialogues and cooperation. For instance, during the BRICS Goa Summit in 2016, Russia and India signed an agreement on cyber security cooperation and sought to launch high-level dialogues in cyber security and thus strengthen their cooperation in combating cybercrime, reinforcing national defense and tackling terrorism.

8.4 Rules of Cyberspace Governance

The rules of global cyberspace governance are faced with a major principle controversy: to transplant or transform the existing international laws to make them applicable to global cyberspace governance or to design and establish a set of new rules for it in accordance with the endogenous demand of the Internet and IT revolution. There has been no consensus for this controversy, but in practice, pragmatic principles are observed: for problems similar to the cases covered by the original international rule system, the existing rules or laws are transplanted or transformed to be applicable to them; for problems with their characteristics, new rules are designed and established.

8.4.1 Coverage of Rules and Representative Achievements

1. Rules construction facilitated by global international organizations

Some rules of cyberspace governance have been made in the framework of the UN through UNGGE and UN General Assembly. For instance, the necessity for applying cyber security norms in peaceful period has been identified. A State should not conduct or knowingly support ICT activity that intentionally damages or otherwise impairs the use and operation of critical infrastructure. No State should harm the information systems of the authorized emergency response teams of another State or use those teams to engage in malicious international activity.[43] Moreover, the necessity to protect human rights in cyberspace has been emphasized. In December 2013, the UN General Assembly passed No. 68/167 Resolution, which makes it explicit that the people's right in the real world should be protected in the virtual space.[44] ITU has formulated *ITU Convention*, *Constitution of the International Telecommunication Union* and *International Telecommunication Rules* (new version), which define the role of the members and their supporting actions in the telecommunication industry. ISO has formulated over 3000 international standards through its information technology committee. WTO has also released *Information Technology Agreement*, *Telecommunication Annex* and *E-commerce Joint Declaration*, which all focus on the establishment of digital trade norms.

2. Rules construction facilitated by international Internet communities

International Internet communities, such as ICANN, IETF and ISOC, all play an important role in the construction and promotion of Internet technology rules and standards, and have formulated a series of specific rules on IP address allocation, domain name registration management, global domain name resolution root servers, and root zone file and root zone file system management. IETF has done most of the work concerning the formulation of Internet technology standards. It is an international civil organization participated in and managed by experts contributing to the Internet technologies and their development. It develops and promotes Internet standards, including IPv4, IPv6, TCP (Transmission Control Protocol) and UDP (User Datagram Protocol). Especially, IETF has formulated the TCP/IP standards and undertaken the formulation of 90% of global Internet standards.

3. Rules construction facilitated by regional international organizations

In recent years, an increasing number of regional international organizations have begun to promote the formulation of cyberspace cybercrime. The effort made by EU and OECD deserves attention in cyber terrorism combat. In 2001, European Commission formulated *Budapest Cybercrime Convention*, with the aim to set up a multilateral legal framework for cross-national fight against cybercrime. In the *European Agenda on Security* (2015–2020) established in 2015, combating cybercrime is one of the three pillars. OECD is committed to releasing to the world the guidelines and suggestions for all actors on tackling cybercrime to facilitate capacity building of all stakeholders in this area. It released a recommendation information system and cyber security in 2002 and another one in 2015 on digital security risk management, calling on all stakeholders to channel digital security risk

---

[43]United Nations General Assembly. *Group of Governmental Experts on Developments in the Field of Information and Telecommunications in the Context of International Security*. July 22, 2015, http://www.un.org/ga/search/view_doc.asp?symbol=A/70/174.

[44]United Nations General Assembly. Resolution Adopted by the General Assembly on 18 December 2013: 68/167. *The Right to Privacy in the Digital Age*. January 21, 2014, http://www.un.org/en/ga/search/view_doc.asp?symbol=A/RES/68/167.

management into macro-economic and social policy-making.[45] In 2008, it released the *Seoul Declaration for the Future of the Internet Economy*, which is about digital economy, with an aim to provide guidelines and suggestions on facilitating digital economy development through cooperation among multi-stakeholders.[46] The discussion of G20 and APEC on digital economy norms has become a highlight on their agenda. *G20 Digital Economy Development and Cooperation Initiative* was released in 2016 to seek for consensus on digital economy development and cooperation. In April 2017, Roadmap for Digitalization: Policies for a Digital Future was released. APEC focuses on the necessity of e-commerce development for the connectivity in the Asia-Pacific Region. In *APEC Telecommunications and Information Working Group Strategic Action Plan 2016–2020*, APEC proposes promoting a secure, resilient and trusted ICT environment, enhancing the digital economy and facilitating regional economic integration.[47]

4. **Rules formulation advocated by major countries and participated in by multiple stakeholders**

Besides the existing international organizations, some countries influential in cyberspace have begun to try other diplomatic means, advocating and promoting multilateral agendas of global cyberspace governance. In 2011, the United Kingdom launched GCCS (Global Conference on Cyberspace) London Process, which has been the first and an important practice in that aspect. Since 2011, five sessions GCCS have been held successively in London, Budapest, Seoul, Hague and India. The first was held in London, and hence "London Process". The World Internet Conference launched by China soon attracted the attention of all countries. It was first held in 2014 and then was and will be held every year in Wuzhen, Zhejiang Province. At the past sessions, the conference has had focuses on "a community of shared future in cyberspace". It reflects the wide involvement of a large number of participants including so many developing countries and so many non-state actors like businesses and international organizations (See Table 8.6). The High-level Expert Advisory Committee of the World Internet Conference Organizing Committee Secretariat was founded in 2015, co-chaired by Ma Yun (Jack Ma) and Fadi Chehadé, president and CEO of ICANN. The Committee is made up of experts from all stakeholders and they offer opinions and suggestions on the Conference.

"London Process' and the World Internet Conference differ from each other in the route of rule formulation, but they have similarities. Both of them extensively involve multiple actors, like national governments and businesses. On the other

---

[45]OECD. *Digital Security Risk Management for Economic and Social Prosperity: OECD Recommendation and Companion Document.* 2015, http://www.oecd.org/sti/ieconomy/digital-security-risk-management.pdf.

[46]OECD (2008), "The Seoul Declaration for the Future of the Internet Economy", OECD Digital Economy Papers, No. 147, OECD Publishing, Paris. http://dx.doi.org/10.1787/230445718605.

[47]APEC. *APEC Telecommunications and Information Working Group Strategic Action Plan 2016 —2020.* 2016, https://www.apec.org/Meeting-Papers/Sectoral-Ministerial-Meetings/Tele-communications-and-Information/2015_tel.

Table 8.6 Overview of the world internet conference

Session/Year	Theme	Scale
First/2014	An Interconnected World Shared and Governed by All – Building a Community of Shared Future in Cyberspace	Over 100 countries and 51 businesses
Second/2015	An Interconnected World Shared and Governed by All – Building a Community of Shared Future in Cyberspace	Over 120 countries, over 20 international organizations and 258 businesses
Third/2016	Innovation-driven Internet Development for the Benefit of All – Building a Community of Common Future in Cyberspace	Over 110 countries, 16 international organizations and 310 businesses
Fourth/2017	Developing Digital Economy for Openness and Shared Benefits: Building a Community of Common Future in Cyberspace	

Source wuzhenwic.org and other relevant reports

hand, at its beginning, London Process involved few developing countries while the World Internet Conference has been attaching importance to the involvement of developing countries and emerging economies. The topics of London Process include economic development, social welfare, cybercrime, secure and reliable network access and international security. The World Internet conference highlights the innovative concept "a community of shared future in cyberspace", which covers the interconnectivity in terms of infrastructure, online cultural exchange, economic innovative development and common prosperity, collaboration and equitable global Internet governance system, advocating common responsibilities, win-win cooperation, inclusiveness and mutual benefit.

5. **Rules formulation facilitated by think tanks, multinationals and researchers**

The diversity of global cyberspace governance actors is, to some extent, manifested in think tanks, multinationals and research communities, who are active in formulation of cyberspace rules. The cyberspace governance norms they advocate have their own characteristics and can lead the global governance rule formulation in special areas.

East West Institute is a representative think tank, which has made great achievement in advocating the formulation of cyberspace rules. In 2010, it launched the Global Cyberspace Cooperation Summit, seven sessions of which have been held in different cities of the world. The Summit advocates global dialogues and cooperation in cyber security. It has also launched the Global Cooperation in Cyberspace Initiative. Its 2016–2017 action agenda contained the follow goals: (1) enhance deterrence against malicious cyber activities; (2) improve the security of Internet products and services; and (3) maintain efficient information and technology flows across borders consistent with local values.[48] Besides, the Institute and The Hague Centre for Strategic Studies jointly established in February 2017 the Global Commission on the Stability of Cyberspace (GCSC), which will strongly support policy and norms coherence related to the security and stability in and of cyberspace and will launch a three-year policy and recommendation formulation initiative.[49]

Multinationals are becoming increasingly active in facilitating the formulation of cyberspace governance rules. For instance, the *Digital Geneva Convention* drafted and formulated by Microsoft contains some principles and norms. For example, such a convention should commit governments to avoid cyber attacks that target the private sector or critical infrastructure, to assist private sector efforts to detect, contain, respond to and recover from cyber attacks, to report vulnerabilities to vendors rather than stockpile, sell or exploit them and to promise non-proliferation of weapons.[50] It can be seen that *Digital Geneva Convention* was aimed to prevent the governments from launching cyber attacks and guarantee the security and position of technical businesses in cyberspace governance.

In the global cyberspace rule system, international laws have special position. How to ensure strategic stability, regulate national action and formulate norms for possible cyber warfare in the cyberspace has become the focus of legal professionals in discussing the formulation of new rules. NATO Cooperative Cyber Defence Centre of Excellence has carried out some adequate research. It released in 2009 *Tallinn Manual on the International Law Applicable to Cyber Warfare*, whose version 2.0 was published in 2017. The manual is committed to identifying the rules in the existing international law framework that are applicable to cyber warfare and explaining how to interpret the rules against the cyberspace backdrop. In the first version of the Manual, 95 rules were marked, but the number increased to 154 in the second version, covering sovereignty, national responsibilities and human rights in cyber warfare international law category.[51]

---

[48]East West Institute. *Global Cooperation in Cyberspace Initiative: 2016–2017 Action Agenda.* 2015, https://www.eastwest.ngo/sites/default/files/ideas-files/ActionAgenda2016_Spread.pdf.

[49]The Hague Centre for Strategic Studies. The Global Commission on the Stability of Cyberspace. July 18, 2017, https://hcss.nl/news/global-commission-stability-cyberspace.

[50]Brad Smith. *The Need for a Digital Geneva Convention.* Microsoft, February 14, 2017, https://blogs.microsoft.com/on-the-issues/2017/02/14/need-digital-geneva-convention/.

[51]Michael N. Schmitt, ed. *Tallinn Manual on the International Law Applicable to Cyber Warfare*, Cambridge University Press, 2013; Michael N Schmitt and Liis Vihul, eds. Tallinn Manual 2.0 on the International Law Applicable to Cyber Operations, Cambridge University Press, 2017.

8.4.2 *Controversies in the Formulation of Cyberspace Governance Rules*

1. Diversity of issues and complexity of rules

First, there are multiple participants in rules formulation, including governments, international organizations, technical communities, businesses, experts and research institutes. Such diversity leads to the difficulty in forming the unified understanding and reaching consensus. There are not many models or rules widely acknowledged or producing remarkable effects in controlling cyberspace threats and risks. Secondly, cyberspace governance involves diverse fields, including politics, economy, society, culture and military factors. For instance, in essence, cyberattack has obscure boundary with traditional military activities and it is often concerned with economic demand and political situation. In physical space, complex economic and political relations and power structure between countries, and between state and non-state actors affect the formulation of cyberspace rules. Thirdly, technology of cyberspace is progressing fast while rules formulation is unable to catch up with the demand of Internet development and governance because it takes long-time consultation and argumentation to formulate the rules. Fourthly, cyberspace resources and capabilities are distributed unevenly in the world, which inevitably produces influence on rules formulation. Developed countries (early-developing countries) advocate bottom-up paths and community-dominated models to consolidate their own advantages while developing countries (later-developing countries) advocate up-bottom paths and state-dominated model to close the gap between them and developed countries. This difference in position has resulted in the division between rule makers and rule observers, which, in turn, makes rules formulation and observance very complicated.

2. Controversy in guiding principles for cyberspace governance rules formulation

There are two guiding principles for cyberspace governance rules formulation, namely, the "multistakeholder" principle and the "multilateralism" principle. The difference between the two lies in whether to identify with the government-dominated cyberspace rules formulation. Countries like the United Kingdom and the United States hold that the government's strengthening of the sovereign control over cyberspace will impair the applicability of international laws of human rights in cyberspace; instead, they advocate that hard power of ICT and resource dominates the formulation and execution of cyberspace governance rules. But the control authority of the government over cyberspace cannot be ignored. Multilateralism stresses the government's rights to and responsibilities for cyberspace, and thus the formulation and execution of global cyberspace governance rules should be done by all countries and international organizations.

3. Controversy in the consensus on and control effort in promoting the development of digital economy

Digital economy is an important achievement of ICT prosperity, but the cross-border circulation of digital products and services has given rise to problems in supervision and thus the worldwide controversy in digital economy rules. During Obama's presidency, the United States was committed to channeling the new-generation digital trade rules into TPP negotiation, demanding to promote the non-localization of data storage and cancel cybersecurity censorship and thus reduce the non-tariff barriers against digital trade within the region.[52] Some countries including China and India advocate pre-storage localization of data and strengthening of cyberspace censorship, emphasizing managed cross-border data circulation. With their ICT development, emerging countries have increasingly strong wish to promote their domestic rules reform and participate in international rules formulation. Their communication and cooperation with countries who have traditional advantages in information technologies will be expanded.

4. Controversy in consensus on necessity to regulate national action in cyberspace and in militarization of cyberspace

Though many state and non-state actors are advocating the effort for non-proliferation of cyber weapons, the formulation of cyberspace militarization rules has been a controversy, which covers four problems: (1) how to define cyber weapons and cyber attacks, including whether to categorize cyber espionage and network monitoring software into cyberweapons and cyber attacks[53]; (2) necessity to formulate international rules for cyberspace militarization issues including whether to admit the reality and threat of cyber warfare; (3) effectiveness of formulation of international rules of cyberspace militarization issues (Even though they have been formulated, rules of cyber weapon non-proliferation may not be implemented or they may not produce effect.)[54]; (4) how to establish international rules of cyberspace militarization, i.e. whether to directly apply the existing international law frameworks to cyberspace or to formulate special rules and norms of cyber warfare.

---

[52]Lee Branstetter. TPP and Digital Trade. in Jeffrey J. Schott and Cathleen Cimino-Isaacs, eds., *Assessing the Trans-Pacific Partnership*, Volume 2: Innovations in Trading Rules, Peterson Institute for International Economics, 2016, pp. 72–81.

[53]Mike Lennon. Defining and Debating Cyber Warfare. *Security Week*, April 6, 2010, http://www.securityweek.com/content/defining-and-debating-cyber-warfare.

[54]Clay Wilson. Cybersecurity and Cyber Weapons: Is Nonproliferation Possible? in Maurizio Martellini, ed. *Cyber Security: Deterrence and IT Protection for Critical Infrastructures*, Springer, 2013, pp. 11–24.

8.5 Prospect of Cyberspace Governance

Both state and non-state actors have fully realized the strategic values of cyberspace and its importance to their development, security and interest. At present and in some time, the global cyberspace governance is faced with a major conflict, namely, the conflict between the rapid development of technologies and their applications and the insufficiency of governance capacity, platforms and rules.

8.5.1 Increasing Demand for Cyberspace Governance

An increasing number of countries have fully realized the strategic significance and value of cyberspace and the necessity and importance of their participation in global cyberspace governance and proposal of their opinions on it. To only rely on Multi-stakeholder Model cannot meet the demand for cyberspace governance, so there is an immediate need for sovereign countries to get involved in solving key problems of the global cyberspace governance. Secondly, the rapid development of technologies demands the reinforcement of global cyberspace governance. The rapid development of cyber information technologies brings about pressure for cyberspace governance and in particular, cutting-edge technologies like big data, cloud computing, mobile Internet, IoT and AI pose new-challenges to governance capacity. Thus there have emerged a series of problems hard to solve, such as cross-border data flow supervision, jurisdiction over data resources of strategic values, cross-national cybercrime combating and prevention of national cyber weapon proliferation. Therefore, the international community is expected to provide more effective governance mechanisms and rules.

8.5.2 Increasing Enthusiasm of Sovereign Countries About Participating in Global Cyberspace Governance

The security and stability of cyberspace is becoming the key factor influencing all countries' security and development and an increasing number of sovereign countries have fully realized the strategic significance and values of cyberspace and the significance for them to get involved in cyberspace governance and to propose rules in accordance with their demand. President Xi Jinping of the People's Republic of China made a strategic judgement in 2014 that there would be no national security without cyber security and no modernization without informatization. This judgement does not only apply to China but to global cyberspace governance.

On the one hand, all actors participating in global cyberspace governance are reaching their consensus and their cooperation has witnessed prominent progress.

The final achievement document released in 2015 by UN Intergovernmental Working Group of Experts on ICT Security shows the most important part of a new consensus. That is, a new governance structure must be established on the basis of international laws respecting the equality of cyberspace sovereignty. President Xi Jinping has made a creative proposal of the governance model "multilateral participation of multiple parties" to ease the tension between Multilateralism and Multi-stakeholder Model in cyberspace governance. Xi's proposal is an important embodiment of the above-mentioned consensus. On the other hand, major countries involved in global cyberspace governance are lack of strategic mutual trust, and technical communities and sovereign countries, and especially major emerging countries and developing countries are lack of political trust with each other, which is the major barrier against the development of global cyberspace governance. The key reason for the lack of trust lies in the fact that countries with advantages are reluctant to quit the Cold War thinking and national interests of total self-centralism, which may lead countries into risks of security in both the real and virtual world. Affected by the stereotyped image created by European and American countries, technical communities lack political trust with major emerging countries and developing countries following different development roads and models, which hinders effective cooperation between them.

8.5.3 Increasing Importance of the UN as a Platform

Since the end of the Second World War, the UN has been the only multilateral platform of the world in real sense. In the early period of the Internet development and global cyberspace formation, the UN failed to fully play its role, but when the global cyberspace governance has entered the "deep-water zone", the importance of the UN becomes prominent. Especially, the UN is becoming a critical advantageous resource in organizational legality and member countries' authority in terms of ensuring the strategic stability of the cyberspace, safeguarding the cyber security and formulating guiding principles for international cyberspace governance. How to effectively give play to the UN's role is of vitality to the future development of global cyberspace governance. However, due to the lack of trust between major countries and efficiency of the UN's decision-making mechanism lower than that of technical communities' mechanism, the cyberspace governance promoted by the UN framework is lack of drive. The Internet Governance Forum (IGF), founded in the early 21st century, was finally reduced to the "Rome Conference", which covers a variety of topics without any focus. The UN Intergovernmental Working Group of Experts on ICT Security failed to form its achievement document in 2017, which shows more clearly the limit of the UN as a platform.

8.5.4   Increasing Enthusiasm of Non-state Actors About Promoting Cyberspace Governance

Non-state actors and technical communities represented by ICANN are the main force in facilitating the international cyberspace governance from the levels of technologies and norms. In the early period of the Internet development, these actors formed technical communities with businesses and individuals and became one of the forces promoting the governance. Unless they are trusted by sovereign countries and thus they can cooperate with latter, they are faced with limit in resources and capacities for effective international actions. Even though the Internet technical communities have had their technical advantages, sovereign countries still exercise check and balance on them through laws and systems. With the Internet's being integrated into politics, economy, culture and society, it is no longer practicable to only rely on non-state actors in cyberspace governance.

In prospect, to facilitate the international cyberspace governance system reform should become the common mission of all responsible actors, including China and the United States, who are expected to facilitate the building of a community of shared future in cyberspace. They should promote their mutual development, safeguard their mutual security, participate in the governance together and share the achievements, so that they can make the cyberspace more peaceful, secure, open, cooperative, and prosperous.

Afterword

During the compilation, we have realized that today, the information revolution is progressing and the global IT is experiencing a new round of innovation. The Internet is changing the production and life styles and promoting the social development, which has turned the world into a global village and the international community into a community of shared future. To promote mutual sharing and governance of the Internet is an opportunity and a mission. It is hoped that the *World Internet Development Report 2017* (the "Report" hereinafter) comprehensively shows and interprets the global Internet development status and trend from the perspective of China and provides China's solutions to the problems of the Internet development, so that all the countries can promote their mutual development, guarantee their mutual security, participate in the governance together and share the achievements.

The compilation of the Report has won support and guidance from Cyberspace Administration of China, especially guidance from the leaders of the Administration and support in terms of data and material from the departments and institutions of the Administration. China Academy of Cyberspace, as the coordinator, has invited a number of think tanks to participate in the compilation, including China Academy of Information and Communication Technology (CAICT), the First Research Institution of Ministry of Industry and Information Technology, State Information Center, CCID, Institute of National Governance of Tsinghua University, Internet Development Research Center of Peking University, and Cyberspace Governance Research Center and Internet Laboratory of Fudan University.

The publication of the *Report* is also owed to the support and help from society. Due to our limited experience, capability and time for the compilation, there are probably errors in it. We sincerely hope that governmental agencies, international organizations, research institutions, Internet businesses, and social associations will offer their opinions and suggestions and provide more detailed materials so that we can make modifications and improvement to make better contributions to world Internet development.

Chinese Academy of Cyberspace Studies

December 2017

Chinese Academy of Cyberspace Studies (ed.), *World Internet Development Report 2017*,
https://doi.org/10.1007/978-3-662-57524-6

GPSR Compliance
The European Union's (EU) General Product Safety Regulation (GPSR) is a set
of rules that requires consumer products to be safe and our obligations to
ensure this.

If you have any concerns about our products, you can contact us on

ProductSafety@springernature.com

In case Publisher is established outside the EU, the EU authorized
representative is:

Springer Nature Customer Service Center GmbH
Europaplatz 3
69115 Heidelberg, Germany